PRECALCULUS
THE
EASY
WAY

Lawrence S. Leff
Former Assistant Principal, Mathematics Supervision
Franklin D. Roosevelt High School
Brooklyn, New York

BARRON'S

Dedication

To Rhona . . .

For the understanding,

for the sacrifices,

for the love,

. . . and with love.

All inquiries should be addressed to:
Barron's Educational Series, Inc.
250 Wireless Blvd.
Hauppauge, NY 11788
www.barronseduc.com

International Standard Book No.: 0-7641-2892-2

Library of Congress Catalog Card No.: 2004062447

Library of Congress Cataloging-in-Publication Data
Leff, Lawrence S.
   Precalculus the easy way / Lawrence S. Leff.
      p. cm.—(Easy way series)
   Includes index.
   ISBN 0-7641-2892-2
   1. Algebra—Problems, exercises, etc.   I. Title.   II. Easy way.

QA154.3.L43 2005
512—dc22

                              2004062447

PRINTED IN THE UNITED STATES OF AMERICA
9 8 7 6 5 4 3 2 1

CONTENTS

PREFACE

Unlike many other precalculus books, *Precalculus the Easy Way* assumes little, if any, prior knowledge of high-school-level algebra. Instead, the book begins with the concepts and skills typically introduced in a first-year high-school-level algebra course and then gradually progresses to the more advanced topics in algebra and trigonometry needed for the study of calculus. Emphasis is on providing clear, straightforward explanations while avoiding unnecessary details. Because of this approach, adults who have not recently studied mathematics, as well as students enrolled in high-school-level mathematics courses that precede calculus, will find this book especially helpful and easy to follow.

The book is modern in its approach, with special consideration given to graphing-calculator approaches. Although it is assumed that you have access to a graphing calculator, the development of a solid foundation in algebraic operations and mathematical reasoning remains a primary goal throughout the book.

Here are some additional features of the book that you should know about:

- The topic organization and simple lesson format break down the subject matter into manageable learning modules that will help you organize your study time.
- The clear explanations, helpful illustrations, and numerous step-by-step examples make the book ideal for self-study and rapid learning.
- Special math "tips" that anticipate and help resolve potential student difficulties are strategically located throughout the book.
- Each chapter ends with a comprehensive set of practice exercises designed to reinforce and extend key skills and concepts. These checkup exercises, together with the answers or solution hints included at the back of the book, will help you assess your understanding and monitor your progress.

STUDY UNIT I: ALGEBRA AND GRAPHING METHODS

Chapter 1

BASIC ALGEBRAIC METHODS

OVERVIEW

This chapter reviews basic algebraic terms and skills that you may have forgotten or never learned. If you have already mastered the algebraic skills reflected in any of the lessons listed below, you may want to skip that lesson and move ahead to the next one.

Lesson 1-1   Real Numbers, Variables, and Exponents

KEY IDEAS

Algebra is generalized arithmetic that uses both letters and numbers. These letters, called **variables**, are placeholders for unknown numbers. By using variables rather than specific numbers, the language of algebra allows general statements to be made about numbers. We know, for example, that $2 + 3 = 3 + 2$. To generalize that the order in which *any* two numbers are added together does not matter, we can write $a + b = b + a$, where the variables a and b stand for *any* two numbers.

Ordering Real Numbers on a Number Line

Positive 3 and negative 3 are *opposites* in the same sense that winning three games (+3) and losing three games (–3) are opposite situations. The size relationship between positive and negative numbers can be represented by using a ruler-like diagram called a **number line**. A number line extends indefinitely in opposite directions on either side of the point labeled 0, which is called the **origin**. On a horizontal number line, positive numbers are labeled in increasing order to the right of 0, and negative numbers in decreasing order to the left of 0. The opposite of each positive number lies on the left side of the origin and at the same distance from the origin as its positive counterpart, as shown in Figure 1.1.

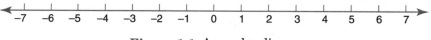

Figure 1.1 A number line

Any number on the number line is greater than any number to its left. For example, –2 is greater than –3, –1 is greater than –2, 0 is greater than –1, and 1 is greater than 0.

Classifying Numbers

The set of all points on the number line corresponds to the set of **real numbers**. The set of real numbers is comprised of the following sets.

- **Integers**. The set of integers consists of the counting numbers, their opposites, and 0. Thus, the set of integers is {. . . , –4, –3, –2, –1, 0, 1, 2, 3, 4, . . .}. The three periods at each end indicate that the pattern continues without ceasing.
- **Rational numbers**. The set of rational numbers includes every number that can be expressed as a fraction with an integer in the numerator and a nonzero integer in the denominator, as in $\frac{-3}{5}$. Decimal numbers that terminate, such as $0.25 \left(= \frac{1}{4} \right)$ and

$0.113 \left(= \dfrac{113}{1000} \right)$, are rational. Decimal numbers that endlessly repeat the same set of digits, such as $0.33333\ldots \left(= \dfrac{1}{3} \right)$ and $0.45454545\ldots \left(= \dfrac{5}{11} \right)$, are rational since they can be rewritten as the quotients of two integers. Since every integer can be written as a fraction with 1 as its denominator, the set of integers is a subset of the set of rational numbers.

- **Irrational numbers**. The set of irrational numbers includes numbers that have no rational equivalents, such as $\pi = 3.1415926\ldots$ and $\sqrt{2} = 1.4142136\ldots$, where the trailing three dots mean that the decimal digits never end. Thus, the set of real numbers is the union of the set of rational numbers (and its subsets) and the set of irrational numbers. A real number is either rational or irrational, but cannot be both.

Absolute Value

The distance that a number, x, is from 0 on a number line is called the **absolute value** of x, written as $\left| x \right|$. Since –3 and +3 are each a distance of 3 units from the origin, $\left| +3 \right| = 3$ and $\left| -3 \right| = 3$. The absolute value of a real number is always nonnegative.

Using Symbols to Make Comparisons

The symbol = is read as "is equal to," while the symbol ≠ means "is not equal to." For example, $1 + 1 = 2$ and $1 + 1 \neq 3$. If two numbers are not equal, then one number must be greater than the other. The symbol for "is greater than" is >, while the symbol < means "is less than."

EXAMPLE: $5 > 0$ **EXAMPLE:** $-3 < -2$

An equal sign may be combined with an inequality symbol. For instance:

EXAMPLE: $x \leq 3$ is read as "x is less than *or* equal to 3," so x is *at most* 3.

EXAMPLE: $y \geq 6$ is read as "y is greater than *or* equal to 6," so y is *at least* 6.

Representing One Quantity in Terms of Another

Sometimes it is necessary to compare two quantities by representing one in terms of the other. Here are some examples:

EXAMPLE: Tim's weight exceeds Sue's weight by 13 pounds. If Sue weighs x pounds, then Tim weighs $x + 13$ pounds.

EXAMPLE: The number of dimes in Chris's bank exceeds 3 times the number of pennies by 2. If there are x pennies, then there are $3x + 2$ dimes.

EXAMPLE: Bill has 7 fewer dollars than twice the number that Kim has. If Kim has x dollars, then Bill has $2x - 7$ dollars.

EXAMPLE: The sum of two numbers is 25. If x represents one of these numbers, then $25 - x$ represents the other number.

EXAMPLE: Vincent has x compact discs, and Joe has 4 compact discs. If the number of compact discs that Ellen has is 3 less than twice the sum of the number that Vincent and Joe have, then Ellen has $2(x + 4) - 3$ compact discs.

Laws of Integer Exponents

Repeated multiplication of the same quantity can be abbreviated by using a shorthand notation in which the number of times the quantity is being used in multiplication is placed to the right of the term and one-half line above it. For example:

$$32 = \underbrace{2 \times 2 \times 2 \times 2 \times 2}_{\text{2 appears 5 times.}} = \underset{\text{Base}}{2}^{5} \quad \leftarrow \text{Exponent}$$

In 2^5, 2 is the **base** and 5 is the **exponent**. The expression 2^5 is read as "2 raised to the fifth power."

- **Multiplication law of exponents: $a^x \times a^y = a^{x+y}$**
  To multiply powers with the same base, *add* their exponents and keep the same base. For example:

  $$5^7 \times 5^3 = 5^{7+3} = 5^{10} \quad \text{and} \quad x \cdot x^7 = x^1 \cdot x^7 = x^8.$$

- **Quotient law of exponents: $a^x \div a^y = a^{x-y}$**
  To divide powers of the same base, *subtract* their exponents and keep the same base. For example:

  $$\frac{5^7}{5^3} = 5^{7-3} = 5^4 \quad \text{and} \quad \frac{x^8}{x^7} = x^{8-7} = x^1 = x.$$

- **Power of a power law of exponents: $\left(a^x\right)^y = a^{xy}$**

  To simplify a power of a power, *multiply* the exponents. For example:

  $$\left(2^4\right)^2 = 2^{4 \times 2} = 2^8 \quad \text{and} \quad \left(mn^5\right)^2 = m^{1 \times 2} n^{5 \times 2} = m^2 n^{10}.$$

- **Zero exponent law: $a^0 = 1$ (provided that $a \neq 0$)**
  Any nonzero quantity raised to the 0 power is 1. For example:

$$4^0 = 1 \quad \text{and} \quad \left(3x\right)^0 = 1 \quad \text{and} \quad 2y^0 = 2 \cdot 1 = 2.$$

- **Negative exponent law**: $a^{-x} = \dfrac{1}{a^x}$ **and** $\dfrac{1}{a^{-x}} = a^x$

  To change from a negative to a positive exponent, invert the base and make the exponent positive. For example:

$$3^{-2} = \frac{1}{3^2} = \frac{1}{9} \quad \text{and} \quad \frac{1}{x^{-3}} = x^3.$$

  If a fraction has a negative exponent, switch the top and bottom numbers of the fraction and make the exponent positive. For example:

$$\left(\frac{3}{5}\right)^{-2} = \left(\frac{5}{3}\right)^2 = \frac{25}{9}.$$

Solving Linear Equations

KEY IDEAS

An **equation** is a statement that two quantities have the same value:

$$1 + 1 = 2 \text{ (true)} \quad \text{and} \quad 1 + 2 = 12 \text{ (false)}.$$

An equation that contains variables cannot be judged true or false until the variables are replaced with specific numbers. For example, the equation $x - 2 = 5$ is neither true nor false. If x is replaced by 3, the equation becomes $3 - 2 = 5$, which is false; if x is replaced by 7, the equation becomes $7 - 2 = 5$, which is true. Much of algebra is concerned with solving different types of equations.

Some Terms Related to Solving Equations

Here are some terms that you should know related to solving equations:

- The **domain** or **replacement set** of a variable is the set of all possible values that the variable can have. Unless otherwise stated, the domain is always the largest possible set of real numbers.
- **Solving an equation** means finding all values in the domain of the variable for which the equation is true. Each of the values is a **solution** or **root** of the equation. The collection of all roots of an equation is the **solution set**.

- **Equivalent equations** are equations that have the same solution set. The equations $x - 1 = 2$, $2x = 6$, and $x = 3$ are equivalent since 3 is the solution of each equation.
- A **linear equation** in variable x is an equation that can be written in the form $ax + b = 0$, where a and b are constants that stand for real numbers with $a \neq 0$. The equation $3x + 7 = 2$ is a linear equation since, if 2 is subtracted from both sides of the equation, the result is $3x + 5 = 0$, which has the form $ax + b = 0$, where $a = 3$ and $b = 5$. The exponent of the variable in a linear equation is 1.

General Strategy for Solving a Linear Equation

An equivalent equation is created whenever the same arithmetic operation is performed on both sides of an equation. If $x - 2 = 5$, then an equivalent equation results when 2 is added to both sides of the equation: $x - 2 + 2 = 5 + 2$. This equation simplifies to another equivalent equation, $x = 7$. To solve a linear equation, transform the original equation into one or more equivalent equations until the variable is isolated on one side of the equal symbol with a coefficient of 1. The solution can then be read from the other side, as in $x = 7$. Unless otherwise indicated, a variable in an equation represents a real number.

Solving a Linear Equation Using Inverse Operations

In the equation $3x + 7 = -5$, 3 and x are connected by multiplication while $3x$ and 7 are connected by addition. You can undo an arithmetic operation that is connected to a variable by performing the inverse operation on both sides of the equation. Addition and subtraction are inverse operations, as are multiplication and division. When solving equations involving two arithmetic operations, isolate the variable by undoing any addition or subtraction *before* undoing any multiplication or division. To solve $3x + 7 = -5$:

1. Subtract 7 from each side of the equation:

$$(3x + 7) - 7 = (-5) - 7$$
$$3x = -12$$

2. Divide each side of the equation by 3:

$$\frac{3x}{3} = \frac{-12}{3}$$
$$x = -4$$

3. Check $x = -4$ in the original equation:

$$3(-4) + 7 = -5$$
$$-12 + 7 = -5$$
$$-5 = -5$$

Solving a Proportion

A **proportion** is an equation that states that two fractions are equal: $\frac{9}{12} = \frac{3}{4}$. The products 12×3 and 9×4 are called *cross-products* and, in a true proportion, must be equal. To solve $\frac{2}{3} = \frac{y+9}{21}$ for y, eliminate the fraction by cross-multiplying:

$$\frac{2}{3} \quad\diagdown\hspace{-0.5em}\diagup\quad \frac{y+9}{21}$$

$$3(y+9) = 2 \times 21$$

1. Remove the parentheses: $\qquad\qquad 3y + 27 = 42$

2. Subtract 27 from each side: $\quad (3y+27) - 27 = (42) - 27$
$$3y \ = 15$$

3. Divide each side by 3: $\qquad\qquad \dfrac{3y}{3} = \dfrac{15}{3}$
$$y = 5$$

Solving an Equation with Like Terms

Terms such as $4x$ and $5x$ are called **like terms** since they differ *only* in their numerical coefficients, 4 and 5. To combine like terms, combine their numerical coefficients and keep the variable:

$$4x + 5x = 9x \quad \text{and} \quad 7x - x = 7x - 1x = 6x.$$

To solve an equation in which the variable appears in more than one term, first combine like terms.

EXAMPLE: If $5x + 3x = -24$, then $8x = -24$, so $x = -\dfrac{24}{8} = -3$.

EXAMPLE: If $3(2y + 5) = 1 + 8y$, remove the parentheses by "distributing" the 3 over $2y$ and 5. In other words, multiply each term inside the parentheses by 3; the result is $6y + 15 = 1 + 8y$. Collect like variable terms on the left side of the equation and the number terms on the right side:

$$6y - 8y = 1 - 15, \text{ so } -2y = -14 \text{ and } y = \dfrac{-14}{-2} = 7.$$

Exercise 1   Solving a Word Problem Involving Like Terms

A geologist collected 13 rocks that were found to have exactly the same weight. Nine of these rocks with an additional 5-ounce weight at one end of a balance scale can balance the remaining rocks and a 23-ounce weight at the opposite end of the scale. What is the number of ounces in the weight of one rock?

Solution: Since the unknown quantity is the weight of one of the rocks, let x = weight of one rock. Translate the conditions of the problem into an equation:

$$\underbrace{\overset{\substack{\text{Weight of 9 rocks +}\\\text{5-ounce weight}}}{9x + 5}}\ =\ \underbrace{\overset{\substack{\text{Weight of } 4 \left(= 13 - 9\right) \text{ remaining}\\\text{rocks + 23-ounce weight}}}{4x + 23}}$$

$$9x - 4x = 23 - 5$$
$$5x = 18$$
$$x = \frac{18}{5} = 3.6$$

Each rock weighs **3.6** ounces.

Exercise 2   Solving a Word Problem Involving Like Terms

Tickets for a concert cost $4.00 for a balcony seat and $7.50 for an orchestra seat. If ticket sales totaled $1585 for 300 tickets sold, how many tickets for balcony seats were sold?

Solution: If x = number of tickets sold for balcony seats, then $300 - x$ = number of tickets sold for orchestra seats. Since ticket sales totaled $1585:

$$4.00x + 7.50\left(300 - x\right) = 1585$$
$$4.00x + 2250 - 7.50x = 1585$$
$$-3.50x = 1585 - 2250$$
$$x = \frac{-665}{-3.50} = 190$$

190 tickets for balcony seats were sold.

Exercise 3   Solving a Consecutive Integer Problem

A set of three consecutive odd integers has the property that twice the sum of the second and third integers is 43 more than 3 times the first integer. What is the smallest of the three integers?

Solution: Consecutive odd integers, such as 1, 3, 5, and 7, differ by 2. If x is the first odd integer in the set, then $x + 2$ and $x + 4$ are the next two consecutive odd integers.

Twice the sum of the second and third integers ⏜ $\overset{\text{is}}{\sim}$ 43 more than 3 times the first integer.

$$2\big[(x+2)+(x+4)\big] = 3x+43$$

$$2(2x+6)=3x+43$$

$$4x+12=3x+43$$

$$4x-3x=43-12$$

$$x=31$$

The smallest odd integer of the set is **31**.

Solving Linear Inequalities

KEY IDEAS

A linear equation becomes a **linear inequality** when the equal symbol is replaced by an inequality symbol. A linear inequality is solved in much the same way that a linear equation is solved.

Writing Equivalent Inequalities

Arithmetic operations are performed with inequalities in much the same way as with equations, except when multiplying or dividing by a negative number.

- **Addition and subtraction property of inequalities**
  An equivalent inequality results when the same quantity is added or subtracted on each side of the inequality. For example, if $4 < 8$, it is also true that

  $$\underset{7}{4+\boxed{3}} < \underset{11}{8+\boxed{3}}.$$

- **Multiplication and division property of inequalities**
  An equivalent inequality results when each side of an inequality is multiplied or divided by the same *positive* quantity. For example, if $4 < 8$, it is also true that

  $$\underset{12}{4\times3} < \underset{24}{8\times3}.$$

An equivalent inequality results when each side of an inequality is multiplied or divided by the same *negative* quantity *and* the direction of the inequality sign is reversed. For example, if $-2x < 6$, then

$$\frac{-2x}{-2} \;\boxed{>}\; \frac{6}{-2}, \text{ so } x > -3.$$

Solving a Linear Inequality

You can solve a linear inequality as you would an equation *except* that, when both sides of an inequality are multiplied or divided by the same *negative* number, you must reverse the sign of the inequality. To find the solution set of $1 - 2x \le x + 13$:

1. Subtract 1 from each side:

$$\left(1 - 2x\right) - 1 \le \left(x + 13\right) - 1$$
$$-2x \le x + 12$$

2. Subtract x from each side:

$$\left(-2x\right) - x \le \left(x + 12\right) - x$$
$$-3x \le 12$$

3. Divide each side by –3 *and* reverse the inequality:

$$\downarrow$$
$$\frac{-3x}{-3} \boxed{\ge} \frac{12}{-3}, \text{ so } x \ge -4$$

Interval Notation

The set of all numbers that belong to a continuous portion of the number line can be indicated using a shorthand notation in which the left and right boundary values of that interval are written as an ordered pair, as shown in Figure 1.2. If an interval does not include a boundary value, then the interval is "open" on that side and an open circle is placed around that endpoint on its graph. If, however, an interval includes a boundary value, then the interval is "closed" on that side and a solid (filled) circle is placed around that endpoint on its graph.

Inequality	Interval	Example	Graph
$a < x < b$	(a,b)	$-2 < x < 3$ *or* $(-2,3)$	
$a \le x < b$	$[a,b)$	$-2 \le x < 3$ *or* $[-2,3)$	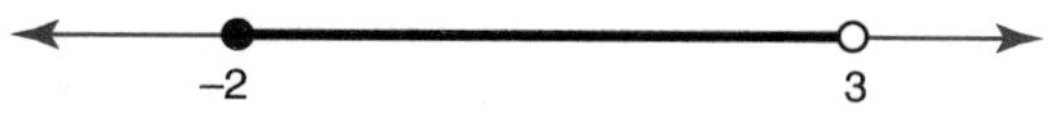
$a < x \le b$	$(a,b]$	$-2 < x \le 3$ *or* $(-2,3]$	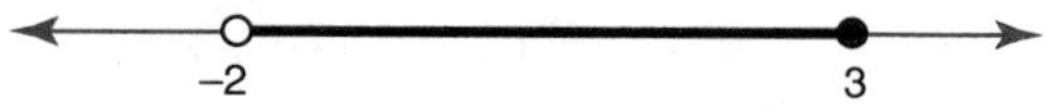
$a \le x \le b$	$[a,b]$	$-2 \le x \le 3$ *or* $[-2,3]$	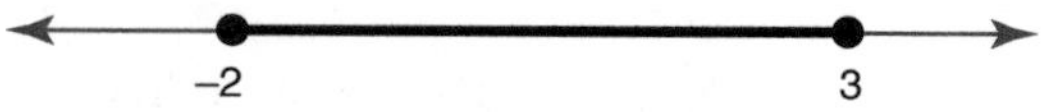

Figure 1.2 Intervals from a to b, where $a < b$

Unbounded Intervals

An interval that extends without an upper or lower limit is "unbounded." An interval may be unbounded on either side or both sides. The symbol for infinity, ∞, is used to indicate that a bound does not exist. The entire set of real numbers, in interval notation, is $(-\infty, +\infty)$.

- (a,∞): the set of all real x such that $x > a$ with no upper bound. In interval notation, $x > 3$ corresponds to $(3,\infty)$.
- $[a,\infty)$: the set of all real x such that $x \geq a$ with no upper bound. In interval notation, $x \geq -2$ corresponds to $[-2,\infty)$.
- $(-\infty,a)$: the set of all real x such that $x < a$ with no lower bound. In interval notation, $x < 4$ corresponds to $(-\infty,4)$.
- $(-\infty,a]$: the set of all real x such that $x \leq a$ with no lower bound. In interval notation, $x \leq -1$ corresponds to $(-\infty,-1]$.

Solving a "Sandwich" Inequality

If the smallest value of x is 2 and the greatest value is 5, as illustrated in Figure 1.3, this fact can be expressed by writing the double or **"sandwich" inequality** $2 \leq x \leq 5$, in which the variable term is in the middle. The inequality $2 \leq x \leq 5$ is read as "2 is less than or equal to x and x is less than or equal to 5."

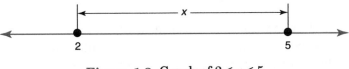

Figure 1.3 Graph of $2 \leq x \leq 5$

To solve an inequality such as $-5 < 3x + 7 \leq 28$, isolate the variable by performing the same operation on the left, middle, and right members of the inequality:

1. Subtract 7 from each member of the inequality:

$$-5 < 3x + 7 \leq 28$$
$$\underline{-7 \qquad -7 \quad -7}$$
$$-12 < \quad 3x \quad \leq 21$$

2. Divide each member by 3:

$$\frac{-12}{3} < \frac{3x}{3} \leq \frac{21}{3}$$

3. Simplify:

$$-4 < \quad x \quad \leq 7$$

The solution set may also be represented, using interval notation, as $(-4, 7]$.

Operations with Polynomials

KEY IDEAS

Algebraic terms such as $7x$, $-3xy$, and $\frac{1}{2}ab^2$ are called *monomials*. A **monomial** is a single number, a variable, or the *product* of one or more numbers and variables with positive integer exponents. A **polynomial** is a monomial or the sum of monomials, as in $3x^2 + 5x - 1$. Since polynomials represent real numbers, arithmetic operations can be performed with polynomials.

Like Monomials

The monomials $3xy^2$ and $4xy^2$ are **like monomials** since they differ only in their numerical coefficients. Like monomials can be added or subtracted by combining their numerical coefficients:

$$7b^2 - 4b^2 = \left(7 - 4\right)b^2 = 3b^2 \quad \text{and} \quad 2xy^2 + 3xy^2 = \left(2 + 3\right)xy^2 = 5xy^2.$$

Polynomials in One Variable

A polynomial in one variable is in **standard form** when its monomial terms are arranged so that the exponents decrease from left to right. The polynomial $2x - 5x^2 + 3$ can be put into standard form by rearranging its terms and writing it as $-5x^2 + 2x + 3$.

The **degree of a polynomial** in one variable is the greatest exponent of the variable. Thus, $-5x^2 + 7x + 3$ is a second-degree polynomial since the greatest exponent is 2. The polynomial $y^3 + 3y^2 - 4y + 9$ is a third-degree polynomial.

Naming Polynomials

Sometimes it is convenient to refer to a polynomial by the number of terms it contains:

Example	Degree	Number of Terms	Name of Polynomial
$4x^3$	3	One	Monomial
$7x - 2y$	1	Two	Binomial
$5n^2 - 3n - 4$	2	Three	Trinomial

Adding Polynomials

To add two polynomials such as $2x^2 - 5x - 1$ and $x^2 - 3x - 7$, write one polynomial underneath the other, being careful to align like terms in the same column. Then combine the like terms in each column:

$$2x^2 - 5x - 1$$
$$+$$
$$\underline{x^2 + 3x - 7}$$
$$3x^2 - 2x - 8 \quad \leftarrow \text{Sum}$$

Subtracting Polynomials

To subtract a polynomial from another polynomial, add the *opposite* of each term of the polynomial that is being subtracted. For example, to subtract $3a - 2b - 9c$ from $5a + 7b - 4c$, first arrange the polynomials either horizontally or vertically. Then:

- To subtract horizontally, write the two polynomials on the same line, the larger one first, enclosing each polynomial in parentheses. Then change the sign between the parentheses to plus, write the second polynomial by taking the opposite of each term inside the parentheses, and combine like terms:

$$
\begin{aligned}
\left(5a + 7b - 4c\right) - \left(3a - 2b - 9c\right) &= \left(5a + 7b - 4c\right) + \left(-3a + 2b + 9c\right) \\
&= \left(5a - 3a\right) + \left(7b + 2b\right) + \left(-4c + 9c\right) \\
&= \quad 2a \quad + \quad 9b \quad + \quad 5c
\end{aligned}
$$

- To subtract vertically, write the polynomial being subtracted on the line underneath the first polynomial, aligning like terms in the same vertical column. Change to an addition example by replacing the sign of each term in the polynomial on the second line with its opposite side. Then add like terms.

Original subtraction example	$\longrightarrow$	Equivalent addition example

$$
\begin{array}{r}
5a + 7b - 4c \\
- \qquad\qquad \\
\underline{3a - 2b - 9c} \\
\end{array}
\qquad\qquad
\begin{array}{r}
5a + 7b - 4c \\
+ \qquad\qquad \\
\underline{-3a + 2b + 9c} \\
2a + 9b + 5c \\
\end{array}
$$

Multiplying Monomials

To multiply two monomials, group like factors. Then multiply the like factors together. For example:

$$\left(6x^3y\right)\left(\frac{1}{3}xy^2\right)=\left(6\cdot\frac{1}{3}\right)\left(x^3\cdot x\right)\left(y\cdot y^2\right)=2x^4y^3.$$

Dividing Monomials

To divide a monomial by another monomial, group the quotients of like factors. Then divide the like factors. For example:

$$\frac{-12a^6b}{3a^4b^2}=\left(\frac{-12}{3}\right)\left(\frac{a^6}{a^4}\right)\left(\frac{b}{b^2}\right)$$
$$=-4a^2b^{-1}$$
$$=\frac{4a^2}{b}$$

Multiplying a Polynomial by a Monomial

To multiply a polynomial by a monomial, multiply each term of the polynomial by the monomial. For example:

$$3a^2\left(a^2-4a+5\right)=3a^2\left(a^2\right)+3a^2\left(-4a\right)+3a^2\left(5\right)$$
$$=3a^4\quad-12a^3\quad+15a^2$$

Dividing a Polynomial by a Monomial

To divide a polynomial by a monomial, divide each term of the polynomial by the monomial. For example:

$$\frac{72x^3-32x^2+8x}{8x}=\frac{72x^3}{8x}+\frac{-32x^2}{8x}+\frac{8x}{8x}$$
$$=9x^{3-1}\quad-\quad4x^{2-1}\quad+\quad1$$
$$=9x^2\quad-\quad4x\quad+\quad1$$

Multiplying a Polynomial by a Polynomial

To multiply a polynomial by another polynomial, write the second polynomial underneath the first one. Then multiply each term of the first polynomial by each term of the polynomial below it, just as two multidigit numbers are multiplied. For example:

$$x^2 - 2x + 8$$
$$x + 3$$

$$x\left(x^2 - 2x + 8\right) = x^3 - 2x^2 + 8x$$
$$3\left(x^2 - 2x + 8\right) = \quad 3x^2 - 6x + 24$$

Add like terms in each column: $x^3 + x^2 + 2x + 24$ ← Final product

Multiplying Binomials Using *"FOIL"*

To multiply two binomials such as $\left(2x + 7\right)$ and $\left(x + 3\right)$, write the binomials side by side and then form the sum of these four products:

- Product of **F**irst terms: $\left(\underline{2x} + 7\right)\left(\underline{x} + 3\right) = \boxed{2x^2} + \cdots$

- Product of **O**uter terms: $\left(\underline{2x} + 7\right)\left(x + \underline{3}\right) = 2x^2 + \boxed{6x} + \cdots$

- Product of **I**nner terms: $\left(2x + \underline{7}\right)\left(\underline{x} + 3\right) = 2x^2 + 6x + \boxed{7x} + \cdots$

- Product of **L**ast terms: $\left(2x + \underline{7}\right)\left(x + \underline{3}\right) = 2x^2 + 6x + 7x + \boxed{21} = 2x^2 + 13x + 21$

The letters of the word *"FOIL"* tell which pairs of terms to multiply.

A Geometric Model of *FOIL*

The process of multiplying two binomials can be represented geometrically. For example, to find $(2x + 7)(x + 3)$, represent the area of the rectangle whose length is $2x + 7$ and whose width is $x + 3$, as shown in Figure 1.4.

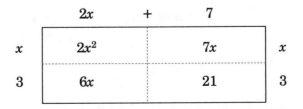

Figure 1.4 Finding $(2x + 7)(x + 3)$ geometrically

The area of the largest rectangle must be equal to the sum of the areas of the four smaller rectangles:

$$\overbrace{\text{Area of largest rectangle}}^{\left(2x+7\right)\left(x+3\right)} \quad = \quad \overbrace{\text{Sum of areas of four small rectangles}}^{2x^2 + 6x + 7x + 21}$$

EXAMPLE:
$$\left(x-6\right)\left(x+2\right) = \overbrace{x\cdot x}^{F} + \overbrace{\left(2\right)\left(x\right)}^{O} + \overbrace{\left(-6\right)\left(x\right)}^{I} + \overbrace{\left(-6\right)\left(+2\right)}^{L}$$
$$= x^2 + [\left(2x\right) + \left(-6x\right)] - 12$$
$$= x^2 - 4x - 12$$

EXAMPLE:
$$\left(3x-4\right)\left(3x+4\right) = \overbrace{3x\cdot 3x}^{F} + \overbrace{\left(3x\right)\left(4\right)}^{O} + \overbrace{\left(-4\right)\left(3x\right)}^{I} + \overbrace{\left(-4\right)\left(+4\right)}^{L}$$
$$= 9x^2 + \left[\left(12x\right) + \left(-12x\right)\right] - 16$$
$$= 9x^2 + 0 - 16$$
$$= 9x^2 - 16$$

Factoring Polynomials

Lesson 1-5

KEY IDEAS

Sometimes you know the result of multiplying two quantities together, but you would like to know also which two quantities were multiplied together to give that result. The process of reversing multiplication is called **factoring**. To *factor an expression* means to rewrite it as a product.

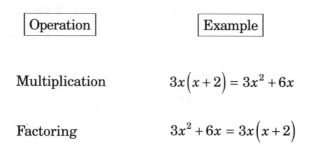

Operation	Example
Multiplication	$3x\left(x+2\right) = 3x^2 + 6x$
Factoring	$3x^2 + 6x = 3x\left(x+2\right)$

Finding the *Greatest* Common *Factor* (GCF)

The **greatest common factor (GCF)** of a polynomial is the greatest monomial that divides evenly into *each* term of the polynomial. For example, the GCF of $21a^5 + 14a^3$ is $7a^3$ since 7 is the greatest integer that divides evenly into 21 *and* 14, and a^3 is the greatest power of a that is contained in a^5 *and* a^3.

Factoring a Polynomial by Removing the GCF

If you write $30 = 5 \times 6$, you have *factored* 30 as the product of 5 and 6. **Factoring** a polynomial means writing the polynomial as the product of two or more other polynomials, each of which is called a **factor** of the original polynomial. If you know the GCF of a polynomial, you can find the other factor by dividing the polynomial by the GCF. For example, since 5 is a factor of 30, you can obtain the corresponding factor by dividing 30 by 5; the result is 6. To factor a polynomial such as $21a^5 + 14a^3$:

1. Determine the GCF. The GCF of $21a^5 + 14a^3$ is $7a^3$.
2. Find the corresponding factor by dividing the polynomial by the GCF:

$$\frac{21a^5}{7a^3} + \frac{14a^3}{7a^3} = 3a^2 + 2.$$

3. Write $21a^5 + 14a^3$ in factored form: $21a^5 + 14a^3 = 7a^3(3a^2 + 2)$.
4. Check by multiplying the two factors together to make sure the result is the original polynomial:

$$7a^3\left(3a^2 + 2\right) = 7a^3 \cdot 3a^2 + 7a^3 \cdot 2 = 21a^5 + 14a^3.$$

It's not hard to think of a polynomial that cannot be factored. A polynomial, such as $3x^2 + 5$, that cannot be factored except by writing it as the product of itself and 1 (or as the product of its opposite and –1) is called a **prime polynomial**.

Exercise 1 Factoring by Removing the GCF

Factor $9x^4 - 3x^3 + 12x$.

Solution: The GCF of $9x^4 - 3x^3 + 12x$ is $3x$, and the corresponding factor is

$$\frac{9x^4}{3x} - \frac{3x^3}{3x} + \frac{12x}{3x} = 3x^3 - x^2 + 4.$$

Hence, $9x^4 - 3x^3 + 12x = \mathbf{3x(3x^3 - x^2 + 4)}$.

Exercise 2   Factoring by Removing the GCF

Factor $6a^2b - 21ab$.

Solution: The GCF of $6a^2b - 21ab$ is $3ab$, and the corresponding factor is

$$\frac{6a^2b}{3ab} - \frac{21ab}{3ab} = 2a - 7.$$

Hence, $6a^2b - 21ab = \mathbf{3ab(2a - 7)}$.

Exercise 3   Using Factoring to Solve an Equation

If $xz = y - x$, solve for x in terms of y and z.

Solution:

- The given equation is: $\qquad\qquad xz = y - x$
- Add x to each side: $\qquad\qquad xz + x = y$
- Factor out x: $\qquad\qquad\quad\; x(z + 1) = y$
- Divide each side by $z + 1$: $\qquad x = \dfrac{y}{z+1}$

Factoring by Grouping

Some polynomials that have four terms, such as $x^3 + 3x^2 + 4x + 12$, can be factored in stages:

1. Group pairs of terms together: $\qquad x^3 + 3x^2 + 4x + 12 = \left(x^3 + 3x^2\right) + \left(4x + 12\right)$

2. Factor out the greatest common
   monomial factor from each pair of terms: $\qquad = x^2\left(x + 3\right) + 4\left(x + 3\right)$

3. Factor out the common *binomial* factor: $\qquad = \left(x + 3\right)\left(x^2 + 4\right)$

Exercise 4   Factoring by Grouping

Factor $8mx + 4px - 6m - 3p$.

Solution: Since the polynomial has four terms, first group pairs of terms together:

$$8mx + 4px - 6m - 3p = \left(8mx + 4px\right) + \left(-6m - 3p\right)$$
$$= 4x\left(2m + p\right) - 3\left(2m + p\right)$$
$$= \mathbf{\left(2m + p\right)\left(4x - 3\right)}$$

Factoring Quadratic Trinomials

KEY IDEAS

To factor a quadratic trinomial into the product of two binomials, use the reverse of FOIL.

Factoring $x^2 + bx + c$

Since $(x + 2)(x + 5) = x^2 + 7x + 10$, you know that $x^2 + 7x + 10 = (x + 2)(x + 5)$. Notice that the binomial factors contain 2 and 5 since these are the only two integers that, when multiplied together, give 10, the last term in $x^2 + 7x + 10$, *and*, when added together, equal 7, the coefficient of the x-term in $x^2 + 7x + 10$. If you start with $x^2 + 7x + 10$ and want to know its factors, work in reverse:

1. Write the general form of the binomial factors:

$$x^2 + 7x + 10 = \left(x + \boxed{?}\right)\left(x + \boxed{?}\right).$$

2. Fill in the missing terms in the binomial factors with the two numbers whose product is the last number term of the quadratic trinomial $\left(x^2 + 7x + \boxed{10}\right)$ and whose sum is the numerical coefficient of the middle x-term $\left(x^2 + \boxed{7}x + 10\right)$. Since $2 \times 5 = 10$ and $2 + 5 = 7$:

$$x^2 + 7x + 10 = \left(x + 2\right)\left(x + 5\right).$$

3. Check your work by multiplying the two binomial factors together and then comparing the product to the original quadratic trinomial.

Exercise 1   Factoring a Quadratic Trinomial

Factor $y^2 - 7y + 12$ as the product of two binomials.

Solution: Find the two numbers whose product is $+12$ *and* whose sum is 7. Because the product of the two numbers you are seeking is positive ($+12$) and their sum is negative (-7), both numbers must be negative. Since

$$\left(-3\right) \times \left(-4\right) = +12 \text{ and } \left(-3\right) + \left(-4\right) = -7,$$

the correct factors of $+12$ are -3 and -4. Hence:

$$y^2 - 7y + 12 = \left(\boldsymbol{y - 3}\right)\left(\boldsymbol{y - 4}\right).$$

Exercise 2   **Factoring a Quadratic Trinomial**

Factor $n^2 - 5n - 14$ as the product of two binomials.

Solution: Find the two numbers whose product is –14 *and* whose sum is –5. The two factors of –14 must have opposite signs. Since

$$(+2) \times (-7) = -14 \text{ and } (+2) + (-7) = -5,$$

the correct factors of –14 are +2 and –7.

Hence:

$$n^2 - 5n - 14 = (n + 2)(n - 7).$$

Factoring $ax^2 + bx + c$ $(a > 1)$

As you can imagine, factoring a quadratic trinomial becomes more difficult when the numerical coefficient of the x^2-term is different from 1 since there are more possibilities to consider. To factor $3x^2 + 10x + 8$:

1.   Factor the x^2-term, and set up the binomial factors:

$$3x^2 + 10x + 8 = (3x + \text{?})(x + \text{?}).$$

2.   Identify possibilities for the missing numerical terms in the binomial factors. The missing terms are the two integers whose product is +8, the last term of $3x^2 + 10x + 8$, and that make the sum of the outer and inner products of the terms of the binomial factors equal to +10x. The two factors of +8 must have the same sign. Because the coefficient of the x-term in $3x^2 + 10x + 8$ is positive, the two factors of 8 are both positive. It follows that the possible integer factors of 8 must be either 1 and 8 or 2 and 4.

3.   Use trial and elimination to find the correct factors of 8 and their proper placements in the binomial factors:

Outer product = +6x
$$3x^2 + 10x + 8 = (3x + 4)(x + 2)$$
Sum = $\boxed{+10x}$
Inner product = +4x

4. Check that the factors work. Keep in mind that the placement of the factors of 8 matters. Although the binomial factors $(3x + 2)(x + 4)$ contain the correct factors of 8, the factors are not placed correctly since the sum of the outer and inner products is $12x + 2x = 14x$ rather than $10x$.

Factoring a Quadratic Trinomial by Linear Decomposition

If you have difficulty factoring a quadratic trinomial whose leading coefficient is greater than 1, you can factor by decomposing the linear (middle) term. This method eliminates some of the guess-and-check work. To factor $3x^2 + 10x + 8$ using this method, break $10x$ down into the sum of two x-terms. Although there are many possibilities, such as $x + 9x$, $2x + 8x$, $3x + 7x$, and $4x + 6x$, choose the pair of x-terms whose numerical coefficients have the same product as the product of the leading coefficient and the constant term of the quadratic trinomial:

$$\overbrace{3 \times 8 = 24}$$
$$\boxed{3}\,x^2 + 10x + \boxed{8}$$

Because $\underline{6} \times \underline{4} = 24$, $10x$ can be decomposed into $6x + 4x$.
To factor $3x^2 + 10x + 8$:

1. Replace $10x$ with $6x + 4x$: $3x^2 + 10x + 8 = 3x^2 + \overbrace{6x + 4x}^{10x} + 8$

2. Group the first and last pairs of terms: $= \left(3x^2 + 6x\right) + \left(4x + 8\right)$

3. Factor out the GCF of each pair of terms: $= 3x\underbrace{\left(x + 2\right)} + 4\underbrace{\left(x + 2\right)}$

4. Factor out the common binomial factors: $= \left(3x + 4\right)\left(x + 2\right)$

Lesson 1-7

Special Products and Factoring Patterns

KEY IDEAS

Because $(A + B)(A - B) = A^2 - B^2$, any polynomial that has the form $A^2 - B^2$ can be factored as the product of the sum and difference of the terms that are being squared. The binomial $4y^2 - 9$ has this form, where

$$4y^2 = \underbrace{\left(2y\right)^2}_{A} \quad \text{and} \quad 9 = \underbrace{\left(3\right)^2}_{B}.$$

Hence, $4y^2 - 9$ can be factored as $(2y + 3)(2y - 3)$.
There are also special factoring patterns involving $A^3 + B^3$ and $A^3 - B^3$.

Multiplying Conjugate Binomials and Factoring Their Products

Can you think of two binomials whose product is another binomial? Here is a simple example:

$$(x+4)(x-4) = x^2 + (-4x+4x) - 16 = x^2 - 16a.$$

As a further illustration, study this example:

$$
\begin{aligned}
(2y+3)(2y-3) &= \overbrace{2y \cdot 2y}^{\text{F}} + \overbrace{2y \cdot (-3)}^{\text{O}} + \overbrace{(3)(2y)}^{\text{I}} + \overbrace{(3)(-3)}^{\text{L}} \\
&= 4y^2 \;+\; [-6y \;+\; 6y] \;-\; 9 \\
&= 4y^2 - 9
\end{aligned}
$$

Notice that, when the sum and the difference of the same two terms are multiplied together, as in $(2y + 3)(2y - 3)$, the products of the outer and inner terms cancel each other; that is, their sum is 0. The result is a *binomial* that can be formed by simply taking the difference of the *squares* of the two terms that are being added and subtracted:

$$(2y+3)(2y-3) = (2y)^2 - (3)^2 = 4y^2 - 9.$$

In general:

- When two binomials have the form $A + B$ and $A - B$, the binomials are called **conjugate binomials**. Conjugate binomials can be multiplied together quickly by following a simple rule.

Rule for Multiplying Conjugate Binomials

$$(A+B)(A-B) = A^2 - B^2.$$

EXAMPLE: $(3y-4)(3y+4) = 9y^2 - 16$.

- Whenever you recognize that a binomial has the form $A^2 - B^2$, you can reverse the multiplication by factoring the product as $(A + B)(A - B)$. For instance:

$$9y^2 - 64 = \underbrace{(3y)^2}_{A} - \underbrace{(8)^2}_{B} = \underbrace{(3y+8)}_{A+B}\underbrace{(3y-8)}_{A-B}.$$

EXAMPLE: $n^2 - \dfrac{49}{100} = \left(n\right)^2 - \left(\dfrac{7}{10}\right)^2 = \left(n + \dfrac{7}{10}\right)\left(n - \dfrac{7}{10}\right).$

EXAMPLE: $4a^2 - 25b^2 = \left(2a\right)^2 - \left(5b\right)^2 = \left(2a + 5b\right)\left(2a - 5b\right).$

EXAMPLE: $0.16y^4 - 0.09 = \left(0.4y^2\right)^2 - \left(0.3\right)^2 = \left(0.4y^2 + 0.3\right)\left(0.4y^2 - 0.3\right).$

Factoring the Sum and the Difference of Cubes

With a little effort you can verify that

$$\left(A + B\right)\left(A^2 - AB + B^2\right) = A^3 + B^3,$$

$$\left(A - B\right)\left(A^2 + AB + B^2\right) = A^3 - B^3.$$

Reading each of these products from right to left gives you a rule for factoring the sum or the difference of two cubes, as illustrated further in the accompanying table.

Factoring the Sum and the Difference of Two Cubes

Factoring Formula	Example
FACTORING THE SUM OF TWO CUBES $$A^3 + B^3 = \left(A + B\right)\left(A^2 - AB + B^2\right)$$	$x^3 + 27 = \left(x\right)^3 + \left(3\right)^3$ Think of x as A and 3 as B: $$x^3 + 27 = \left(x + 3\right)\left(x^2 - x \cdot 3 + 3^2\right)$$ $$= \left(x + 3\right)\left(x^2 - 3x + 9\right)$$
FACTORING THE DIFFERENCE OF TWO CUBES $$A^3 - B^3 = \left(A - B\right)\left(A^2 + AB + B^2\right)$$	$8x^3 - y^3 = \left(2x\right)^3 - \left(y\right)^3$ Think of $2x$ as A and y as B: $$8x^3 - y^3 = \left(2x - y\right)\left(\left(2x\right)^2 + 2xy + y^2\right)$$ $$= \left(2x - y\right)\left(4x^2 + 2xy + y^2\right)$$

TIP

The signs of the terms of the factors of the sum or the difference of two cubes depend on the sign between A^3 and B^3:

Factoring Completely

A polynomial is **factored completely** when each of its factors cannot be factored further. When factoring a polynomial completely, you may need to use more than one factoring method.

EXAMPLE: $2x^3 - 50x = 2x\left(x^2 - 25\right) = 2x\left(x+5\right)\left(x-5\right)$

EXAMPLE: $3t^3 + 18t^2 - 48t = 3t\left(t^2 + 6t - 16\right) = 3t\left(t+8\right)\left(t-2\right)$

EXAMPLE: $a^2b + 2a^2 - b - 2 = \left(a^2b + 2a^2\right) + \left(-b - 2\right)$

Factor out the GCF from each pair of terms: $= a^2\left(b+2\right) - 1\left(b+2\right)$

Factor out $\left(b+2\right)$: $= \left(b+2\right)\left(a^2 - 1\right)$

Factor $a^2 - 1$: $= \left(b+2\right)\left(a+1\right)\left(a-1\right)$

<table>
<tr><td>**Chapter**
1</td><td colspan="2">## CHECKUP EXERCISES</td></tr>
</table>

1—18. Solve for the variable.

1. $x - 7 = 3x - 13$

7. $9w = 2w - 3(8 - w)$

13. $8 - 3(3 - x) \leq 1$

2. $2n - 5n + 1 = -26$

8. $0.54 - 0.07y = 0.2y$

14. $0.1x - 0.02x \leq 0.24$

3. $4(2x - 1) = 5x + 17$

9. $1 - 2x \leq x + 13$

15. $\dfrac{10 - x}{5} = \dfrac{7 - x}{2}$

4. $2y + 1 > 5y - 8$

10. $b^x \cdot b^{2x-1} = b^{11}$

16. $-3 < \dfrac{2x - 1}{4} \leq 5$

5. $3p + 2(8 - p) = 3$

11. $6\left(\dfrac{x}{2} + 1\right) = -9$

17. $35 - z < z + 1.5z$

6. $5(6 - q) = -3(q + 2)$

12. $11 > \dfrac{2x}{3} - 5$

18. $\dfrac{3}{x + 2} = \dfrac{4}{7 - 2x}$

19—26. Perform the indicated operation, and write the answer in simplest form.

19. $(x - 3y)^2$

23. $\left(4x^3 - 9x^2 + 5x + 1\right) - \left(-x^3 + 7x^2 - 2x - 1\right)$

20. $(2x - 1)(x + 7)$

24. $\dfrac{30y^6 + 5y^3 - 10y^2}{-5y^2}$

21. $(4y + 3)(4y - 3)$

25. $\dfrac{h^3k^2 + h^2k^3 - 7hk}{hk}$

22. $(9x^2 - x + 5)(4x - 3)$

26. $\dfrac{0.14a^3 - 1.05a^2b}{0.7a}$

27—41. Factor completely.

27. $y^2 - .09$

32. $-y^2 + 10y + 11$

37. $3x^2 + 5x - 2$

28. $x^4 - 16$

33. $r^2t - s^4t$

38. $2y^2 - 13y - 45$

29. $x^2 - 4x - 21$

34. $w^3 - 8v^3$

39. $a^4 - a^2b^2 - 9a^2 + 9b^2$

30. $n^3 - n^2 - 12n$

35. $64h^3 + k^3$

40. $4x^2 - 4x - 15$

31. $8m^5n^2 - 12m^3n^4$

36. $x^2y^2 - x^2 - 4y^2 + 4$

41. $x^5 + 2x^3 - x^2 - 2$

42—45. Solve for the indicated variable.

42. Solve for w: $P = 2\ell + 2w$.

44. Solve for x: $a(x + b) = c$.

43. Solve for t: $xt = x - st$.

45. Solve for F: $C = \dfrac{5}{9}(F - 32)$.

46 and 47. Simplify.

46. $\dfrac{\left(b^{2n+1}\right)^3}{b^n \cdot b^{4n+3}}$

47. $\dfrac{\left(3a^{3n} \cdot b^n\right)^2}{\left(a^4 \cdot b^3\right)^n}$

48. If $y = \dfrac{1}{4}x^2 - \dfrac{2\left(x^2 - x^0\right)}{x + 1} + 3x^{-1}$, what is the value of y when $x = 6$?

49. Carlos has \$365 in savings and saves \$20 each week while not spending any of his savings. His brother has \$590 in savings and spends \$25 of his savings each week while not saving any additional money. After how many weeks will Carlos and his brother have the same amount in savings?

50. At a movie theater, a cashier sold 250 more adults' tickets than children's tickets. The adults' tickets were \$6.50 each, and the children's tickets were \$3.50 each. What is the *least* number of children's tickets that the cashier had to sell for the total cash receipts to be at least \$2750?

51. The denominator of a fraction is 4 less than twice the numerator. If 3 is added to both the numerator and the denominator, the new fraction has a value of $\frac{2}{3}$. What was the original fraction?

52. A postal clerk sold 50 postage stamps for $15.84. Some were 37-cent stamps, and the rest were 23-cent stamps. Find the number of 37-cent stamps that were sold.

53. The lengths of the sides of a triangle are consecutive even integers. The perimeter of the triangle is equal to the perimeter of a square whose side measures 5 less than the shortest side of the triangle. Find the length of the longest side of the triangle.

54. Three numbers are in the ratio of 2:3:5. If the smallest number is multiplied by 8, the result is 32 more than the sum of the second and third numbers. What is the smallest of the three numbers?

55. Tickets to a concert that were purchased in advance cost $4.50 each, and tickets purchased at the box office on the day of the concert cost $8.00 each. The total amount of money collected in ticket sales was the same as if every ticket purchased had cost $6.00. If 180 tickets were purchased in advance, what was the total number of tickets purchased at the box office?

56. George is twice as old as Edward, and Edward's age exceeds Robert's age by 4 years. If the sum of the three ages is at least 56 years, what is Robert's minimum age?

57. A portion of a wire 70 inches in length is bent to form a rectangle having the greatest possible area such that the length of the rectangle exceeds three times its width by 2 inches, and the dimensions of the rectangle are whole numbers. Find the length of the wire that is *not* used to form the rectangle.

Chapter 2

RATIONAL AND IRRATIONAL EXPRESSIONS

OVERVIEW

*R*ational expression is another name for an algebraic fraction that has a polynomial numerator and a polynomial denominator. We always assume that, whenever a variable appears in the denominator of a fraction, the variable cannot have a value that makes the denominator evaluate to 0. For example, x in the rational expression $\dfrac{x^2-9}{x+1}$ cannot equal –1 since this value of x makes the denominator of the fraction, $x + 1$, evaluate to 0. The rules for working with rational expressions are the same as those for handling fractions in arithmetic.

Lessons in Chapter 2
Lesson 2-1: Operations with Rational Expressions
Lesson 2-2: Simplifying Complex Fractions
Lesson 2-3: Radicals and Fractional Exponents
Lesson 2-4: Operations with Radicals

Operations with Rational Expressions

KEY IDEAS

Fractions in arithmetic can be written in lowest terms, multiplied, divided, added, and subtracted. These operations can be performed with rational expressions in much the same way.

Writing a Rational Expression in Lowest Terms

A fraction is in **lowest terms** when its numerator and denominator have no factors in common other than 1 or –1. To write a rational expression in lowest terms, first divide out any factors that are common to both the numerator and the denominator since the quotient of such factors is 1, as in

$$\frac{21}{28} = \frac{3 \times \cancel{7}}{4 \times \cancel{7}} = \frac{3}{4}.$$

To write $\dfrac{4x+12}{x^2+3x}$ in lowest terms:

1. Factor the numerator and the denominator:
$$\frac{4x+12}{x^2+2x} = \frac{4(x+3)}{x(x+3)}$$

2. Divide out (cancel) the common factor:
$$= \frac{4\,\cancel{(x+3)}^{\,1}}{x\,\cancel{(x+3)}}$$

3. Multiply the remaining factors in the numerator together, and multiply the remaining factors in the denominator together:
$$= \frac{4}{x}$$

TIPS

- Only *factors* of a *product* common to both the numerator and the denominator of a fraction can be divided out. The example at the left below illustrates a common mistake. This cancellation is *wrong* because the numerator $3a$ is being *added* to b^2, so $3a$ is not a factor of the numerator, as it is in the example at the right:

$$Wrong: \quad \frac{\overset{1}{\cancel{3a}} + b^2}{\cancel{3a}} = 1 + b^2 \qquad Correct: \quad \frac{\overset{1}{\cancel{3a}} \cdot b^2}{\cancel{3a}} = b^2$$

- The order in which terms are subtracted can be switched if a negative sign is placed in front of the revised difference, as is done in the numerator of the following fraction:

$$\frac{b-a}{a^2-b^2} = \frac{-(a-b)}{a^2-b^2} = \frac{-\overset{1}{\cancel{(a-b)}}}{\cancel{(a-b)}(a+b)} = \frac{-1}{(a+b)}.$$

Exercise 1   Writing an Algebraic Fraction in Lowest Terms

Write $\dfrac{3-3y^2}{y^2+4y-5}$ in lowest terms.

Solution:

- Factor both the numerator and the denominator completely:

$$\frac{3-3y^2}{y^2+4y-5} = \frac{3(1-y^2)}{(y-1)(y+5)} = \frac{3(1-y)(1+y)}{(y-1)(y+5)}$$

- Rewrite $1 - y$ as $-(y - 1)$:

$$= \frac{-3(y-1)(1+y)}{(y-1)(y+5)}$$

- Divide out any pairs of common factors:

$$= \frac{-3\,\overset{1}{\cancel{(y-1)}}(1+y)}{\cancel{(y-1)}(y+5)}$$

- Multiply the remaining factors together:

$$= \frac{-3(1+y)}{y+5}$$

Exercise 2   Writing an Algebraic Fraction in Lowest Terms

Write $\dfrac{18a^4 - 30a^3}{3a^2}$ in lowest terms.

Solution: Factor the numerator, and then divide out any factors common to both the numerator and the denominator.

$$\frac{18a^4 - 30a^3}{3a^2} = \frac{6a^3\left(3a-5\right)}{3a^2} = \frac{\overset{2}{\cancel{6}}\,a^3}{\cancel{3}\,a^2} \cdot \left(3a-5\right)$$

$$= 2a^{3-2} \cdot \left(3a-5\right)$$

$$= \mathbf{2a\left(3a-5\right)} \;\; or \;\; \mathbf{6a^2-10a}$$

If the original fraction looks to you like a division example, you are right. You can get the same answer by dividing each term in the numerator by the denominator:

$$\frac{18a^4 - 30a^3}{3a^2} = \frac{18a^4}{3a^2} - \frac{30a^3}{3a^2} = 6a^2 - 10a.$$

Multiplying Rational Expressions

To prevent the algebra from getting too ugly, factor and then divide out any matching pairs of factors in the numerators and in the denominators *before* multiplying two rational expressions together.

To find the product $\dfrac{12y^2}{x^2+7x} \cdot \dfrac{x^2-49}{2y^5}$ in lowest terms:

1. Factor where possible:
$$\frac{12y^2}{x^2+7x} \cdot \frac{x^2-49}{2y^5} = \frac{12y^2}{x\left(x+7\right)} \cdot \frac{\left(x+7\right)\left(x-7\right)}{2y^5}$$

2. Divide out pairs of common factors:
$$= \frac{\overset{6}{\cancel{12}}\,\cancel{y^2}}{x\left(\cancel{x+7}\right)} \cdot \frac{\overset{1}{\cancel{(x+7)}}\left(x-7\right)}{\underset{y^3}{\cancel{2}\,\cancel{y^5}}}$$

3. Multiply the remaining factors together:
$$= \frac{6\left(x-7\right)}{xy^3}$$

Dividing Rational Expressions

If you know how to multiply rational expressions, then you also know how to divide these expressions: invert the second fraction and then multiply, just as in arithmetic.

To find the quotient $\dfrac{8m^2}{3} \div \dfrac{6m^3}{3m-12}$:

1. Change to a multiplication example:
$$\frac{8m^2}{3} \div \frac{6m^3}{3m-12} = \frac{8m^2}{3} \times \frac{3m-12}{6m^3}$$

2. Factor:
$$= \frac{8m^2}{3} \times \frac{3(m-4)}{6m^3}$$

3. Divide out pairs of common factors:
$$= \frac{\overset{4}{\cancel{8}}\,\cancel{m^2}}{\cancel{3}} \times \frac{\overset{1}{\cancel{3}}(m-4)}{\underset{3m}{\cancel{6}\,\cancel{m^3}}}$$

4. Multiply the remaining factors together:
$$= \frac{4(m-4)}{3m}$$

Exercise 3   **Dividing Rational Expressions**

Write the quotient in lowest terms: $\dfrac{x^2-2x-8}{x^2-25} \div \dfrac{x^2-4}{2x+10}$.

Solution: Invert the second fraction (the divisor) and multiply.

- Change to a multiplication example:
$$\frac{x^2-2x-8}{x^2-25} \div \frac{x^2-4}{2x+10} = \frac{x^2-2x-8}{x^2-25} \cdot \frac{2x+10}{x^2-4}$$

- Factor where possible:
$$= \frac{(x-4)(x+2)}{(x-5)(x+5)} \cdot \frac{2(x+5)}{(x+2)(x-2)}$$

- Divide out pairs of common factors:
$$= \frac{(x-4)\cancel{(x+2)}}{(x-5)\cancel{(x+5)}} \cdot \frac{2\cancel{(x+5)}}{\cancel{(x+2)}(x-2)}$$

- Simplify:
$$= \frac{2(x-4)}{(x-5)(x-2)}$$

Combining Rational Expressions with Like Denominators

To add or subtract rational expressions with the same denominators, write the sum or difference of the numerators over the common denominator.

To find the difference $\dfrac{5a+b}{10ab} - \dfrac{3a-b}{10ab}$:

- Write the difference of the numerators over the common denominator:
$$\frac{5a+b}{10ab} - \frac{3a-b}{10ab} = \frac{5a+b-\left(3a-b\right)}{10ab}$$

- Combine like terms in the numerator:
$$= \frac{5a+b-3a+b}{10ab}$$

$$= \frac{2a+2b}{10ab}$$

- Write the fraction in lowest terms:
$$= \frac{\overset{1}{\cancel{2}}\left(a+b\right)}{\underset{5}{\cancel{10}}\,ab} = \frac{a+b}{5ab}$$

Combining Rational Expressions with Unlike Denominators

Fractions are easy to combine once they have the same denominator. When adding or subtracting rational expressions with different denominators, first change the expressions into equivalent fractions with the same denominator, usually the lowest common multiple of the original denominators. In arithmetic this is called the **lowest common denominator (LCD)**.

To write $\dfrac{3}{10x} + \dfrac{4}{15x}$ as a single fraction:

1. Determine the LCD. Since $30x$ is the smallest expression into which the denominators $10x$ and $15x$ both divide evenly, the LCD is $30x$.

2. Change each fraction into an equivalent fraction that has $30x$ as its denominator. Multiply the first fraction by 1 in the form of $\dfrac{3}{3}$, and multiply the second fraction by 1 in the form of $\dfrac{2}{2}$. Thus:

$$\frac{3}{10x} + \frac{4}{15x} = \frac{3}{3}\cdot\left(\frac{3}{10x}\right) + \frac{2}{2}\cdot\left(\frac{4}{15x}\right)$$

$$= \frac{9}{30x} + \frac{8}{30x}$$

3. Add the like fractions:
$$= \frac{9+8}{30x} = \frac{17}{30x}$$

Exercise 4   ## Combining Rational Expressions with Unlike Monomial Denominations

Write $\dfrac{2x+1}{6y} + \dfrac{3x-5}{9y} + \dfrac{1}{18y}$ in simplest form.

Solution: The LCD of $6y$, $9y$, and $18y$ is $18y$ since this is the smallest expression into which $6y$, $9y$, and $18y$ divide evenly. Change the first two fractions into equivalent fractions that have $18y$ as their denominators by multiplying the first fraction by 1 in the form of $\dfrac{3}{3}$ and multiplying the second fraction by 1 in the form of $\dfrac{2}{2}$. Thus:

$$\frac{2x+1}{6y} + \frac{3x-5}{9y} = \frac{3}{3}\cdot\left(\frac{2x+1}{6y}\right) + \frac{2}{2}\left(\frac{3x-5}{9y}\right) + \frac{1}{18y}$$

$$= \frac{3(2x+1)+2(3x-5)+1}{18y}$$

$$= \frac{12x-6}{18y}$$

$$= \frac{\cancel{6}(2x-1)}{\underset{3}{\cancel{18}y}}$$

$$= \boldsymbol{\frac{2x-1}{3y}}$$

When combining rational expressions with unlike polynomial denominators, you may need to begin by factoring the denominators to help determine the LCD.

| Exercise 5 | Combining Rational Expressions with Unlike Polynomial Denominators |

When $\dfrac{4}{x^2-2x-3}$ is subtracted from $\dfrac{x+3}{x^2-1}$, what is the difference, expressed as a single fraction in lowest terms?

Solution:

- Write the difference with the denominators in factored form:

$$\frac{x+3}{x^2-1}-\frac{4}{x^2-2x-3}=\frac{x+3}{\left(x+1\right)\left(x-1\right)}-\frac{4}{\left(x-3\right)\left(x+1\right)}$$

- Determine the LCD. The LCD is $(x-3)\,(x+1)\,(x-1)$ because this product is the smallest expression into which each of the denominators divides evenly. Change each fraction into an equivalent fraction with the LCD as its denominator by multiplying the first fraction by 1 in the form of $\dfrac{x-3}{x-3}$ and multiplying the second fraction by 1 in the form of $\dfrac{x-1}{x-1}$:

$$\frac{x+3}{x^2-1}-\frac{4}{x^2-2x-3}=\frac{x+3}{\left(x+1\right)\left(x-1\right)}\cdot\frac{\left(x-3\right)}{\left(x-3\right)}-\frac{4}{\left(x-3\right)\left(x+1\right)}\cdot\frac{\left(x-1\right)}{\left(x-1\right)}$$

- Combine the fractions:

$$=\frac{\left(x+3\right)\left(x-3\right)-\,4\left(x-1\right)}{\left(x+1\right)\left(x-1\right)\left(x-3\right)}$$

$$=\frac{x^2-4x-5}{\left(x+3\right)\left(x+1\right)\left(x-1\right)}$$

- Write the fraction in lowest terms: $=\dfrac{\left(x-5\right)\overset{1}{\cancel{\left(x+1\right)}}}{\left(x+3\right)\cancel{\left(x+1\right)}\left(x-1\right)}=\dfrac{x-5}{\left(x+3\right)\left(x-1\right)}$

Simplifying Complex Fractions

KEY IDEAS

A **complex fraction** is a fraction in which the numerator, denominator, or both contain other fractions. *Simplifying a complex fraction* means eliminating the fractions from the numerator and the denominator of the "big" fraction.

Simplifying a Complex Fraction Using Division

To simplify the complex fraction $\dfrac{1+\dfrac{2}{x}}{1-\dfrac{4}{x^2}}$:

1. Combine terms in the numerator and in the denominator:

$$\frac{1+\dfrac{2}{x}}{1-\dfrac{4}{x^2}} = \frac{\dfrac{x}{x}+\dfrac{2}{x}}{\dfrac{x^2}{x^2}-\dfrac{4}{x^2}} = \frac{\dfrac{x+2}{x}}{\dfrac{x^2-4}{x^2}}.$$

2. Divide the numerator by the denominator:

$$\overbrace{\left(\frac{x+2}{x}\right)}^{\text{Numerator}} \div \underbrace{\left(\frac{x^2-4}{x^2}\right)}_{\text{Denominator}} = \left(\frac{x+2}{x}\right) \cdot \left(\frac{x^2}{x^2-4}\right)$$

$$= \frac{(x+2)\,x^{\cancel{2}}}{\cancel{x}\,(x-2)(x+2)}$$

$$= \frac{x}{x-2}$$

Simplifying a Complex Fraction by Clearing Fractions

You may find it easier to eliminate the fractions contained in a complex fraction by multiplying the numerator and the denominator by the lowest common multiple (LCM) of all of its denominators. For the complex fraction considered on the previous page, the LCM of the denominators of $\dfrac{2}{x}$ and $\dfrac{4}{x^2}$ is x^2:

$$\frac{1+\dfrac{2}{x}}{1-\dfrac{4}{x^2}} = \frac{x^2\left(1+\dfrac{2}{x}\right)}{x^2\left(1-\dfrac{4}{x^2}\right)}$$

$$= \frac{x^2+2x}{x^2-4}$$

$$= \frac{x\left(x+2\right)}{\left(x+2\right)\left(x-2\right)}$$

$$= \frac{x}{x-2}$$

| Lesson 2-3 | Radicals and Fractional Exponents |

KEY IDEAS

You know that $4^2 = 16$. To undo the squaring, take the *principal square root* of 16, written as $\sqrt{16} = 4$. The symbol $\sqrt{}$ is a **radical sign**. The number underneath the radical sign, 16 in this case, is the **radicand**. Squaring a number and taking the square root of a number are inverse operations in much the same way that multiplication and division are inverse operations. Because $10^3 = 1000$, 10 is the cube root of 1000. This fact can be expressed in either radical or exponent form:

$$\sqrt[3]{1000} = 1000^{\frac{1}{3}} = 10.$$

It can be generalized that, if k is a positive integer greater than 1 and $b^k = x$, then b is a kth root of x.

Square Roots and Perfect Squares

A **square root** of a nonnegative number is one of two identical factors of the number. A square root of 9 is 3 since $3 \times 3 = 9$. The other square root of 9 is –3 since $-3 \times -3 = 9$. A square root of $\frac{4}{25}$ is $\frac{2}{5}$ since $\frac{2}{5} \times \frac{2}{5} = \frac{4}{25}$.

Numbers that have rational square roots are called **perfect squares**. For example, 9 and $\frac{4}{25}$ are perfect squares, but 13 is not a perfect square.

The kth Root of a Number

You know that $10^3 = 1000$, so 10 is a cube root of 1000. Because $3^4 = 81$, 3 is a fourth root of 81. Furthermore, $2^5 = 32$, so 2 is a fifth root of 32.

Definition of the kth Root of a Number

Let b and x represent real numbers. If $b^k = x$ and k is a positive integer greater than 1, then b is a kth root of x.

The Principal Root of a Number

Because $4^2 = 16$ and $(-4)^2 = 16$, both 4 and –4 are square roots of 16. The notation $\sqrt{16}$, however, refers only to the square root of 16 that is nonnegative, namely 4, which is called the **principal square root**. The principal cube root of –8, written as $\sqrt[3]{-8}$, is –2. The principal fourth root of 81, written as $\sqrt[4]{81}$, is 3.

The notation $\sqrt[k]{x}$ always represents the **principal kth root of x** and, by definition, has the same sign as x. The number k that indicates what root of x is to be taken is the **index** of the radical. When the index is not indicated, as in $\sqrt{9}$, it is understood to be 2. Thus, $\sqrt{9} = 3$.

$x^{\frac{1}{k}}$ Means the kth Root of x

By extending the Laws of Integer Exponents to include rational exponents it follows that

$$x^{\frac{1}{2}} \cdot x^{\frac{1}{2}} = x^{\frac{1}{2} + \frac{1}{2}} = x, \qquad x^{\frac{1}{3}} \cdot x^{\frac{1}{3}} \cdot x^{\frac{1}{3}} = x^{\frac{1}{3} + \frac{1}{3} + \frac{1}{3}} = x, \quad \text{and so on.}$$

A reasonable generalization is that, since $x^{\frac{1}{k}}$ is one of k identical factors of x, it represents the kth root of x.

Definition of a Fractional Exponent

If k is a positive integer greater than 1, then

$$x^{\frac{1}{k}} = \sqrt[k]{x},$$

provided that $\sqrt[k]{x}$ is a real number. For example, $8^{\frac{1}{3}} = \sqrt[3]{8} = 2$.

TIPS

- Since the product of two identical numbers cannot be negative, $\sqrt{-64}$ is *not* a real number; the even root of a negative number never represents a real number. The odd root of a negative number does, however, represent a real number, as in $\sqrt[3]{-64} = -4$.
- Because $x^{\frac{1}{k}}$ is defined as $\sqrt[k]{x}$, $x^{\frac{1}{k}}$ represents the principal kth root of x.

Pythagorean Theorem

The hypotenuse of a right triangle is the side opposite the right angle and is the longest side of the right triangle. According to the Pythagorean theorem, the square of the length of the hypotenuse is equal to the sum of the squares of the lengths of the legs.

Exercise 1   Reviewing the Pythagorean Theorem

Find the length of the hypotenuse of a right triangle whose legs measure 8 inches and 15 inches, as shown in the accompanying diagram.

Solution: If x represents the length of the hypotenuse of the right triangle, then:

$$x^2 = 8^2 + 15^2$$
$$= 64 + 225$$
$$= 289$$

Since $x^2 = 289$, $x = \sqrt{289} = \mathbf{17}$.

Evaluating $x^{\frac{n}{k}}$

You know that the denominator of a fractional exponent tells what root to take of the base. Because

$$\left(x^{\frac{1}{k}}\right)^n = x^{\frac{n}{k}} = x^{\frac{\text{power}}{\text{root}}},$$

the numerator of the exponent of $x^{\frac{n}{k}}$ tells what power of the base to take, while the denominator of the exponent indicates what root is to be taken. For example, $8^{\frac{2}{3}}$ can be evaluated in either of the following two ways:

- Take the cube root of 8 and then raise the result to the second power:

$$8^{\frac{2}{3}} = \left(\sqrt[3]{8}\right)^2 = \left(2\right)^2 = 4.$$

- Raise 8 to the second power and then take the cube root of the result:

$$8^{\frac{2}{3}} = \sqrt[3]{8^2} = \sqrt[3]{64} = 4.$$

TIP

Although $x^{\frac{n}{k}}$ can be evaluated by first taking the power or the root of the base, always find the root first. As a result, you will work with smaller numbers:

$$27^{\frac{4}{3}} = \left(\sqrt[3]{27}\right)^4 = (3)^4 = 81 \quad \text{rather than} \quad 27^{\frac{4}{3}} = \sqrt[3]{(27)^4} = \sqrt[3]{531,441} = 81.$$

Evaluating Roots Using a Calculator

Your graphing calculator can be used to evaluate powers and roots. If you have the Texas Instruments TI-83+/TI-84+ or a similar graphing calculator, quit the current screen and enter the Home Screen by pressing the 2nd key followed by the MODE key, which has the QUIT command printed above it.

- To evaluate 3^7, press 3 ∧ 7 ENTER, which produces the answer 2187, shown on the right side of the next line.

- To evaluate $\sqrt[3]{4}$, write the cube-root radical in exponential form as $4^{\frac{1}{3}}$. Then press $\boxed{4}\ \boxed{\wedge}\ \boxed{(}\ \boxed{1}\ \boxed{\div}\ \boxed{3}\ \boxed{)}\ \boxed{\text{ENTER}}$. You should verify that $\sqrt[3]{4} \approx 1.587401052$, where the symbol $\approx$ is read as "is approximately equal to."

- To evaluate 5^{-2}, use this sequence of keystrokes:

$$\boxed{5}\ \boxed{\wedge}\ \boxed{(-)}\ \boxed{2}\ \boxed{\text{ENTER}}.$$

Verify that the answer displayed is .04 since $5^{-2} = \dfrac{1}{5^2} = \dfrac{1}{25}$.

Operations with Radicals

KEY IDEAS

Radicals may be multiplied and divided, provided that they have the same index. Radicals that have the same index and the same radicand can be added or subtracted.

Simplifying Radicals

When simplifying square-root radicals, look for the perfect-square factors of the radicands. For example, to simplify $\sqrt{75}$, first factor 75 so that one of the factors is a perfect square:

$$\sqrt{75} = \sqrt{25 \cdot 3} = \sqrt{25} \cdot \sqrt{3} = 5\sqrt{3}.$$

To simplify $\dfrac{1}{2}\sqrt{72}$, you could factor 72 as 8×9. Since, however, 8 also contains a perfect-square factor, it in turn would need to be simplified later on. Try factoring 72 so that one of the factors is the *greatest* perfect-square factor of 72:

$$\frac{1}{2}\sqrt{72} = \frac{1}{2}\sqrt{36 \cdot 2} = \frac{1}{2}\sqrt{36} \cdot \sqrt{2} = \frac{1}{2} \cdot 6 \cdot \sqrt{2} = 3\sqrt{2}.$$

A radicand may also include a variable factor:

$$\sqrt{18x^3} = \sqrt{9x^2 \cdot 2x} = \sqrt{9x^2} \cdot \sqrt{2x} = 3x\sqrt{2x}, \text{ provided that } x \geq 0.$$

You can apply the same procedure to simplifying cube-root radicals, except that you look for perfect *cubes* in the radicand rather than perfect squares:

$$\sqrt[3]{54x^6} = \sqrt[3]{27x^6 \cdot 2} = \sqrt[3]{27x^6} \cdot \sqrt[3]{2} = 3x^2\sqrt[3]{2}.$$

Some radicals can be simplified by using the properties of exponents. For example:

$$\sqrt[4]{81x^{12}y^8z^2} = \left(81x^{12}y^8z^2\right)^{\frac{1}{4}}$$

$$= \left(81^{\frac{1}{4}}\right)\left(x^{12}\right)^{\frac{1}{4}}\left(y^8\right)^{\frac{1}{4}}\left(z^2\right)^{\frac{1}{4}}$$

$$= 3x^{\frac{12}{4}}\,y^{\frac{8}{4}}\,z^{\frac{2}{4}}$$

$$= 3x^3y^2z^{\frac{1}{2}} \quad or \quad 3x^3y^2\sqrt{z}$$

Multiplying Radicals

To multiply two radicals with the same index, write the product of the two radicands underneath the radical sign and, if possible, simplify. Assume $x \geq 0$; then:

$$\sqrt{8x} \cdot \sqrt{6x^3} = \sqrt{48x^4}.$$

Now try to simplify the product by finding the greatest perfect-square factor of $48x^4$:

$$\sqrt{48x^4} = \sqrt{16x^4} \cdot \sqrt{3} = 4x^2\sqrt{3}.$$

TIP

The product of two identical square-root radicals is the radicand. For example:

$$\sqrt{7x} \cdot \sqrt{7x} = 7x, \text{ provided that } x \geq 0.$$

Dividing Radicals

To divide two radicals with the same index, write the quotient of the two radicands underneath the radical sign and, if possible, simplify. Assume $x \geq 0$; then:

$$\frac{\sqrt{80x^{15}}}{\sqrt{5x^3}} = \sqrt{\frac{80}{5} \cdot \frac{x^{15}}{x^3}} = \sqrt{16x^{12}}$$

$$= \sqrt{16} \cdot \sqrt{x^6 \cdot x^6}$$

$$= 4x^6$$

Combining Radicals

If radicals have the same index and the same radicand, they can be combined in the same way that like algebraic terms are combined.

EXAMPLE: $3\sqrt{5} + 4\sqrt{5} = 7\sqrt{5}$

EXAMPLE: $8\sqrt{2x} - 3\sqrt{2x} = 5\sqrt{2x}$, provided that $x \geq 0$

To combine radicals with the same index but different radicands, rewrite the radicals, if possible, so that they have the same radicand. Then add or subtract the like radicals.

EXAMPLE: $\sqrt{48} + \sqrt{27} = \left(\sqrt{16} \cdot \sqrt{3}\right) + \left(\sqrt{9} \cdot \sqrt{3}\right)$

$$= 4\sqrt{3} + 3\sqrt{3}$$

$$= 7\sqrt{3}$$

EXAMPLE: $\sqrt{200y} - \sqrt{32y} = \left(\sqrt{100} \cdot \sqrt{2y}\right) - \left(\sqrt{16} \cdot \sqrt{2y}\right)$

$$= 10\sqrt{2y} - 4\sqrt{2y}$$

$$= 6\sqrt{2y}, \text{ provided that } y \geq 0$$

Exercise 1   Combining Cube-Root Radicals

Write the difference $4\sqrt[3]{16} - 5\sqrt[3]{54}$ as a single radical.

Solution: Factor the radicands of $\sqrt[3]{16}$ and $\sqrt[3]{54}$ so that one of the factors is a perfect *cube*. Then simplify:

$$
\begin{aligned}
4\sqrt[3]{16} - 5\sqrt[3]{54} &= 4\sqrt[3]{8}\cdot\sqrt[3]{2} - 5\sqrt[3]{27}\cdot\sqrt[3]{2} \\
&= 4\cdot 2\cdot\sqrt[3]{2} - 5\cdot 3\cdot\sqrt[3]{2} \\
&= 8\sqrt[3]{2} - 15\sqrt[3]{2} \\
&= \mathbf{-7\sqrt[3]{2}}
\end{aligned}
$$

Multiplying Radical Expressions

Radical expressions that have the form of binomials can be multiplied using *FOIL*.

EXAMPLE: $\left(4 - 2\sqrt{3}\right)^2 = \left(4 - 2\sqrt{3}\right)\left(4 - 2\sqrt{3}\right)$

$$
\begin{aligned}
&= \overbrace{4\cdot 4}^{F} + \overbrace{4\left(-2\sqrt{3}\right)}^{O} + \overbrace{\left(-2\sqrt{3}\right)4}^{I} + \overbrace{\left(-2\sqrt{3}\right)\left(-2\sqrt{3}\right)}^{L} \\
&= 16 + \left[\left(-8\sqrt{3}\right) + \left(-8\sqrt{3}\right)\right] + (4)\left(\sqrt{3}\right)^2 \\
&= 16 \qquad\qquad -16\sqrt{3} \qquad + 12 \\
&= 28 \quad - 16\sqrt{3}
\end{aligned}
$$

Multiplying Conjugate Radical Expressions

Two radical expressions having the form of binominals that take the sum and the difference of the same two terms, such as $3 + \sqrt{2}$ and $3 - \sqrt{2}$, are called **conjugate radical expressions**. The product of conjugate square roots does not contain a radical and is always rational:

$$
\begin{aligned}
\left(3 + \sqrt{2}\right)\left(3 - \sqrt{2}\right) &= 9 \quad \overbrace{-3\sqrt{2} + 3\sqrt{2}}^{= 0} - \left(\sqrt{2}\cdot\sqrt{2}\right) \\
&= 9 \qquad\qquad\qquad - 2 \\
&= 7
\end{aligned}
$$

TIPS

- Conjugate square-root expressions can be quickly multiplied together by using the familiar pattern $(A+B)(A-B) = A^2 - B^2$:

$$\left(3+\sqrt{2}\right)\left(3-\sqrt{2}\right) = \left(3\right)^2 - \left(\sqrt{2}\right)^2 = 9 - 2 = 7$$

- When a square-root expression is written in simplest form, no radicand should contain a fraction or a perfect-square factor other than 1.

Exercise 2   Multiplying Conjugate Square Roots

Find the product $\left(\sqrt{10}+5\sqrt{3}\right)\left(\sqrt{10}-5\sqrt{3}\right)$.

Solution: $\left(\sqrt{10}+5\sqrt{3}\right)\left(\sqrt{10}-5\sqrt{3}\right) = \left(\sqrt{10}\right)^2 - \left(5\sqrt{3}\right)^2$

$$= \;\; 10 \;\;\; - \;\; 75$$

$$= \mathbf{-65}$$

Rationalizing Denominators

To simplify $\dfrac{3}{7-3\sqrt{5}}$ so that the fraction is transformed into an equivalent fraction without a radical in the denominator, multiply the numerator and the denominator of the fraction by the conjugate of the denominator:

$$\frac{3}{7-3\sqrt{5}} = \frac{3}{7-3\sqrt{5}} \cdot \overset{1}{\overbrace{\left(\frac{7+3\sqrt{5}}{7+3\sqrt{5}}\right)}}$$

$$= \frac{21+9\sqrt{5}}{7^2 - 3^2 \cdot 5}$$

$$= \frac{21+9\sqrt{5}}{4}$$

Factoring Over the Set of Real Numbers

Factoring is performed, unless otherwise indicated, over the set of rational numbers. Therefore, the factored expressions may contain only rational numbers. Thus, $x^2 - 3$ cannot be factored over the set of rational numbers. However, if factoring over the set of real numbers were allowed, the factored expression could then contain irrational numbers. For example:

$$x^2 - 3 = \left(x\right)^2 - \left(\sqrt{3}\right)^2 = \left(x+\sqrt{3}\right)\left(x-\sqrt{3}\right).$$

Chapter 2	CHECKUP EXERCISES

1–5. Multiple Choice

1. The expression $\dfrac{2+\sqrt{3}}{2-\sqrt{3}}$ is equivalent to

   (1) $11\sqrt{3}$ (2) $7-4\sqrt{3}$ (3) $7+4\sqrt{3}$ (4) $\dfrac{7+4\sqrt{3}}{7}$

2. The expression $\sqrt{\dfrac{4}{3}} - \sqrt{\dfrac{3}{4}}$ is equivalent to

   (1) $\dfrac{2-\sqrt{3}}{2\sqrt{3}}$ (2) $\sqrt{\dfrac{7}{12}}$ (3) $\dfrac{\sqrt{3}}{6}$ (4) $2\sqrt{3}$

3. What is the product $(y+1)\left(\dfrac{y}{1-y^2}\right)$ expressed in lowest terms?

   (1) $\dfrac{y}{y-1}$ (2) $\dfrac{y}{1-y}$ (3) $\dfrac{y+1}{y-1}$ (4) -1

4. The expression $\dfrac{\sqrt{7}+\sqrt{2}}{\sqrt{7}-\sqrt{2}}$ is equivalent to

   (1) $\dfrac{9}{5}$ (2) -1 (3) $\dfrac{9+2\sqrt{14}}{5}$ (4) $\dfrac{11+\sqrt{2}}{14}$

5. When resistors R_1 and R_2 are connected in a parallel electric circuit, the total resistance is $\dfrac{1}{\dfrac{1}{R_1}+\dfrac{1}{R_2}}$. This complex fraction is equivalent to

   (1) R_1+R_2 (2) $\dfrac{R_1+R_2}{R_1R_2}$ (3) $\dfrac{R_1}{R_2}+\dfrac{R_2}{R_1}$ (4) $\dfrac{R_1R_2}{R_1+R_2}$

6–11. Write each fraction in lowest terms.

6. $\dfrac{3x+6}{4-x^2}$

8. $\dfrac{ab-b^2}{5ab-5a^2}$

10. $\dfrac{3x^3-27xy^2}{12x^2+36xy}$

7. $\dfrac{10a^2-15ab}{4a^2-9b^2}$

9. $\dfrac{2x^2-50}{2x^2+5x-25}$

11. $\dfrac{x^2-4y^2}{x^2+4xy+4y^2}$

12–26. Perform the indicated operation, and write the answer in simplest form.

12. $2\sqrt{27}+3\sqrt{108}$

17. $\dfrac{2x+y}{x+2y}+\dfrac{x+5y}{x+2y}$

22. $\dfrac{15}{\sqrt{20}}-\sqrt{45}$

13. $x\sqrt{48}-2\sqrt{75x^2}$

18. $\dfrac{a^2+1}{a^2-1}-\dfrac{a}{a+1}$

23. $\dfrac{40}{\sqrt{8}}-\sqrt{50}$

14. $\left(8-2\sqrt{5}\right)\left(8+2\sqrt{5}\right)$

19. $\dfrac{a^2-5}{a-b}+\dfrac{b^2-5}{b-a}$

24. $\dfrac{81-x^2}{6x-54}\div\dfrac{x^2+9x}{3x}$

15. $\left(1+\sqrt{3}\right)\left(2-\sqrt{12}\right)$

20. $5\sqrt[3]{2y^3}-\sqrt[3]{16y^3}$

25. $\dfrac{2x^2-32}{x^2+x-12}\div\dfrac{20-5x}{2x^2-5x-3}$

16. $\left(3\sqrt{2}-2\sqrt{3}\right)^2$

21. $2\sqrt[3]{8y^4}-3y\sqrt[3]{y}$

26. $\dfrac{x+2}{x^2-9}+\dfrac{2}{x^2+x-6}$

27–30. Express each fraction as an equivalent fraction in simplest form with a rational denominator. In Exercise 28, x represents an integer greater than 1.

27. $\dfrac{\sqrt{20}}{1-2\sqrt{5}}$

28. $\dfrac{\sqrt{x}}{x-\sqrt{x}}$

29. $\dfrac{\sqrt{8}}{\sqrt{6}-\sqrt{2}}$

30. $\dfrac{\sqrt{5}+3\sqrt{2}}{\sqrt{5}-3\sqrt{2}}$

31–36. Simplify each complex fraction.

31. $\dfrac{\dfrac{3}{x^2}+\dfrac{1}{x}}{1-\dfrac{9}{x^2}}$

33. $\dfrac{1-\dfrac{1}{w}}{\dfrac{1}{w^2}-\dfrac{1}{w}}$

35. $\dfrac{m-\dfrac{1}{m}}{m-2+\dfrac{1}{m}}$

32. $\dfrac{\dfrac{x-y}{y}}{y^{-1}-x^{-1}}$

34. $\dfrac{x+y^{-1}}{y+x^{-1}}$

36. $\dfrac{\dfrac{b}{b-3}+\dfrac{4}{b}}{1+\dfrac{1}{b-3}}$

37–40. Perform each of the indicated operations, and write the answer in simplest form.

37. $\dfrac{3}{x^2-4}+\dfrac{2}{x^2+5x+6}$

39. $\dfrac{x}{x^2+3x-4}-\dfrac{x+1}{2x^2-2}$

38. $\dfrac{x^3-9x}{\left(xy\right)^2-8y}\cdot\dfrac{x^3y-8x}{x^2-6x+9}$

40. $\dfrac{t^2-1}{t^2-4}\div\dfrac{9t+9}{4t+12}\cdot\dfrac{2-t}{t^2+2t-3}$

41. During the design of a rectangular box, its dimensions are represented as

$$\dfrac{2x^2+2x-24}{4x^2+x}\ \text{by}\ \dfrac{x^2+x-6}{x+4}\ \text{by}\ \dfrac{8x^2+2x}{x^2-9}\ .$$

Express the volume of the box in simplest form in terms of x.

Chapter 3

GRAPHING AND SYSTEMS OF EQUATIONS

OVERVIEW

Drawing horizontal and vertical number lines creates a **coordinate plane** in which a point is located by using an ordered pair of real numbers (x,y). The *horizontal* number line is the x-axis, and the *vertical* number line is the y-axis. Collectively, these axes are referred to as the **coordinate axes**. The point at which the coordinate axes intersect, $(0,0)$, is the **origin**.

The set of all points that satisfy a linear equation in two variables can be represented graphically as a straight line. The solution of a system of two linear equations is the ordered pair of numbers, if any, that make both equations true at the same time. The solution may be obtained graphically by determining the point at which the two lines intersect or algebraically by eliminating one of the two variables.

Lesson 3-1   Graphing Points and Linear Equations

KEY IDEAS

The location of a point in the coordinate plane is described by an ordered pair of numbers (x,y), where x is the directed distance of the point from the vertical y-axis and y is the directed distance of the point from the horizontal x-axis. Relative to the origin:

- $(+4,+3)$ is 4 units to the right and 3 units up.
- $(-4,+3)$ is 4 units to the left and 3 units up.
- $(-4,-3)$ is 4 units to the left and 3 units down.
- $(+4,-3)$ is 4 units to the right and 3 units down.

An equation that has the form $ax + by = c$ is a **linear equation** in two variables. The graph of such an equation is a line. A minimum of two points is needed to graph a linear equation.

The Coordinate Plane

The x- and y-axes divide the coordinate plane into four quadrants that are numbered in the counterclockwise direction using Roman numerals. If the coordinates of point P are $(-4,3)$, the x-coordinate is -4, the y-coordinate is $+3$, and the point is located in Quadrant II, as shown in Figure 3.1.

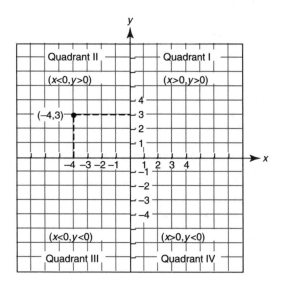

Figure 3.1 The coordinate plane

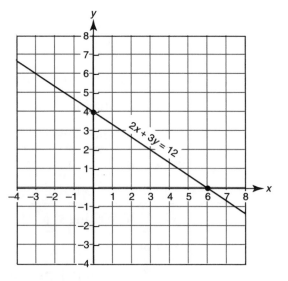

Figure 3.2 Graph of $2x + 3y = 12$

Using Intercepts to Graph a Line

An easy way to graph a line is to locate the points where the line cuts the coordinate axes.

- An **x-intercept** of a graph is a point at which the graph meets the x-axis. At an x-intercept, y is always 0.
- A **y-intercept** of a graph is a point at which the graph meets the y-axis. At a y-intercept, x is always 0.

To graph the line $2x + 3y = 12$:

1. Find the x-intercept by letting $y = 0$ and solving for x:

$$2x + 3(0) = 12$$
$$2x = 12$$
$$x = \frac{12}{2} = 6.$$

    Hence, the x-intercept is (6,0).

2. Find the y-intercept by letting $x = 0$ and solving for y:

$$2(0) + 3y = 12$$
$$3y = 12$$
$$y = \frac{12}{3} = 4.$$

    Hence, the y-intercept is (0,4).

3. Plot (6,0) and (0,4), and then draw a line to connect these points, as shown in Figure 3.2.

Of course, any two points that satisfy the equation could have been chosen.

 **Lesson 3-2**   **Midpoint and Distance Formulas**

KEY IDEAS

If you know the coordinates of points A and B, you can use formulas to find the midpoint and length of $\overline{AB}$.

Midpoint Formula

The **midpoint** of a line segment is the point on the segment that divides it into two parts of equal length. The **coordinates** of the midpoint of a line segment are the averages of the corresponding x- and y-coordinates of its endpoints. In Figure 3.3, the midpoint of $\overline{AB}$ is $\left(\dfrac{x_A + x_B}{2}, \dfrac{y_A + y_B}{2} \right)$. If the endpoints are $A(3,2)$ and $B(11,8)$, the midpoint is at $\left(\overline{x}, \overline{y} \right)$, where

$$\overline{x} = \frac{3+11}{2} = \frac{14}{2} = 7$$

and

$$\overline{y} = \frac{2+8}{2} = \frac{10}{2} = 5.$$

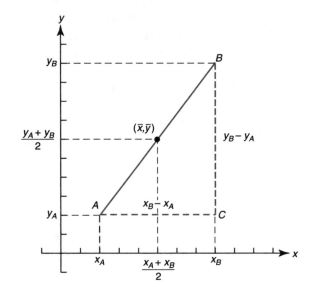

Figure 3.3 Midpoint and distance formulas

Distance Formula

For the right triangle in Figure 3.3, the length of the vertical leg is $y_B - y_A$ and the length of the horizontal leg is $x_B - x_A$. The distance from point A to point B is measured by the length of hypotenuse $\overline{AB}$ of right triangle ABC. By the Pythagorean theorem:

$$\left(AB \right)^2 = \left(x_B - x_A \right)^2 + \left(y_B - y_A \right)^2 , \text{ so } AB = \sqrt{ \left(x_B - x_A \right)^2 + \left(y_B - y_A \right)^2 } .$$

Sometimes the Greek letter Δ (delta) is used as a shorthand notation for "change in," as in

$$\Delta x = x_B - x_A \quad \text{and} \quad \Delta y = y_B - y_A.$$

For example, if the endpoints of line segment AB are $A(3,2)$ and $B(11,8)$, then $\Delta x = 11 - 3 = 8$ and $\Delta y = 8 - 2 = 6$, so

$$AB = \sqrt{ \left(\Delta x \right)^2 + \left(\Delta y \right)^2 } = \sqrt{ 8^2 + 6^2 } = \sqrt{100} = 10.$$

Midpoint and Distance Formulas

If $A(x_A, y_A)$ and $B(x_B, y_B)$ are the endpoints of line segment AB, then

- Midpoint of $\overline{AB} = \left(\dfrac{x_A + x_B}{2}, \dfrac{y_A + y_B}{2} \right)$.

- $AB = \sqrt{\left(\Delta x\right)^2 + \left(\Delta y\right)^2}$, where $\Delta x = x_B - x_A$ and $\Delta y = y_B - y_A$.

Exercise 1   Finding the Distance Between Two Points

What is the distance from $(6, -9)$ to $(-3, 4)$?

Solution: Since $\Delta x = -3 - 6 = -9$ and $\Delta y = 4 - (-9) = 13$:

$$\text{Distance} = \sqrt{\left(-9\right)^2 + \left(13\right)^2}$$
$$= \sqrt{250}$$
$$= \sqrt{25} \cdot \sqrt{10} = \mathbf{5\sqrt{10}}$$

Equation of a Circle

The distance formula is sometimes used to obtain other formulas. A **circle** is the set of all points in the plane that are the same distance, called the **radius**, from a fixed point that is the **center**. If the center of a circle is at the origin and $P(x,y)$ is any point on the circle, then the distance from the origin to P must be r, the radius, as illustrated in Figure 3.4. The distance from the origin to point $P(x,y)$ is

$$\sqrt{\left(x - 0\right)^2 + \left(y - 0\right)^2} = r$$

or, equivalently, $x^2 + y^2 = r^2$. For example, if a circle whose center is at the origin has a radius of 5, then an equation of the circle is $x^2 + y^2 = 25$.

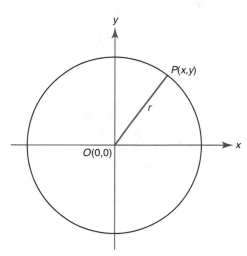

Figure 3.4 Equation of a circle: $x^2 + y^2 = r^2$

The Slope of a Line

KEY IDEAS

If you think of a line in the coordinate plane as a hill, then the slope of the line is a number that measures the steepness of the hill. "Walking down" the hill from left to right has a negative slope, and "walking up" the hill from left to right has a positive slope. If two lines have positive slopes, the steeper line is the line with the greater slope.

Finding the Slope of a Line

A line that is slanted so that it is not parallel to a coordinate axis is called an **oblique line**. When moving from one point to another on an oblique line, there is a change in vertical distance, called the *rise*, and a change in horizontal distance, called the *run*. The **slope** of a line is defined as the ratio of these changes, $\dfrac{\text{rise}}{\text{run}}$. If you know the coordinates of any two points on a line, you can calculate the slope of that line.

The Slope Formula

- The slope, m, of the line that contains points $A(x_A, y_A)$ and $B(x_B, y_B)$ is

$$m = \frac{\Delta y}{\Delta x} = \frac{y_B - y_A}{x_B - x_A}.$$

- The order in which the points are taken when subtracting in the denominator must be the same as the order taken when subtracting in the numerator.

Six Facts About Slope

The slope m, of a line may be positive, negative, zero, or undefined.
In Figure 3.5a:

- Line *p* *rises*. To find the slope of line *p*, choose any two convenient points on the line, such as (4,5) and (2,1):

$$\text{Slope} \; = \; m = \frac{\text{rise}}{\text{run}} = \frac{\Delta y}{\Delta x} = \frac{1-5}{2-4} = \frac{-4}{-2} = +2 \, .$$

Fact 1: The slope of a line that rises as x increases is always positive.

- Line q *falls*. To find the slope of line q, choose any two convenient points on the line, such as $(4,-2)$ and $(1,0)$:

$$\text{Slope} = m = \frac{\text{rise}}{\text{run}} = \frac{\Delta y}{\Delta x} = \frac{0-(-2)}{1-4} = \frac{2}{-3} = -\frac{2}{3}.$$

Fact 2: The slope of a line that falls as x increases is always negative.

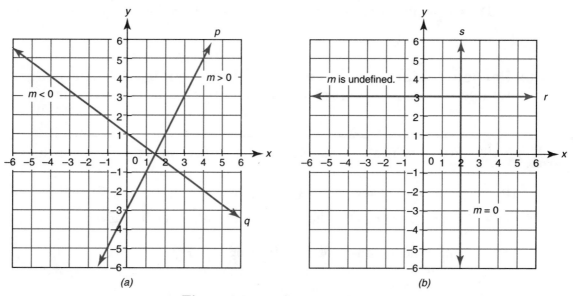

Figure 3.5 Possible values for slope

In Figure 3.5*b*:

- Line r is a *horizontal* line. To find the slope of line r, choose any two convenient points on the line, such as $(0,3)$ and $(4,3)$:

$$\text{Slope} = m = \frac{\text{rise}}{\text{run}} = \frac{\Delta y}{\Delta x} = \frac{3-3}{4-0} = \frac{0}{4} = 0.$$

Fact 3: The slope of a horizontal line is 0.

- Line s is a *vertical* line. To find the slope of line s, choose any two convenient points on the line, such as $(2,0)$ and $(2,5)$:

$$\text{Slope} = m = \frac{\text{rise}}{\text{run}} = \frac{\Delta y}{\Delta x} = \frac{5-0}{2-2} = \frac{5}{0} \quad \leftarrow \text{Undefined!}$$

Fact 4: The slope of a vertical line is undefined.

Pairs of lines that are parallel or perpendicular have special slope relationships.

- When two nonvertical lines have *different* slopes, the lines intersect. Parallel lines do not intersect, so their slopes must be the same.

Fact 5: Parallel lines have equal slopes; lines that have equal slopes are parallel.

- Pairs of numbers such as $\dfrac{3}{4}$ and $-\dfrac{4}{3}$ are negative reciprocals because their product is -1.

Fact 6: Perpendicular lines have slopes that are negative reciprocals; lines whose slopes are negative reciprocals are perpendicular.

Exercise 2   Proving That Lines Meet at a Right Angle

If the coordinates of the vertices of $\triangle PQR$ are $P(-1,-1)$, $Q(1,-2)$, and $R(3,2)$, prove that $\triangle PQR$ is a right triangle by showing that $\overline{PQ}$ is perpendicular to $\overline{QR}$.

Solution: Find and then compare the slopes of $\overline{PQ}$ and $\overline{QR}$.

- For $\overline{PQ}$, $\Delta y = -2-\left(-1\right)=-2+1=-1$ and $\Delta x = 1-\left(-1\right)=1+1=2$. Hence:

$$\text{Slope of } \overline{PQ} = \frac{\Delta y}{\Delta x} = -\frac{1}{2}.$$

- For $\overline{QR}$, $\Delta y = 2-\left(-2\right)=2+2=4$ and $\Delta x = 3-1=2$. Hence:

$$\text{Slope of } \overline{QR} = \frac{\Delta y}{\Delta x} = \frac{4}{2} = 2.$$

- Since $\left(-\dfrac{1}{2}\right)(2)=-1$, the slopes of $\overline{PQ}$ and $\overline{QR}$ are negative reciprocals. Hence, $\overline{PQ} \perp \overline{QR}$, so $\triangle PQR$ is a right triangle with right angle PQR.

Graphing with a Calculator

KEY IDEAS

If you know the equation of a line, a graphing calculator can be used to display the graph of the line. Because of its popularity, wide availability, and ease of use, the discussions related to graphing calculators are based on the Texas Instruments TI-83+/84+ family of graphing calculators. If you are using a different model, you may have to make minor adjustments in the calculator procedures described throughout this book.

The Viewing Window

The viewing window of a graphing calculator shows only a small part of the coordinate plane. In a **Standard window** the positive and negative coordinate axes each have 10 tic marks, as shown in Figure 3.6. The current values of the window variables X min, X max, Y min, and Y max determine the number of tic marks on the coordinate axes. To display the current window variables, press ⎡WINDOW⎤. Figure 3.7 shows the values of the window variables for a Standard window.

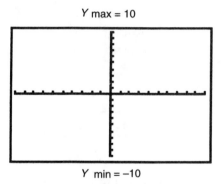

Figure 3.6 Standard viewing window

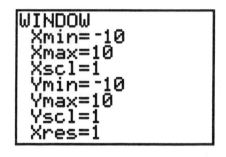

Figure 3.7 Screen variables for a Standard window

Changing the Window Variables

To change the value of any of the window variables, use one of the four cursor arrow keys to move the blinking cursor to the value of the screen variable that you want to change. Then overwrite the old value with the new one.

TIPS

- An arithmetic expression can be entered as the value of a screen variable. For example, to change the value of X min from -10 to -20, on the line that reads X min $= -10$ enter $\times 2$ after -10. Pressing $\boxed{\text{ENTER}}$, or moving to another line, automatically sets X min $= -20$.
- The distance between consecutive tic marks on an axis can be set to a number other than 1 by changing the value of Xscl (Xscale) or Yscl (Yscale). For example, setting Xscl $= 5$ scales the x-axis so that tic marks are 5 units instead of 1 unit apart.

The ZOOM Menu

The **zoom menu** includes options that allow you to quickly resize the viewing window to preset values. When you enter the number of a menu option, the graph is immediately plotted in the resized window.

- Pressing $\boxed{\text{ZOOM}}$ $\boxed{6}$ creates a **Standard window** in which each coordinate axis has 10 tic marks on either side of 0.

- Pressing $\boxed{\text{ZOOM}}$ $\boxed{4}$ creates a basic **Decimal** window (see Figure 3.8) in which the x-coordinate of the cursor position changes in user-friendly steps of 0.1. The screen dimensions for a Decimal window are shown in Figure 3.9. Windows in which the x-coordinates of the cursor locations change in steps of 0.1, or a multiple of 0.1, are called **friendly windows**.

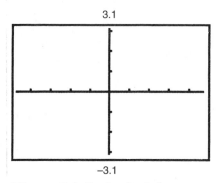

Figure 3.8 Decimal window

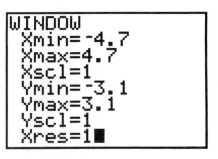

Figure 3.9 Decimal window variables

Decimal Windows

In a **Decimal window** the ratio of (X max $- X$ min) to (Y min $- Y$ max) is the same as the ratio of the pixel width to the pixel height of the viewing rectangle. For many graphing calculators this ratio is approximately 3 to 2. Graphs viewed in a Decimal window maintain their true geometric proportions. In a Decimal window, a circle looks like a circle rather than an oval.

TIPS

- You can create other Decimal windows by using positive-integer multiples of the settings for [X min,X max] and [Y min,Y max] in Figure 3.9. If you change the screen variables so that X min $= -4.7 \times 2$ and X max $= 4.7 \times 2$, the x-coordinate of the cursor will change in friendly steps of 0.2.
- Setting X max and X min so that their difference is a multiple of 9.4 will produce a friendly x-coordinate readout. For example, if X min $= 0$ and X max $= 9.4$, the x-coordinate of the cursor will change in friendly steps of 0.1.

Finding an Appropriate Viewing Window

The window size may need to be adjusted so that the basic shape of a graph and all of its important features can be seen. For example, when graphing a slanted line, you may need to adjust the size of the viewing window so that the x-intercept falls within the interval from X min to X max and the y-intercept falls within the interval from Y min to Y max.

TIPS

- To find an appropriate viewing window, some experimenting may be required. Start with a Standard window when you suspect that the key part of a graph will fall within $-10 \leq x \leq 10$. If the graph doesn't fit within this window, change the values of the screen variables in the WINDOW editor as needed.
- If the graph fits easily within a Standard viewing window, try a basic Decimal window or a multiple of it.

Indicating Window Size

When a graph drawn by a calculator is recreated on paper, it is helpful to provide the [X min, X max] and [Y min, Y max] values of the display window in which the graph is being viewed. For example, the notation [$-4.7,+4.7$] $\times$ [$-6.2,+6.2$] under a graph means that the graph is being viewed in a rectangular window that is sized so that $-4.7 \leq x \leq 4.7$ and $-6.2 \leq y \leq 6.2$.

Graphing a Linear Equation

To graph $y - 2x = 3$, first solve for y: $y = 2x + 3$. Then:

1. Enter the equation. Press $\boxed{Y =}$ to open the $Y =$ editor. Set Y_1 equal to $2x + 3$ by pressing:

   $\boxed{2}$ $\boxed{x, T, \theta, n}$ $\boxed{+}$ $\boxed{3}$.

2. Select a viewing window. Press ZOOM 6 to draw the graph in a Standard window, as shown in Figure 3.10.

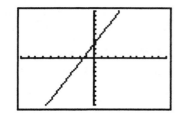

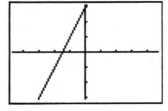

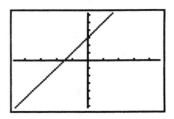

Figure 3.10 Standard window   **Figure 3.11** Decimal window   **Figure 3.12** Multiple of a Decimal window

3. Look at the graph. Press ZOOM 4 to view the graph in a Decimal window, as shown in Figure 3.11. The y-intercept of the graph now appears to be cut off. Multiplying the Decimal window values for Y min and Y max by 2 produces the graph in Figure 3.12.

TIPS

To graph an equation, solve for y in terms of x, if necessary. Then follow this procedure:

- Enter the equation in the $Y =$ editor. If the coefficient of x is negative, as in $y = -2x$, press (−) rather than − .

- Display the graph by pressing either ZOOM 4 to see the graph in a Decimal window or ZOOM 6 to view the graph in a Standard window.

- Adjust the size of the viewing window, if necessary, so that the graph fits. You can increase the window variables by pressing WINDOW and then multiplying one or more of the window variables by a number that will allow you to see more of the graph.

2nd Function Keys and Bracket Notation

Most keys on your graphing calculator have two functions. The second function is printed on top of the key. To activate the second function, first press 2nd and then press the key that has the desired function printed above it. For example, notice that ON has the label OFF printed above it. To turn the calculator off, press 2nd [OFF] in succession; the brackets refer to the key that has OFF *above* it.

The Table-Building Feature

Suppose that a cable television service offers two plans. Plan A costs \$11 per month plus \$7 for each premium channel. Plan B costs \$27 per month plus \$3 for each premium channel. For what number of premium channels will the two plans cost the same? You can compare the costs of the two plans by using your graphing calculator to create a table of values.

Let y represent the cost of ordering x premium channels under each service plan.

1. Set $Y_1 = 11 + 7x$ for plan A and set $Y_2 = 27 + 3x$ for plan B.
2. Enter the TABLE SETUP mode by pressing $\boxed{\text{2nd}}$ [TBLSET]. Set TblStart = 1, as shown in Figure 3.13, so that 1 is the starting value of X when you view the table. The consecutive values of X in the table will increase by 1 when ΔTbl = 1.

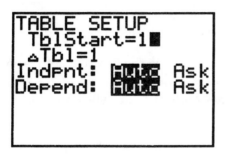

Figure 3.13 Table setup

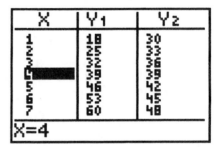

Figure 3.14 Table representation of the plans

3. Display the table by pressing $\boxed{\text{2nd}}$ $\boxed{\text{TABLE}}$. The plans will have the same cost for the value of X that makes $Y_1 = Y_2$. According to the table in Figure 3.14, when $X = 4$ premium channels, $Y_1 = Y_2 = \$39$.

The two plans have the same cost (\$39) for 4 premium channels.

Lesson 3-5   **Graphing a Linear Inequality**

KEY IDEAS

The graph of $y = 2x + 3$ serves as a boundary line between two distinct regions:

- The set of points in the region *above* the line satisfies the inequality $y > 2x + 3$.
- The set of points in the region *below* the line satisfies the inequality $y < 2x + 3$.

Graphing a Linear Inequality

To graph a linear inequality such as $y - 2x > 3$:

1. Replace the inequality sign with an equal sign, and graph the resulting equation as a broken line since points on the line are not included in the solution set of $y - 2x > 3$, as shown in Figure 3.15. If the inequality relation is $\leq$ or $\geq$, draw a solid line to indicate that the solution set includes the points on the line.

2. Decide which side of the boundary line, above or below it, represents the solution set. If you are not sure, pick a convenient test point not on the boundary line, such as $(0,0)$. Determine whether $(0,0)$ satisfies the inequality:

$$y - 2x > 3$$
$$0 > 2(0) + 3 \quad ?$$
$$0 > 0 + 3 \quad\quad ?$$
$$0 > 3 \quad\quad\quad \text{No!}$$

Test point $(0,0)$ does *not* satisfy the inequality, so the solution set is the region *above* the boundary line since that region does *not* include $(0,0)$.

3. Shade in the region that represents the solution set, as shown in Figure 3.16.

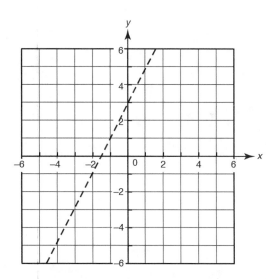

Figure 3.15 Boundary line for $y - 2x > 3$

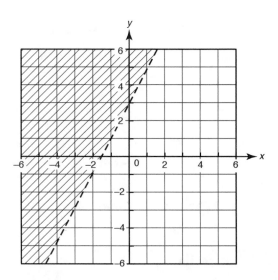

Figure 3.16 Graph of $y - 2x > 3$

TIPS

You can use your graphing calculator to graph a linear inequality. In the $Y =$ editor, place the cursor over the diagonal that comes before Y_1.

- If the inequality relation is $>$ (greater than), press $\boxed{\text{ENTER}}$ until you get the solid right triangle shown below at the left. This triangular symbol indicates that the area above the graph will be shaded. After entering the right side of the inequality, display the graph, as shown below at the right.

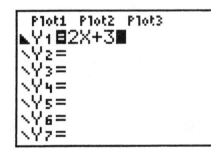
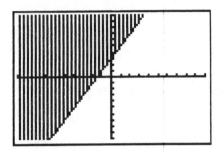

- If the inequality relation is $<$ (less than), press $\boxed{\text{ENTER}}$ until you get the solid right triangle shown below at the left. This triangular symbol indicates that the region below the line will be shaded. Display the graph, as shown below at the right.

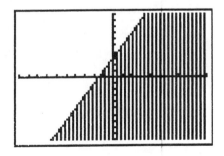

Lesson 3-6

Writing Equations of Lines

KEY IDEAS

An equation of a line expresses the general relationship between the x- and the y-coordinates of each point on that line. To be able to write an equation of a line, you need to know two facts about the line, such as its slope and y-intercept, or its slope and a point on the line, or two different points on the line.

Equation of a Line: Slope-Intercept Form

If in an equation of a line y is solved for in terms of x, as in $y = 2x - 3$, then the equation of a line is said to be in **slope-intercept form** because the coefficient of x is the slope of the line and the constant term represents the y-intercept. The slope of the line $y = 2x - 3$ is 2, and its y-intercept is (0,3). If the slope of a line is –4 and its y-intercept is 1, then an equation of the line is $y = -4x + 1$.

> ### Slope-Intercept Form: $y = mx + b$
>
> If the slope of a line is m and the y-intercept is (0,b), then an equation of this line is $y = mx + b$.

Exercise 1   Finding the Slope and y-Intercept of a Line from Its Equation

What are the slope and the y-intercept of the line whose equation is $4x - 3y - 8 = 0$?

Solution: Solve the given equation for y. If $4x - 3y - 8 = 0$, then

$$-3y = -4x + 8$$

$$y = \frac{-4x}{-3} + \frac{8}{-3}$$

$$= \frac{4}{3}x - \frac{8}{3}$$

Since $m = \frac{4}{3}$ and $b = -\frac{8}{3}$, the slope of the line is $\frac{4}{3}$ and the y-intercept is $\left(0, -\frac{8}{3}\right)$.

Exercise 2   Writing an Equation of a Line

The y-intercept of line q is (0,1). If line q is parallel to the line $2y = x - 6$, what is an equation of line q?

Solution: If $2y = x - 6$, then $y = \frac{1}{2}x - 3$, so the slope of this line is $\frac{1}{2}$. Because parallel lines have the same slope, the slope of line q is also $\frac{1}{2}$. It is given that the y-intercept of line q is (0,1), so you now know that, for line q, $m = \frac{1}{2}$ and $b = 1$. An equation of line q is

$$y = \frac{1}{2}x + 1.$$

Equation of a Line: Point-Slope Form

If you know the slope and a point on a line, or two points on a line, using the $y = mx + b$ form of the equation may not be the easiest way to obtain the equation of the line. In these situations, use the **point-slope form** of the equation of a line.

> ### Point-Slope Form: $y - k = m(x - h)$
>
> If (h,k) is a point on a line whose slope is m, then an equation of this line is $y - k = m(x - h)$.

EXAMPLE: If the slope of a line is 2 and the line passes through (4,3), then $m = 2$, $h = 4$, and $k = 3$, so an equation of this line is $y - 3 = 2(x - 4)$. If needed, this equation can be put into $y = mx + b$ form by solving for y:

$$y - 3 = 2(x - 4)$$
$$= 2x - 8$$
$$y = 2x - 8 + 3$$
$$= 2x - 5$$

Exercise 3   Writing an Equation of a Line, Given Two Points

Line AB contains points $(-1,0)$ and $(1,6)$.

   a.   Write an equation of line AB.
   b.   If line p is perpendicular to line AB and contains point $C(-4,1)$, write an equation of line p.

Solutions:

a.   Calculate the slope of line AB using the given points. Then use either point to write an equation of the line using the point-slope form.

- Find the slope, m, of $\overleftrightarrow{AB}$. Because $\Delta y = 6 - 0 = 6$ and $\Delta x = 1 - (-1) = 2$:

$$m = \frac{\Delta y}{\Delta x} = \frac{6}{2} = 3.$$

- An equation of line AB has the form $y - k = 3(x - h)$. Replace the values of h and k with the corresponding coordinates of either point A or point B. If you choose point A, then $h = -1$ and $k = 0$, so an equation of line AB is $\boldsymbol{y - 0 = 3(x - (-1))}$ or, equivalently, $\boldsymbol{y = 3x + 3}$.

b. Perpendicular lines have slopes that are negative reciprocals. Because the slope of line AB is 3, the slope of line p is $-\dfrac{1}{3}$. It is also given that line p contains point $C(-4,1)$. Use the point-slope form of the equation of a line where $m = -\dfrac{1}{3}, h = -6$, and $k = 1$. An equation of line p is $\boldsymbol{y - 1 = -\dfrac{1}{3}(x + 6)}$.

Equations of Horizontal and Vertical Lines

If the *y*-intercept of a *horizontal line* is $(0,b)$, then the *y*-coordinate of each point on the line is *b*, so the equation of the line is $y = b$. See Figure 3.17.

Similarly, if the *x*–intercept of a *vertical line* is $(a,0)$, then the *x*-coordinate of each point on the line is *a*, so the equation of the line is $x = a$. See Figure 3.18.

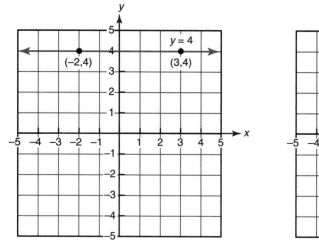

Figure 3.17 Vertical line: $y = b$

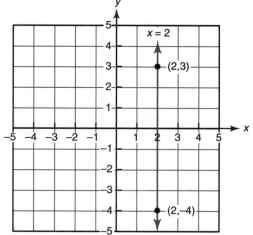

Figure 3.18 Horizontal line: $x = a$

Fitting a Line to Data

A researcher may suspect that two real-world variables are related in such a way that the dependent variable changes at a constant rate. To find an equation that describes this linear relationship, the researcher must rely on the raw data and tools borrowed from statistics.

If the data collected by the researcher are plotted as a set of ordered pairs, called a **scatter plot**, it will probably not be possible to draw the same line through all or even most of the data points. Nevertheless, the data may still appear to trace out a path that is approximately linear. This is the case in Figure 3.19, in which the height and weight measurements of a group of people are plotted. The line that comes "closest" to the set of data points is called the **line of best fit**. Although drawing different lines will appear to the eye to fit the data points equally well, the statistics features of a graphing calculator can be used to find an equation of *the* line of best fit. Because of the way the equation of the line of best fit is calculated, it is sometimes referred to as the **least-squares regression line**.

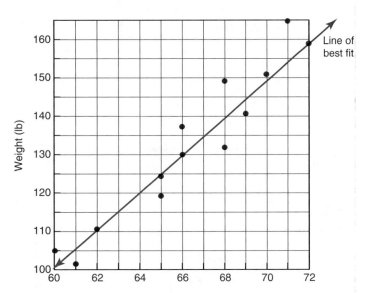

Figure 3.19 Scatter plot of (height, weight) measurements

Finding a Regression Line

The accompanying table shows the numbers of applications for admission received by a college for certain years from 1991 to 2000.

College Applications for Certain Years

YEAR	APPLICATIONS
1991	297
1993	331
1995	409
1996	482
1999	647
2000	615

Figure 3.20 Storing the data as lists

Since the number of applicants depends on the year, let x represent the year number and y represent the number of applicants for that year. If you suspect that the number of applicants is increasing at a constant rate, then your goal is to find an equation of the form $y = ax + b$ that best fits the data presented in the table. Proceed as follows:

1. Store the data. To make the data easier to work with, let $x = 1$ represent 199<u>1</u>, $x = 3$ represent 199<u>3</u>, and so forth, with the year 2000 represented as $x = 10$.

    Press $\boxed{\text{STAT}}$ $\boxed{\text{ENTER}}$. Enter each x-value as a one- or two-digit number in list L1 and the corresponding y-value in list L2, as shown in Figure 3.20.

2.   Calculate the regression line. Press $\boxed{\text{STAT}}$ $\boxed{\triangleright}$ $\boxed{4}$ to choose the LinReg(ax + b) option.

3.   Store the regression equation as Y_1. Press $\boxed{\text{VARS}}$ $\boxed{\triangleright}$ $\boxed{1}$ $\boxed{1}$.

4.   Display the results. Press $\boxed{\text{ENTER}}$ to get the display in Figure 3.21, where a is the slope of the regression line and b is the y-intercept. The equation of the regression line is approximately $y = 41.140x + 230.371$.

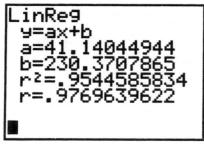

Figure 3.21   Regression line and correlation coefficient

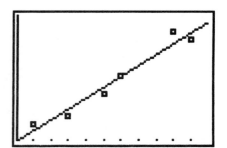

Figure 3.22   Regression line with the scatter plot

Press $\boxed{\text{ZOOM}}$ $\boxed{9}$ to see the regression line with a scatter plot of the data, as shown in Figure 3.22.

The r-value that appears in Figure 3.21 is the **coefficient of linear correlation**; it measures how close the regression line comes to passing through all of the actual data points. It can range from –1 to +1. The closer the absolute value of r is to 1, the closer the regression line fits the data. Because the r-value here is approximately 0.977, the calculated regression line fits the data very closely.

Using a Regression Line to Make Predictions

When there is strong linear correlation ($r \approx \pm 1$), the regression equation becomes a useful model for predicting y by using values of x that were not available or could not readily be observed when the original set of data was collected. This is one of the most important reasons for fitting lines to data.

- Estimating *within* the range of observed measurements for x is called **interpolation**. For example, to estimate or *interpolate* the number of college applications received for the year 1997, evaluate the regression equation when $x = 7$. If the regression equation for these data is stored as Y_1, then the easiest way of finding the predicted y-value when $x = 7$ is to create a table of values and then to scroll down to the line on which $x = 7$, as shown in Figure 3.23.

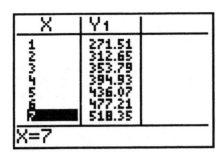

Figure 3.23 Interpolation   Figure 3.24 Extrapolation

- **Extrapolation** is estimation *outside* the range of observed measurements for x. To predict or *extrapolate* the number of college applications that will be received for the year 2009, use the same table and scroll down to $x = 19$, when the predicted y-value is 1012 applications, as shown in Figure 3.24. It is assumed, of course, that the same linear relationship between x and y exists beyond the original set of data measurements. Sometimes, unfortunately, this assumption is not warranted and can lead to faulty conclusions.

Lesson 3-7   Solving Linear Systems Graphically

KEY IDEAS

Solving a system of equations or inequalities means finding the set of all ordered pairs of numbers that make both equations or inequalities true at the same time.

Solving a System of Linear Equations by Graphing

To solve the system

$$2x + y = 5$$
$$y - x = 2,$$

graph both equations on the same set of axes.

1. Graph each equation by using its intercepts. The intercepts of $2x + y = 5$ are $\left(\dfrac{5}{2}, 0\right)$ and $(0, 5)$. The intercepts of $y - x = 2$ are $(-2, 0)$ and $(0, 2)$.

2. As shown in Figure 3.25, the solution of the system of equations is (1,3), the point at which the two lines intersect.
3. Check algebraically that (1,3) is the solution point by verifying that the solution $x = 3$ and $y = -1$ satisfies both of the original equations.

The check is left for you.

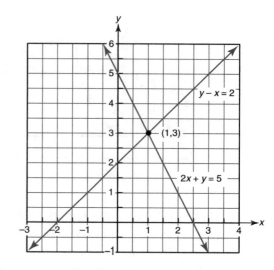

Figure 3.25 Graphs of $2x + y = 5$ and $y - x = 2$

Using the Intersect Feature of a Graphing Calculator

To solve the system

$$2x + y = 5$$
$$y - x = -4$$

using a graphing calculator, set $Y_1 = -2x + 5$ and $Y_2 = x - 4$. Then display the graph in a Decimal window. To find the point of intersection:

1. Press TRACE . Then use the cursor keys to move the cursor very close to the point of intersection of the two lines.
2. Press 2nd [CALC] to access the CALC menu. Select **5: intersect** by pressing 5 .
3. The graphs are displayed with the prompt "First curve?," as shown in Figure 3.26. Press ENTER .

Figure 3.26 First curve?

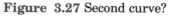

Figure 3.27 Second curve?

4. The graphs are now displayed with the prompt "Second curve?," as shown in Figure 3.27. Press ENTER .

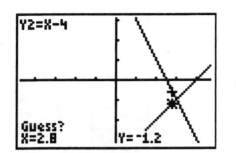

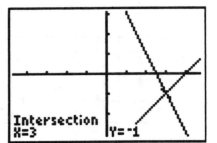

Figure 3.28 Guess?   Figure 3.29 The solution

5. The prompt "Guess?" now appears, as shown in Figure 3.28. Press $\boxed{\text{ENTER}}$ to get the display in Figure 3.29.

Solving a System of Linear Inequalities by Graphing

To find the set of all ordered pairs that satisfy two linear inequalities simultaneously, graph each inequality on the same set of axes. Then determine the region in which the two solution sets overlap.

To solve the system:

$$y > 3x - 4$$

$$x + 2y \le 6,$$

proceed as follows:

1. Graph $y > 3x - 4$.

    - First graph the boundary line, $y = 3x - 4$, using the intercepts $\left(\dfrac{4}{3}, 0\right)$ and $(0,-4)$.

      Instead of using the x-intercept, you may want to choose a point without fractional coordinates. If $x = 1$, then $y = 3 \cdot 1 - 4 = -1$, so $(1,-1)$ is a point on the line. Since the solution set of $y > 3x - 4$ does *not* include the points on the boundary line, draw the graph of $y = 3x - 4$ as a *broken* line, as shown in Figure 3.30.

    - Decide which side of the boundary line, above or below, represents the solution set. If you are not sure, pick a test point not on the boundary line, say $(0,0)$. Since $0 > 3 \cdot 0 - 4$ is true, the region on the side of the boundary line that includes $(0,0)$ represents the solution set of $y > 3x - 4$. Shade in this region, as shown in Figure 3.31.

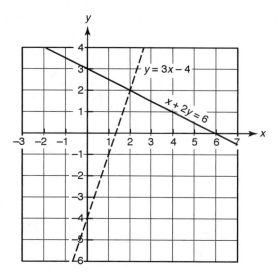

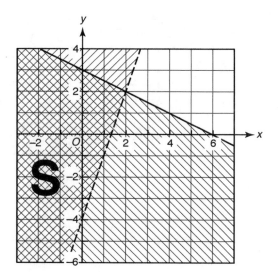

Figure 3.30 Graphing the boundary lines $y = 3x - 4$ and $x + 2y = 6$

Figure 3.31 Identifying the solution set

2. Graph $x + 2y \leq 6$ on the same set of axes.

    - Graph the boundary line, $x + 2y = 6$, as a *solid* line, as shown in Figure 3.30.

    - Use $(0,0)$ as a test point. Since $0 + 2(0) \leq 6$ is true, the region on the side of the boundary line that includes $(0,0)$ represents the solution set of $x + 2y \leq 6$. Shade in this region, as shown in Figure 3.31.

3. Identify the solution set. The solution set is the cross-hatched region in which the individual solution sets overlap. Label the solution set with an "**S**," as shown in Figure 3.31.

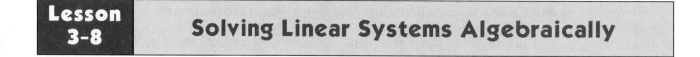

Lesson 3-8   Solving Linear Systems Algebraically

KEY IDEAS

To solve a system of linear equations algebraically, reduce the system to a single equation with only one variable. If the original system of equations is

$$3x + y = 10$$
$$y = 2x,$$

then either:

- Substitute $2x$ for y in the first equation, reducing the original system of two equations to the single equation $3x + 2x = 10$. Then $5x = 10$, so $x = \dfrac{10}{5} = 2$. To find the solution for y, substitute 2 for x in either of the two original equations. Since the second equation is simpler than the first one, replace x with 2 in the equation $y = 2x$; then $y = 2(2) = 4$. The solution for the system is (2,4).

or

- Subtract corresponding sides of the two equations, eliminating y. The result is $3x = 10 - 2x$. Then $5x = 10$, so $x = 2$ and $y = 2x = 2 \cdot 2 = 4$.

Eliminating a Variable by Substitution

Sometimes it is easy to reduce a system of linear equations in two variables to one equation in one unknown by first solving an equation for one variable in terms of the other variable. This equation can then be used to eliminate a variable in the other equation. For example, to solve the system

$$x + 2y = 7$$
$$y - 1 = 2x,$$

begin by eliminating x or y. To eliminate y:

1. Solve the second equation for y: $y = 2x + 1$.
2. Substitute $2x + 1$ for y in the first equation:

$$x + 2\overbrace{(2x + 1)}^{y} = 7$$
$$5x + 2 = 7$$
$$x = \frac{5}{5} = 1$$

3. Find the value of y when $x = 1$ by substituting 1 for x in either of the two original equations. Choosing the first equation gives $1 + 2y = 7$, so $y = \frac{6}{2} = 3$.

Hence, the solution is (1,3). The check is left for you.

Alternatively, you can obtain this solution by first eliminating x. Since $x = 7 - 2y$, x can be eliminated in the second equation by replacing it with $7 - 2y$.

Exercise 1   Solving a Word Problem

The Town Recreation Department ordered a total of 100 balls and bats for the summer baseball camp. Balls cost $4.50 each, and bats cost $20.00 each. The total purchase was $822. How many of each item were ordered?

Solution: Solve this problem either by forming a system of two linear equations in two variables or by using one equation in which both unknowns are expressed in terms of the same variable.

- Solve by using a system of two linear equations in two variables.
- Let x = number of balls ordered, and y = number of bats ordered.
  Translate each condition of the problem into an algebraic equation:

CONDITION 1: The total number of balls and bats is 100. Hence:

$$x + y = 100.$$

CONDITION 2: The total purchase is $822 when one ball costs $4.50 and one bat costs $20.00. Hence:

$$\overbrace{\$4.50x}^{\text{Cost of } x \text{ balls}} + \overbrace{\$20.00y}^{\text{Cost of } y \text{ bats}} = \$822.$$

- Solve the system of two equations in two unknowns:

$$x + y = 100$$
$$4.50x + 20.00y = 822$$

Since $x + y = 100, y = 100 - x$: $4.50x + 20(100 - x) = 822$

Remove the parentheses: $4.50x + 2000 - 20x = 822$

Combine like terms: $-15.50x = 822 - 2000$

Solve for x: $x = \dfrac{-1178}{-15.50} = 76$

Solve for y: $y = 100 - x = 100 - 76 = \mathbf{24}$

Alternatively, solve the system by using one equation in one unknown. If x = number of balls, then the difference $100 - x$ must represent the number of bats, so $\$4.50x + \$20(100 - x) = \$822$. You should verify that the solution is $x = 76$ balls and $100 - x = 100 - 76 = 24$ bats.

You can also use a calculator to solve the system $x + y = 100$ and $4.50x + 20.00y = 822$ by setting $Y1 = 100 - x$ and $Y2 = (-4.50/20)x + 822/20$. There is no need to simplify the equations. Choose an appropriate viewing window, such as $[0, +100] \times [0, 100]$. Then use the intersect feature from the CALC menu to find the coordinates of the point of intersection, as shown below.

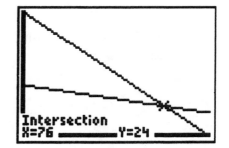

Eliminating a Variable by Adding Equations

Another way of solving a system of linear equations in two variables is to eliminate one of the variables by adding corresponding sides of the two equations. Before combining the two equations, it may be necessary to write the equations in the form $Ax + By = C$ and then multiply one or both equations by a number such that the numerical coefficients of either the x-terms or the y-terms in the two equations will have opposite numerical values.

To solve the system

$$3y = 2x + 1$$
$$5y - 2x = 7,$$

work as follows:

1. Put the first equation in the form $Ax + By = C$, and write it above the second equation with like terms aligned in the same columns:

$$3y - 2x = 1$$
$$5y - 2x = 7$$

2. Multiply both sides of the first (or the second) equation by -1 so that the coefficients of the x–terms will be opposites:

$$[-1] \cdot (3y - 2x) = [-1] \cdot (1) \longrightarrow -3y + 2x = -1$$

$$5y - 2x = 7 \longrightarrow 5y - 2x = 7$$

3. Add corresponding sides of the two equations to eliminate variable x:

$$-3y + 2x = -1$$
$$\underline{5y - 2x = 7}$$
$$2y + 0 \;=\; 6, \quad \text{so} \quad y = \frac{6}{2} = 3$$

4. Find the corresponding value of x by substituting 3 for y in either of the two original equations:

$$5y - 2x \;=\; 7$$
$$5(3) - 2x \;=\; 7$$
$$15 - 2x \;=\; 7$$
$$-2x = -8$$
$$x = \frac{-8}{-2} = 4$$

The solution is **(4,3)**. You should verify that this result works in both equations.

Exercise 2   Solving a Linear System by Combining Equations

Solve for x and y:

$$3x + 4y = 9$$
$$5x + 6y = 13.$$

Solution: Although both equations are already in the form $Ax + By = C$, the coefficients of the x-terms and the y-terms in the two equations are not opposites.

- If you decide to eliminate x, multiply both sides of the first equation by 5 and multiply both sides of the second equation by –3. The coefficient of the x-term of the first equation becomes 15, and the coefficient of the x-term in the second equation becomes –15. Then eliminate x by adding corresponding sides of the two transformed equations:

$$\big[\,5\,\big]\cdot\big(3x + 4y\big) = \big[\,5\,\big]\cdot\big(9\big) \qquad \rightarrow \qquad 15x + 20y = 45$$
$$\big[-3\big]\cdot\big(5x + 6y\big) = \big[-3\big]\cdot\big(13\big) \qquad \rightarrow \qquad \underline{-15x - 18y = -39}$$
$$2y \;=\; 6, \text{ so } y = \frac{6}{2} = 3$$

- To find x, substitute 3 for y in either of the two original equations. Replacing y with 3 in $3x + 4y = 9$ gives $3x + 12 = 9$, so $3x = -3$ and $x = -1$.

The solution is **(–1,3)**. The check is left for you.

Alternatively, you could have chosen to first eliminate y in the original system of equations by multiplying the first equation by 3 and the second equation by –2. The result would be two equivalent equations in which the coefficients of the y-terms are 12 and –12.

Chapter 3	CHECKUP EXERCISES

1—10. Multiple Choice

1. If the coordinates of A are $(-3,2)$ and the coordinates of the midpoint of $\overline{AB}$ are $(-1,5)$, what are the coordinates of B?

   (1) $(1,10)$　　　　(2) $(1,8)$　　　　(3) $(0,7)$　　　　(4) $(-5,8)$

2. The length of the line segment connecting the points whose coordinates are $(3,-1)$ and $(6,5)$ is

   (1) $\sqrt{45}$　　　　(2) 5　　　　(3) 3　　　　(4) $\sqrt{97}$

3. The graph of $x - 3y = 6$ is parallel to the graph of

   (1) $y = -3x + 7$　　(2) $y = -\dfrac{1}{3}x + 5$　　(3) $y = 3x - 8$　　(4) $y = \dfrac{1}{3}x + 8$

4. The graph of which equation is perpendicular to the graph of $y - 3 = \dfrac{1}{2}x$?

   (1) $y = -\dfrac{1}{2}x + 5$　　(2) $2y = x + 3$　　(3) $y = 2x + 5$　　(4) $y = -2x + 3$

5. Which is an equation of the line that is parallel to $y = 2x - 8$ and passes through point $(0,-3)$?

   (1) $y = 2x + 3$　　(2) $y = 2x - 3$　　(3) $y = -\dfrac{1}{2}x + 3$　　(4) $y = -\dfrac{1}{2}x - 3$

6. The point whose coordinates are $(4,-2)$ lies on a line whose slope is $\dfrac{3}{2}$. The coordinates of another point on this line could be

   (1) $(1,0)$　　　　(2) $(2,1)$　　　　(3) $(6,1)$　　　　(4) $(7,0)$

7. Given points $A(0,0)$, $B(3,2)$, and $C(2,3)$, which statement is true?

   (1) $\overline{AB} \parallel \overline{AC}$　　(2) $\overline{AB} \perp \overline{AC}$　　(3) $AB > BC$　　(4) $\overline{BC} \perp \overline{CA}$

8. What is an equation of the line that is parallel to $y - 3x + 5 = 0$ and has the same y-intercept as $y = -2x + 7$?

   (1) $y = 3x - 2$　　(2) $y = -2x - 5$　　(3) $y = 3x + 7$　　(4) $y = -2x - 7$

9. What is the solution for y in the following system of equations?

$$-2x + y = -3$$
$$x + y = 1$$

   (1) $\dfrac{2}{3}$ (2) $-\dfrac{1}{3}$ (3) $\dfrac{4}{3}$ (4) $\dfrac{7}{3}$

10. A cellular telephone company has two plans. Plan A charges \$11 a month and \$0.21 a minute. Plan B charges \$20 a month and \$0.10 a minute. After how much time, to the *nearest minute*, will the cost of plan A be equal to the cost of plan B?

   (1) 1 hr 22 min   (2) 1 hr 36 min   (3) 81 hr 8 min   (4) 81 hr 48 min

11. The coordinates of points A and B are A (–2,–3) and B (2,–5). What is an equation of the line that is perpendicular to $\overline{AB}$ at its midpoint?

12. The equations for two lines are $2x + y + 6 = 0$ and $kx + 4y - 8 = 0$. For what value of k will the two lines be perpendicular?

13. The equations for two lines are $3y - 2x = 6$ and $3x + ky = -7$. For what value of k will the two lines be parallel?

14. Determine an equation of the line that contains the point of intersection of the lines $x + y = 1$ and $2x - 3y = 7$ and is perpendicular to the line $y = \dfrac{1}{2}x$.

15–18. Solve each system of equations or inequalities graphically.

15. $2x + 2y = -10$
     $3x - \dfrac{1}{2}y = 6$

16. $2x = 8 - 5y$
     $x + y = 1$

17. $3y \geq 2x - 6$
     $x + y > 7$

18. $y \leq \dfrac{1}{2}x - 3$
     $y > -2x + 4$

19–24. Solve each system of equations algebraically.

19. $y + 5x = 0$
     $3x - 2y = 26$

20. $\dfrac{y}{4} - x = 0$
     $2y = 3x + 7$

21. $2x = 5y + 8$
     $3x + 2y = 31$

22. $2x + 3y = -6$
     $5x + 2y = 7$

23. $0.3y - 0.2x = 2.1$
     $0.75y = 2.25x$

24. $0.4a + 1.5b = -1$
     $1.2a - b = 8$

25. Five of the same type of pen cost the same as two of the same type of notebook. If one pen and two notebooks costs $4.20, what is the cost of one pen?

26. A club has 15 members consisting of seniors and juniors. After seven more seniors and three more juniors join the club, the ratio of juniors to seniors is 2 to 3. How many juniors are now in the club?

27. The formula to convert degrees Celsius into degrees Fahrenheit is $F = 1.8C + 32$. Determine algebraically the temperature at which the Celsius and Fahrenheit scales show the same value. Confirm your answer graphically and numerically by creating a table in which C changes in steps of 5.

28. Mr. Day and Ms. Knight gave the same test consisting of 20 short-answer questions, but marked their test papers using different grading systems. Mr. Day awarded 5 points for each correct answer, made no deduction for an incorrect or omitted answer, and then added 10 points to the total. Ms. Knight gave 6 points for each correct answer and subtracted 1.5 points for each incorrect or omitted answer. Two students in different classes answered the same number of questions correctly and received the same test grade. Determine, using algebraic methods, the test grade both students received. Confirm your answer graphically or by constructing a table using your graphing calculator.

29. When determining an equation that describes the relationship between the weight, w, in pounds added to a spring and the length, L, in inches of the spring, as shown in the accompanying figure, Cynthia recorded the measurements given in the accompanying table.

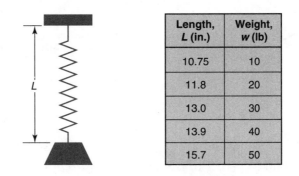

Length, L (in.)	Weight, w (lb)
10.75	10
11.8	20
13.0	30
13.9	40
15.7	50

Find an equation of the line $w = aL + b$ that best fits the data Cynthia recorded.

Estimate a and b correct to the *nearest hundredth*.

Determine the length of the spring to the *nearest tenth of an inch* when $w = 68$ pounds.

30. The accompanying table shows the boiling points of water at different altitudes.

Location	Altitude, h (km)	Boiling point, t (°C)
Wellington, New Zealand	0	100
Banff, Alberta, Canada	1.38	95
Quito, Ecuador	2.85	90
Mt. Logan, Canada	5.95	80

Find an equation of the line $y = ax + b$ that best fits the data.

Estimate a and b correct to the *nearest hundredth*.

Predict to the *nearest tenth of a degree* the boiling point of water at Lhasa, Tibet, where the altitude is 3680 meters.

31. A projector x meters from a screen throws an image on the screen that is y meters in width, as shown in the accompanying diagram. To determine how the width of the image is related to the distance of the screen from the projector, measurements were taken and recorded in the accompanying table.

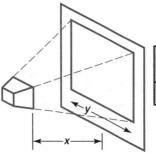

Distance, x, from screen to projector (m)	1.0	1.4	2.7	3.9	5.0
Width, y, of image (m)	0.6	0.9	1.8	2.7	3.5

Find the linear equation $y = ax + b$ that best fits the data.

Estimate a and b correct to the *nearest hundredth*.

Predict to the *nearest tenth of a meter* the width of the image when the projector is 3.0 meters from the screen.

32. The availability of leaded gasoline in New York State is decreasing, as shown in the accompanying table.

Year	1984	1988	1992	1996	2000
Gallons Available (in thousands)	150	124	104	76	50

   a. Determine a linear relationship for x (years) versus y (gallons available) that best fits the data given. Estimate the constants to the *nearest tenth*.

   b. Determine the number of gallons of leaded gasoline that will be available in New York State in the year 2005 if this relationship continues.

   c. Determine the year that leaded gasoline will first become unavailable in New York State if this relationship continues.

Chapter 4

FUNCTIONS AND QUADRATIC EQUATIONS

OVERVIEW

Functions are powerful mathematical tools for describing how one variable quantity depends on another. A function can be represented verbally, numerically, algebraically, or graphically.

Polynomial functions are equations that have the form $y =$ polynomial in x. A polynomial function is classified by its degree. The equation $y = ax + b$, with $a \neq 0$, is a first-degree polynomial function called a **linear function**. The graph of a linear function is a line with constant slope. The second-degree polynomial function $y = ax^2 + bx + c$, with $a \neq 0$, is a **quadratic function**. The graph of a quadratic function is a U-shaped curve, called a **parabola**, whose slope changes along the curve.

This chapter also looks at solving quadratic equations and inequalities, including graphing calculator approaches.

Lessons in Chapter 4

<table>
<tr><td>Lesson
4-1</td><td><h1>Function Concepts</h1></td></tr>
</table>

KEY IDEAS

Functions arise whenever one quantity depends on another. If your grade on your next math test depends on the number of hours you study, your test grade is a *function* of the number of hours studied. In mathematics, however, the term *function* has a narrower meaning. When each possible value of x is paired with only one value of y, y is said to be a function of x.

An Example of a Function

Let set X consist of five teenagers and set Y consist of their possible ages:

$$X = \{\text{Alice, Barbara, Chris, Dennis, Enid}\},$$
$$Y = \{13, 14, 15, 16, 17, 18, 19\}.$$

If each teenager in set X is paired with his or her present age in set Y, the result can be written as a set of ordered pairs:

$$\{(\text{Alice, 17}), (\text{Barbara, 13}), (\text{Chris, 16}), (\text{Dennis, 19}), (\text{Enid, 15})\}.$$

Since each teenager from set X is paired with exactly one age from set Y, the set of ordered pairs is called a **function**.

Definition of a Function

A function is a set of ordered pairs in which no two ordered pairs of the form (x,y) have the same x-value but different y-values.

If

$$f = \{(1,1), (2,3), (3,3), (4,5)\},$$

then f is a function. However, if

$$g = \{(1,2), (2,5), (3,8), (1,4)\},$$

then g is *not* a function since, when $x = 1$, two different y-values are possible.

Function Notation

If f represents a function, then $f(x)$, read as "f of x," represents the value of the function when x takes on the value inside the parentheses.

EXAMPLE: If $f(x) = x^2 + 1$, then $f(-3)$ tells you to find the value of $x^2 + 1$ when $x = -3$:

$$f(-3) = (-3)^2 + 1 = 9 + 1 = 10.$$

EXAMPLE: If function g is defined by the equation $y = x - 2^x$, then $g(5)$ refers to the value of y when $x = 5$:

$$g(5) = 5 - 2^5 = 5 - 32 = -27.$$

EXAMPLE: If $f = \{(1,1), (2,3), (3,0), (4,5)\}$, then $f(3)$ is the y-value in the ordered pair in which $x = 3$:

$$f(3) = 0 \text{ since } (3,0) \text{ belongs to function } f.$$

Exercise 1   **Evaluating a Function**

If $f(x) = x^2 - 1$, what is $f(c + 1)$?

Solution: Replace x with $c + 1$:

$$f\left(x\right) = x^2 - 1$$
$$f\left(c+1\right) = \left(c+1\right)^2 - 1$$
$$= \left(c^2 + 2c + 1\right) - 1$$
$$= \boldsymbol{c^2 + 2c}$$

Exercise 2   **Evaluating $f(-x)$, Given $f(x)$**

If $f(x) = x^2 + \dfrac{1}{x}$, find $f(-x)$.

Solution: Replace x with $-x$:

$$f\left(x\right) = x^2 + \frac{1}{x}$$
$$f\left(-x\right) = \left(-x\right)^2 + \frac{1}{-x}$$
$$= \boldsymbol{x^2 - \frac{1}{x}}$$

Domain and Range of a Function

If you think of function f as a general *rule* for determining y when the value of x is given, the notation $y = f(x)$ expresses this rule as an equation. Since the value of y depends on the choice for x, y is called the *dependent variable* and x is the *independent variable*.

- The **domain** of function f is the set of all possible values of x.
- The **range** of function f is the set of all values that y takes on as x assumes all the values in the domain.

The domain and the range of a function are, unless otherwise indicated, the largest possible sets of real numbers. Therefore, the possible values of x and y for a function may need to be restricted in some way.

EXAMPLE: The domain of $f(x) = \dfrac{1}{x}$ excludes 0 since division by 0 is not defined.

EXAMPLE: The domain of $f(x) = \sqrt{x}$ excludes negative values of x since $f(x)$ is a real number only when $x \geq 0$. The range is also limited to the set of nonnegative real numbers since $y \geq 0$ for each value of x in the domain.

Exercise 3   Finding a Restricted Domain and Range

If function f is defined by $y = \sqrt{x-1}$, find the domain and range of the function.

Solution:

- Any value of x less than 1 makes y equal to the square root of a negative number. Therefore, in order for y to be a real number, $x \geq 1$. The domain of function f is $\{x \mid x \geq 1\}$, read as "the set of all values of x such that x is greater than or equal to 1." The vertical bar inside the braces is translated as "such that." In interval notation, the domain is $[1, \infty)$.
- For any value of x in the domain, y evaluates to either 0 or a positive number. Hence, the range of function f is the set of all values of y such that $y \geq 0$; that is, the range is $\{y \mid y \geq 0\}$ *or* $[0, \infty)$.

Representing Functions

A function can be represented in four different ways:

- **Verbally** by explaining in words how each possible x-value becomes associated with the corresponding y-value.
- **Numerically** by a set of ordered pairs or a table of values.
- **Algebraically** by an equation.
- **Graphically** by a graph. The graph of function f contains point (a,b) if and only if $f(a) = b$.

Determining Whether a Graph Is a Function

Not all sets of ordered pairs, algebraic equations, or graphs represent functions. For example, the equation $x = y^2$ is *not* a function since there are ordered pairs that satisfy the equation, such as (4,2) and (4,–2), in which the same x-value is paired with different y-values. You can tell at a glance whether an equation represents a function by looking at its graph.

> ### Vertical-Line Test
>
> An equation is a function if no vertical line intersects the graph of the equation in more than one point.

The graph in Figure 4.1 represents a function since any vertical line that can be drawn intersects the graph in at most one point. The graph in Figure 4.2 is *not* a function since it fails the vertical-line test.

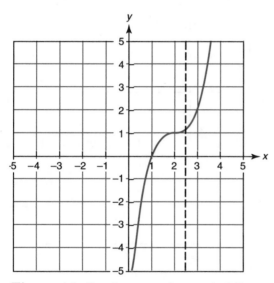

Figure 4.1 Graph passes the vertical-line test

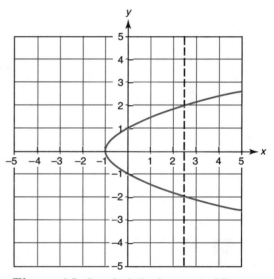

Figure 4.2 Graph *fails* the vertical-line test

Operations on Functions

If f and g represent functions, then a new function, say h, can be formed by performing any of the four arithmetic operations on functions f and g. See the accompanying table, where $f(x) = 3x$ and $g(x) = x^2 - 1$. In each case the domain of function h consists of the set of x-values for which functions f and g are both defined. For $h(x) = \dfrac{f(x)}{g(x)}$, the domain of function h excludes the values of x, if any, for which $g(x) = 0$.

Forming a New Function by Using the Four Arithmetic Operations

Operation	New Function	Domain of Function h
Addition	$h(x) = f(x) + g(x)$ $= 3x + (x^2 - 1)$ $= x^2 + 3x - 1$	{real numbers}
Subtraction	$h(x) = f(x) - g(x)$ $= 3x - (x^2 - 1)$ $= -x^2 + 3x + 1$	{real numbers}
Multiplication	$h(x) = f(x) \cdot g(x)$ $= (3x)(x^2 - 1)$ $= 3x^3 - 3x$	{real numbers}
Division	$h(x) = \dfrac{f(x)}{g(x)}$ $= \dfrac{3\sqrt{x}}{x - 1}$	$\{x \mid x \geq 0 \ \text{and} \ x \neq 1\}$

Exercise 4 Arithmetic Operations with Functions

If $g(x) = 4x$ and $h(x) = \dfrac{1}{2}x^2$ find, in terms of x:

a. $2h(x)$ b. $\left[g(x)\right]^2$ c. $g(x) - x$ d. $\dfrac{1}{2}g(x) \cdot h(x)$ e. $\dfrac{g(x+3)}{g(2)+8}$

Solutions:

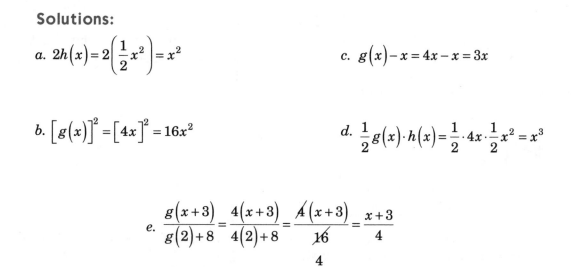

a. $2h(x) = 2\left(\dfrac{1}{2}x^2\right) = x^2$

c. $g(x) - x = 4x - x = 3x$

b. $\left[g(x)\right]^2 = \left[4x\right]^2 = 16x^2$

d. $\dfrac{1}{2}g(x) \cdot h(x) = \dfrac{1}{2} \cdot 4x \cdot \dfrac{1}{2}x^2 = x^3$

e. $\dfrac{g(x+3)}{g(2)+8} = \dfrac{4(x+3)}{4(2)+8} = \dfrac{\cancel{4}(x+3)}{\underset{4}{\cancel{16}}} = \dfrac{x+3}{4}$

Evaluating a Function of Another Function

Two functions may be linked to form a new function by using the output of one function as the input to the other function, as illustrated in Figure 4.3. The new function that is created is called a **composite function**.

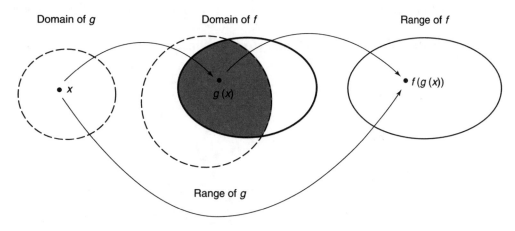

Figure 4.3 Function f composed with function g

EXAMPLE: Function f is defined by $f(x) = x - 1$, and function g is defined by $g(x) = 5x$. Functions f and g can be composed in either order.

- If the output of function g is used as the input to function f, as shown in Figure 4.3, the composite function is denoted as $f \circ g$. The notation $f(g(x))$ represents the value of the composite function for a particular x-value, where g may be referred to as the "inside" function. To evaluate $f(g(2))$, begin by evaluating the inside function for $x = 2$. Because $g(2) = 5 \cdot 2 = 10, f(g(2)) = f(10) = 10 - 1 = 9$.

- If the output of function f is used as the input to function g, the composite function is denoted as $g \circ f$ and its value is given by $g\,(f\,(x))$, where f is now the "inside" function. To evaluate $g\,(f\,(2))$, begin by evaluating the inside function for $x = 2$. Because $f\,(2) = 2 - 1 = 1, g\,(f\,(2)) = g\,(1) = 5 \cdot 1 = 5$.
- Observe that $f\,(g\,(2)) \neq g\,(f\,(2))$.

TIPS

Figure 4.3 illustrates also that composite function $f \circ g$ is defined only for the x-values in the domain of g for which $g\,(x)$ is in the domain of f.

- The values of $f\,(g\,(x))$ and $g\,(f\,(x))$ are usually not equal for the same x. The order in which functions are composed matters.

- The notation $(f \circ g)(x)$ is sometimes used instead of $f\,(g\,(x))$. If $f\,(x) = 4x$ and $g\,(x) = \sqrt{x+1}$, the composite function $f \circ g$ is defined by the equation $(f \circ g)(x) = 4\sqrt{x+1}$.

Exercise 5   Evaluating Composite Functions

If $f\,(x) = 3x + 1$ and $g(x) = \sqrt{x-1}$, find:

a. $f\,(g\,(5))$ b. $g\,(f\,(5))$ c. $g\,(f\,(x))$ d. $f\,(g\,(x + 1))$

Solutions: To evaluate a composite function, first evaluate the "inside" function:

a. To evaluate $f\,(g\,(5))$, first find $g\,(5)$. Since $g\,(5) = \sqrt{5-1} = \sqrt{4} = 2$:

$$f\left(g\left(5\right)\right) = f\left(2\right) = 3 \times 2 + 1 = \mathbf{7}.$$

b. To evaluate $g\,(f\,(5))$, first find $f\,(5)$. Since $f\,(5) = 3(5) + 1 = 16$:

$$g\left(f\left(5\right)\right) = g\left(16\right) = \sqrt{16-1} = \sqrt{\mathbf{15}}.$$

c. Since $f\left(x\right) = 3x+1$, $g\left(f\left(x\right)\right) = g\left(3x+1\right) = \sqrt{\left(3x+1\right)-1} = \sqrt{\mathbf{3x}}$.

d. Since $g\left(x+1\right) = \sqrt{\left(x+1\right)-1} = \sqrt{x}$, $f\left(g\left(x+1\right)\right) = f\left(\sqrt{x}\right) = \mathbf{3\sqrt{x}+1}$.

Quadratic Functions and Their Graphs

KEY IDEAS

The graph of a **linear function**, $f(x) = ax + b$, is a line. The graph of a **quadratic function**, $f(x) = ax^2 + bx + c$, is a special U-shaped curve called a **parabola**.

On your graphing calculator, x^2 is entered in the $Y =$ editor by pressing either $\boxed{\text{x,T,}\theta\text{,n}}$ $\boxed{x^2}$ or $\boxed{\text{x,T,}\theta\text{,n}}$ $\boxed{\wedge}$ $\boxed{2}$.

Terms Related to Parabolas

The graph of a quadratic function of the form $f(x) = ax^2 + bx + c$, with $a \neq 0$, has a vertical **axis of symmetry** that divides the parabola into two parts that, when folded along the axis, exactly coincide. The axis of symmetry intersects a parabola at its turning point, called the **vertex**, as shown in Figure 4.4. The vertex of the parabola is always the lowest (Figure 4.4(a)) or the highest (Figure 4.4(b)) point on the parabola.

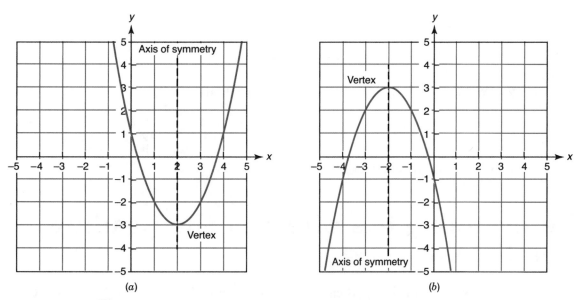

Figure 4.4 Parabolas with minimum and maximum vertex points

Finding the Vertex of a Parabola Without Graphing

The location of the axis of symmetry and the coordinates of the vertex of the graph of $f(x) = ax^2 + bx + c$ depend only on the coefficients a and b.

Parabola Formulas

- Equation of axis of symmetry: $x = -\dfrac{b}{2a}$

- Coordinates of vertex: $\left(-\dfrac{b}{2a}, f\left(-\dfrac{b}{2a} \right) \right)$

To find the coordinates of the vertex of $f(x) = 2x^2 - 8x + 7$:

1. Find the x-coordinate of the vertex by using the formula $x = -\dfrac{b}{2a}$, where $a = 2$ and $b = -8$:

$$x = \frac{-b}{2a} = \frac{-(-8)}{2(2)} = \frac{8}{4} = 2 \,.$$

2. Find the y-coordinate of the vertex by evaluating $f(2)$:

$$f(2) = 2(2)^2 - 8(2) + 7 = -1 \,.$$

Hence, the vertex is at $(2, -1)$.

Determining the Effect of the Leading Coefficient

You can tell from the coefficient of the x^2-term of $f(x) = ax^2 + bx + c$ whether the vertex of a parabola will be a minimum or a maximum point. When $a > 0$, the vertex is the *lowest* point on the graph, and when $a < 0$, the vertex is the *highest* point on the graph, as shown in Figure 4.5.

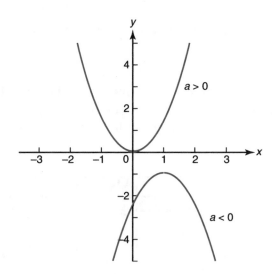

Figure 4.5 Effect of the sign of the coefficient of the x^2-term

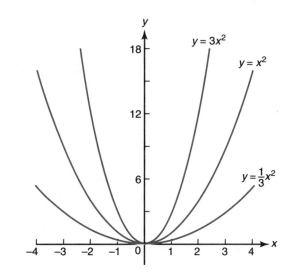

Figure 4.6 Effect of the value of the coefficient of the x^2-term

The width of the parabola depends on $|a|$. Figure 4.6 compares the graphs of $y = \dfrac{1}{3}x^2, y = x^2$, and $y = 3x^2$. Of the three graphs, the parabola of $y = 3x^2$ is the narrowest and the parabola of $y = \dfrac{1}{3}x^2$ is the widest. In general, as $|a|$ decreases, the width of the parabola increases.

Using a Calculator to Find the Vertex of a Parabola

The minimum or maximum feature in the CALC menu of your calculator can be used to find the coordinates of the vertex of a parabola. For example, to find the coordinates of the vertex of $y = 2x^2 - 8x + 7$, set $Y_1 = 2X \char94 2 - 8X + 7$ and display the graph in an appropriate window, such as $[-4.7, 4.7] \times [-3.1, 3.1]$.

1. Open the CALC menu by pressing $\boxed{\text{2nd}}$ [CALC]. Since the vertex of $y = 2x^2 - 8x + 7$ is the lowest point on the parabola, select 3:**minimum**. (If the vertex is the highest point on the parabola, select 4:**maximum**.)

2. Move the cursor slightly to the left of the vertex and press $\boxed{\text{ENTER}}$. Then move the cursor slightly to the right of the vertex and press $\boxed{\text{ENTER}}$ two times.

3. Read the coordinates of the vertex at the bottom of the window, as shown in Figure 4.7. If you do not get the exact value for the x-coordinate, you may need to adjust the window variables so that you are looking at the graph in a friendly window.

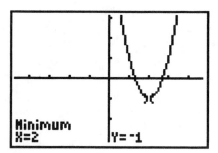

Figure 4.7 Locating the vertex   **Figure 4.8** Table of values

Graphing a Parabola Using a Table of Values

If you need to draw a parabola on graph paper, plot three points on either side of the vertex and connect the points with a smooth curve. To find the coordinates of these points, use the built-in table feature of your graphing calculator. For example, if the equation $y = 2x^2 - 8x + 7$ has already been stored in your calculator, you can obtain the table of values shown in Figure 4.8 by working as follows:

1. Press $\boxed{\text{2nd}}$ [TBLSET].

2. Change the **TblStart** value to –1 since –1 is 3 units less than the x-coordinate of the vertex. If necessary, set ΔTbl = 1 so that x increases in steps of 1 unit.

3. Press $\boxed{\text{2nd}}$ [TABLE]. If you need to look at table entries that are not currently in view, use a cursor key to scroll up or down.

It's easy to tell from the table that the coordinates of the vertex are (2,–1) because corresponding points above and below this point have the same y-coordinate.

Solving Quadratic Equations

KEY IDEAS

If the product of two numbers is 0, you can be sure that at least one of the numbers is 0. This principle, called the **zero-product rule**, applies also to the product of two algebraic expressions. If the left side of the quadratic equation $ax^2 + bx + c = 0$ can be factored, then the zero-product rule can be used to help find the roots of the equation.

Not all quadratic equations can be solved by factoring. If a quadratic equation has real roots but is *not* factorable over the set of rational numbers, the roots are irrational. These irrational roots can be approximated by locating the x-intercepts of the graph of $y = ax^2 + bx + c$.

Solving a Quadratic Equation by Factoring

Before solving a quadratic equation by factoring, it may be necessary to rewrite it so that all of the nonzero terms are on the same side of the equation with 0 on the other side. To solve $x^2 + 4x = 5$, for example, first rewrite the equation as $x^2 + 4x - 5 = 0$, and then as $(x + 5)(x - 1) = 0$. Because of the zero-product rule:

$$x + 5 = 0 \ or \ x - 1 = 0, \ so \ x = -5 \ or \ x = -1.$$

Exercise 1   Solving a Factorable Quadratic Equation

Find the solution set of $28 - x^2 = 3x$.

Solution: If $28 - x^2 = 3x$, then $-x^2 - 3x + 28 = 0$. Make the coefficient of the x^2-term positive by multiplying each member of the equation by -1. This is equivalent to changing the sign of each term to its opposite: $x^2 + 3x - 28 = 0$. Then $(x + 7)(x - 4) = 0$, so $x + 7 = 0$ or $x - 4 = 0$. Since $x = 7$ or $x = -4$, the solution set is $\{-7, 4\}$.

Exercise 2   Finding Values for Which a Fraction Is Undefined

For what values of x is the fraction $\dfrac{x^2 - 1}{2x^2 - 5x - 3}$ undefined?

Solution: A fraction is undefined for any value of a variable that makes the denominator evaluate to 0. To find the values of x for which the fraction $\dfrac{x^2-1}{2x^2-5x-3}$ is undefined, set the denominator equal to 0 and solve for x. If $2x^2 - 5x - 3 = 0$, then $(2x + 1)(x - 3) = 0$, so $2x + 1 = 0$ *or* $x - 3 = 0$. Hence, the fraction is undefined when $x = -\dfrac{1}{2}$ or $x = \mathbf{3}$.

Solving a Quadratic Equation by Graphing

If a quadratic equation $ax^2 + bx + c = 0$ has real roots, these roots are the x-coordinates of the points where the graph of $y = ax^2 + bx + c$ crosses the x-axis since, at each of the x-intercepts, $y = 0$. In Figure 4.9, the graph of $y = x^2 + 2x - 3$ intersects the x-axis at $(-3,0)$ and $(1,0)$. Hence, $x = -3$ and $x = 1$ are the roots of the equation $0 = x^2 + 2x - 3$.

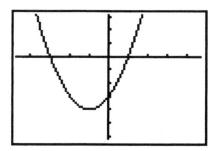　　　　　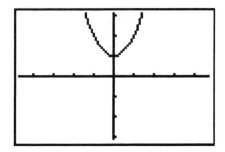

Figure 4.9 Graph of $y = x^2 + 2x - 3$ 　　　 **Figure 4.10** Graph of $y = x^2 + 1$

In general, the real roots of $ax^2 + bx + c = 0$ are the x-intercepts, if any, of the graph of $y = ax^2 + bx + c$. Some parabolas, such as $y = x^2 + 1$ in Figure 4.10, do not have any x-intercepts. This means that $0 = x^2 + 3$ does not have real roots.

Exercise 3   **Approximating the Roots of a Quadratic Equation Graphically**

Solve $x^2 - 3x - 2 = 0$ graphically, and approximate the roots to the *nearest hundredth*.

Solution: Set $Y_1 = X \char`\^ 2 - 3X - 2$, and display the graph in a friendly window such as $[-4.7,4.7] \times [-6.2,6.2]$, as shown in the accompanying graph on the left. Since $x^2 - 3x - 2$ is not factorable, the x-intercepts of the parabola represent irrational roots of $x^2 - 3x - 2 = 0$. From the accompanying graph you know that one root is between 0 and -1, and the other is between 3 and 4.

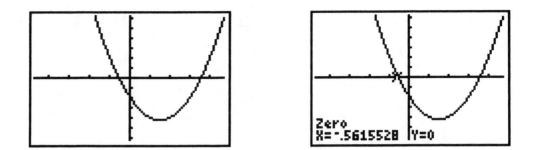

To approximate the negative root, select **2: zero** from the CALC menu.

- Move the cursor to a point on the graph that is slightly to the left of the negative
  x-intercept. Press ENTER .

- Move the cursor to a point on the graph that is slightly to the right of the negative
  x-intercept. Press ENTER two times.

- Read, from the bottom left corner of the viewing window, that the x-intercept is
  –0.5615528, as shown in the accompanying graph on the right.

By moving the cursor slightly to the left of the positive x-intercept and following the
same procedure, you can verify that the positive x-intercept is 3.5615528. Hence, the roots
of $x^2 - 3x - 2 = 0$, to the *nearest hundredth*, are **–0.56** and **3.56**.

Solving a Quadratic Equation with No Middle Term

A quadratic equation that has no linear "x-term" can be solved by writing the equation in
the form $x^2 = k$ so that $x = +\sqrt{k}$ *or* $x = -\sqrt{k}$, which may be abbreviated as $x = \pm\sqrt{k}$.
To solve $2x^2 - 50 = 0$, in which the x-term is missing, solve for the x^2-term. Thus,
$2x^2 - 50 = 0$ becomes $2x^2 = 50$, so $x^2 = \dfrac{50}{2} = 25$. Then take the square root of each side
of the equation: $x = \pm\sqrt{25} = \pm 5$.

Solving Higher Degree Equations by Factoring

Some equations of degree greater than 2 can be solved by factoring.

EXAMPLE: To solve $x^3 + 5x^2 - 6x = 0$, factor the left side completely:

$$x\left(x^2 + 5x - 6\right) = 0$$
$$x\left(x - 1\right)\left(x + 6\right) = 0$$
$$x = 0 \;\; or \;\; x - 1 = 0 \;\; or \;\; x - 6 = 0$$

Hence, the three roots are 0, 1, and 6.

EXAMPLE: To solve $3x^3 + x^2 - 12x - 4 = 0$, factor the left side by grouping pairs of terms together:

$$\left(3x^3 + x^2\right) + \left(-12x - 4\right) = 0$$
$$x^2\left(x + 3\right) - 4\left(x + 3\right) = 0$$
$$\left(x + 3\right)\left(x^2 - 4\right) = 0$$
$$x + 3 = 0 \quad or \quad x^2 - 4 = 0$$
$$x = -3 \qquad\qquad x = \pm 2$$

Hence, the three roots are –3, –2, and 2.

Solving Equations That Are Quadratic in Form

The quadratic equation $x^2 - 5x + 6 = 0$ contains two variable terms, and the greatest exponent is two times the other exponent. The fourth-degree equation $x^4 - 10x^2 + 9 = 0$ has the form of a quadratic equation since it also has two variable terms and the greatest exponent, 4, is also two times the other exponent, 2. To solve $x^4 - 10x^2 + 9 = 0$:

1. Think of x^2 as u, and x^4 as u^2. The original equation now looks like an ordinary quadratic equation:

    $$u^2 - 10u + 9 = 0.$$

2. Solve the quadratic equation by factoring:

    $$\left(u - 1\right)\left(u - 9\right) = 0$$
    $$u - 1 = 0 \;\; or \;\; u - 9 = 0$$
    $$u = 1 \qquad\quad u = 9$$

3. Remember that u stands for x^2:

-   $x^2 = 1$, so $x = \pm\sqrt{1} = \pm 1$;

    or

-   $x^2 = 9$, so $x = \pm\sqrt{9} = \pm 3$.

Hence, the four roots of the equation are –3, –1, 1, and 3. As you gain more experience in solving this type of equation, a temporary "change of variable" may not be necessary. The original equation can be factored as $(x^2 - 1)(x^2 - 9) = 0$ without introducing a new variable.

Exercise 4   **Solving an Equation with a Quadratic Form**

Solve $x + \sqrt{x} - 6 = 0$.

Solution: Since $\sqrt{x} = x^{\frac{1}{2}}$, the equation $x + \sqrt{x} - 6 = 0$ has the form of a quadratic. Make a "u-substitution" by temporarily replacing $\sqrt{x}$ with u, and x with u^2:

$$u^2 + u - 6 = 0$$
$$\left(u + 3\right)\left(u - 2\right) = 0$$
$$u = -3 \;\; or \;\; u = 2$$

Find x, remembering that $u = \sqrt{x}$:

-   If $u = -3$, then $\sqrt{x} = -3$, which is not possible since the square root of a number must be nonnegative. Reject this root.
-   If $u = 2$, then $\sqrt{x} = 2$, so $x = \mathbf{4}$.

You could, of course, have solved the equation directly by factoring.
If $x + \sqrt{x} - 6 = 0$, then

$$\left(\sqrt{x} + 3\right)\left(\sqrt{x} - 2\right) = 0 \text{, so } \sqrt{x} = -3 \;\; or \;\; \sqrt{x} = 2.$$

Reject $\sqrt{x} = -3$. Since $\sqrt{x} = 2$, $x = \mathbf{4}$.

The original equation could also have been solved by isolating the radical:

$$\sqrt{x} = 6 - x$$
$$x = \left(6 - x\right)^2 = 36 - 12x + x^2$$
$$x^2 - 13x + 36 = 0$$
$$\left(x - 4\right)\left(x - 9\right) = 0, \text{ so } x = 4 \;\; or \;\; x = 9$$

You should verify that $x = 4$ works in the original equation, but $x = 9$ fails and is rejected.

| # Solving a Linear-Quadratic System

KEY IDEAS

A linear-quadratic system can be solved graphically or algebraically by using the linear equation to help eliminate one of the two variables in the quadratic equation.

Solving a Linear-Quadratic System Graphically

A linear-quadratic system of equations can be solved by using your calculator to find the points of intersection of a line and a parabola. To solve the system $y = -x^2 + 4x - 3$ and $x + y = 1$ using a calculator, graph the two equations in a friendly window such as $[-4.7, 4.7] \times [-6.2, 6.2]$. Then determine the coordinates of the points of intersection as follows:

1. Select **5: intersect** from the CALC menu.
2. Move the cursor so that it is close to the first point of intersection of the two graphs.
3. Press $\boxed{\text{ENTER}}$ three times. The coordinates of the point of intersection, (1,0), will appear at the bottom of the screen, as shown in Figure 4.11(a).

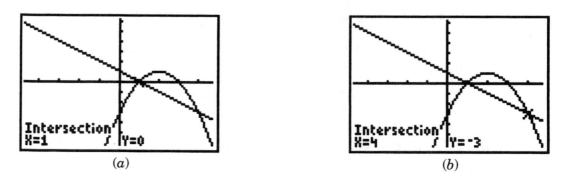

(a)   (b)

Figure 4.11 Using the intersect function in the CALC menu to find the points of intersection of $y = -x^2 + 4x - 3$ and $x + y = 1$

4. Repeat the first three steps for the second point of intersection. The second point of intersection, (4,–3), is shown in Figure 4.11(b).

T I P S

The x-coordinates of the points of intersection of the graphs in Figure 4.11 represent the roots of the equation $f(x) = g(x)$, where $f(x) = -x^2 + 4x - 3$ and $g(x) = 1 - x$. The same approach can be used to find the real roots, provided that they exist, of any equation that has the form $f(x) = g(x)$.

- Set $Y_1 = f(x)$ and $Y_2 = g(x)$:

$$\overbrace{-x^2 + 4x - 3}^{Y_1} = \overbrace{1 - x}^{Y_2}.$$

- Graph the functions in an appropriate viewing window so that their point or points of intersection are visible.
- Use the **intersect** option from the CALC menu to find the x-coordinates of any points of intersection.

Solving a Linear-Quadratic System Algebraically

You can solve the system $y = -x^2 + 4x - 3$ and $x + y = 1$ algebraically by solving the linear equation for y and then replacing y with its equivalent in the quadratic equation. Since $x + y = 1$, $y = 1 - x$. Then $y = 1 - x = -x^2 + 4x - 3$, which, after simplifying, becomes $x^2 - 5x + 4 = 0$. Thus, $(x - 1)(x - 4) = 0$, so $x = 1$ or $x = 4$.

- If $x = 1$, then $1 + y = 1$, so $y = 0$. Hence, $(1,0)$ is a solution.
- If $x = 4$, then $4 + y = 1$, so $y = -3$. Hence, $(4,-3)$ is a solution.

Exercise 1   Solving a Linear-Quadratic System Graphically

Solve $y = 28 - x^2$ and $y = 3x$ graphically. Confirm your answer algebraically.

Solution: Set $Y_1 = 28 - X \wedge 2$ and $Y_2 = 3X$. You should verify that the graphs of Y_1 and Y_2 intersect at $(-7,-21)$ and $(4,12)$, as shown in the accompanying figures.

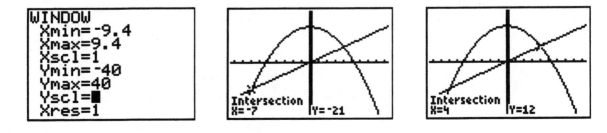

Confirm algebraically. Replace y in the quadratic equation with $3x$. Then $3x = 28 - x^2$ or, equivalently, $x^2 + 3x - 28 = 0$, which can be solved by factoring:

$$(x+7)(x-4)=0$$
$$(x+7)=0 \quad \text{or} \quad (x-4)=0$$
$$x=-7 \qquad\qquad x=4$$

- If $x = -7$, then $y = 3x = 3(-7) = -21$, so **(–7,–21)** is a solution.
- If $x = 4$, then $y = 3 \cdot 4 = 12$, so **(4,12)** is a solution.

Applying Quadratic Equations

KEY IDEAS

Often it is possible to find the maximum or minimum value of a real-world quantity, y, by expressing y as a quadratic function of a related variable, x, and then finding the y-coordinate of the vertex of its graph.

Projectile Motion

When an object such as a ball is tossed in the air, its height as a function of time can be described by a parabola in which the "x-coordinate" represents how much time, t, has elapsed since the object was launched and the "y-coordinate" represents the height, h, at time t, as shown in Figure 4.12. The y-coordinate of the vertex of the parabola represents the maximum height of the object.

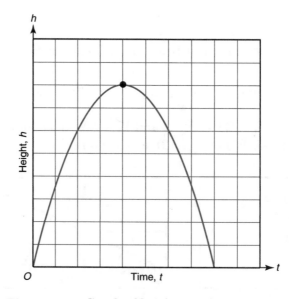

Figure 4.12 Graph of height as a function of time

Exercise 1   Projectile Motion

A ball is tossed straight up in the air from ground level and after t seconds reaches a height of h feet, where $h(t) = 88t - 16t^2$. How many seconds after the ball is tossed will it return to the ground?

Solution: When the ball returns to the ground, $h = 0$. Hence, find the value of t that makes $h(t) = 0$ when $t > 0$:

$$h\left(t\right) = 88t - 16t^2$$
$$0 = 88t - 16t^2$$
$$0 = 8t\left(11 - 2t\right)$$

Therefore, either:

- $8t = 0$, so $t = 0$, which represents the time at which the ball is tossed up in the air;

- $11 - 2t = 0$, so $t = \dfrac{11}{2} = \mathbf{5.5}$, which represents the number of seconds required for the ball to return to the ground.

You can also solve this problem graphically by first graphing $Y_1 = 88X - 16X \char94 2$ in an appropriate viewing window. The values of the window variables for one such window are shown in the accompanying figure at the left. Then use the TRACE feature to move along the curve until $y = 0$, as shown in the figure on the right.

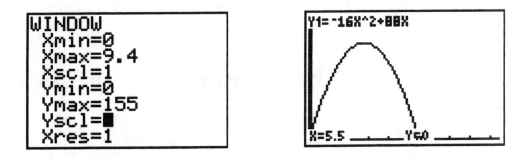

Exercise 2   Projectile Motion

A model rocket is launched from ground level. At t seconds after it is launched, it is h meters above the ground, where $h(t) = -4.9t^2 + 68.6t$. What is the maximum height, to the *nearest meter*, attained by the model rocket?

Solution: Since the coefficient of the t^2-term is negative, the vertex of the graph of $h(t) = -4.9t^2 + 68.6t$ is the highest point on the curve. To find the maximum height, find the y-coordinate of the vertex of the parabola.

- The x-coordinate of the vertex is $t = -\dfrac{b}{2a} = -\dfrac{68.6}{2(-4.9)} = -\dfrac{68.6}{-9.8} = 7$.

- To find the y-coordinate of the vertex, evaluate $h(7)$:

$$h(7) = -4.9(7)^2 + 68.6(7)$$
$$= -240.1 \quad + 480.2$$
$$= 240.1$$

- To the *nearest meter*, the maximum height is **240**.

TIPS

For Exercise 2:

- The value of $h(7)$ can be obtained by using your graphing calculator to display a table of values for $Y_1 = -4.9X \; ^\wedge \; 2 + 68.6X$ and then locating the line on which $x = 7$. The corresponding value of y is $h(7)$.
- A graphing solution can be used. Using your graphing calculator, display the graph of $Y_1 = -4.9X \; ^\wedge \; 2 + 68.6X$ in a viewing window in which the graph fits, such as [–15,15] by [–100,300]. Then use the maximum feature from the CALC menu of your graphing calculator to obtain the y-coordinate of the vertex.

Maximizing the Area of a Rectangle

If a fixed amount of fencing is available to enclose a rectangular region, the area of the enclosed rectangle can be expressed as a quadratic function of the length (or width) of the rectangle.

Exercise 3   Maximizing the Area of a Rectangle

Stacy has 30 yards of fencing with which to enclose a rectangular garden. If all of the fencing is used, what is the maximum area of the garden that can be enclosed?

Solution: Let x represent the length of the enclosed rectangular garden and w represent its width, as shown in the accompanying diagram. Since all 30 yards of fencing must be used, $2x + 2w = 30$, so $x + w = 15$ and $w = 15 - x$. Let $A(x)$ represent the area of the enclosed rectangle as a function of x.

- Because $A = xw$ and $w = 15 - x$,
  $A(x) = x(15 - x) = -x^2 + 15x$.
- The maximum value of $A(x)$ occurs
  at the vertex of the graph of the
  quadratic area function:

$$x = -\frac{b}{2a} = -\frac{15}{2(-1)} = 7.5.$$

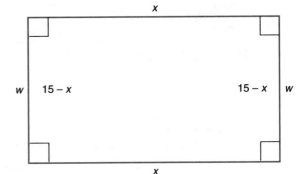

- At $x = 7.5$:

$$A(7.5) = -7.5^2 + (15)(7.5) = -56.25 + 112.5 = \mathbf{56.25 \ yd^2}.$$

Solving a Business Problem

Some types of business problems lead to quadratic functions.

Exercise 4   Maximizing Revenue

The marketing department at Sports Stuff found that approximately 600 pairs of running shoes are sold monthly when the average price of each pair is $90. It was also observed that, for each $5 reduction in price, an additional 50 pairs of running shoes are sold monthly. What price per pair will maximize the store's monthly revenue from the sale of running shoes?

Solution: If x represents the number of $5 reductions in the price of a pair of running shoes, then $50x$ represents the number of *additional* pairs of running shoes that will be sold during the month. Hence, $600 + 50x$ is the total number of pairs of running shoes that will be sold at the reduced price of $90 - 5x$ for each pair.

- The store's monthly revenue, R, from the sale of running shoes is the price of each pair times the number of pairs sold:

$$R(x) = (90 - 5x)(600 + 50x) = 54,000 + 1500x - 250x^2.$$

- The maximum value of $R(x)$ occurs at the vertex of the graph of the quadratic revenue function:

$$x = -\frac{b}{2a} = -\frac{1500}{2(-250)} = 3.$$

- When $x = 3$, the reduced price of a pair of running shoes is:

$$90 - 5x = 90 - (5 \times 3) = \mathbf{\$75}.$$

This price maximizes the monthly revenue from the sale of running shoes.

Solving Quadratic Inequalities

KEY IDEAS

Quadratic inequalities such as $x^2 - 2x - 3 < 0$ and $x^2 - 2x - 3 > 0$ can be solved graphically or algebraically by solving the corresponding quadratic equations.

Solving Quadratic Inequalities Graphically

Use your graphing calculator to obtain the graph of $y = x^2 - 2x - 3$, as shown in Figure 4.13.

- To solve $x^2 - 2x - 3 < 0$, find the interval of x for which the graph is below the x–axis. Since $y < 0$ when x is between -1 and 3, the solution is $-1 < x < 3$.
- To solve $x^2 - 2x - 3 > 0$, find the interval of x for which the graph is above the x–axis. The solution is $x < -1$ *or* $x > 3$.

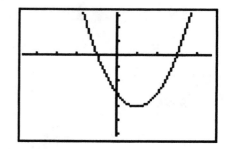

Figure 4.13 Graph of $y = x^2 - 2x - 3$

Solving Quadratic Inequalities Algebraically

From the graphical solutions of $x^2 - 2x - 3 < 0$ and $x^2 - 2x - 3 > 0$, you know that the roots of the related quadratic equation are the boundary points of the solution intervals. Therefore, you can generalize that, if r and R are unequal roots of $ax^2 + bx + c = 0$, where a is positive and $r < R$, the values of x that satisfy the related quadratic inequalities are contained in the intervals shown in Figure 4.14.

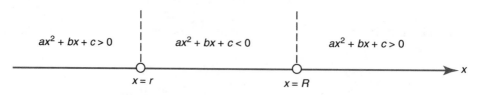

Figure 4.14 Locating the roots of a quadratic inequality on a number line

For example, to solve $x^2 - 3x - 10 < 0$ algebraically, first solve $x^2 - 3x - 10 = 0$ by factoring the left side of the equation:

$$(x + 2)(x - 5) = 0, \text{ so } x = -2 \text{ } or \text{ } x = 5.$$

Then plot –2 and 5 on a number line, as shown in Figure 4.15. Since the middle interval is the solution of the quadratic inequality in the "less than" case, the solution is $-2 < x < 5$.

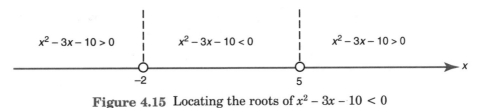

Figure 4.15 Locating the roots of $x^2 - 3x - 10 < 0$

TIP

If the leading coefficient of the quadratic inequality is negative, rewrite the inequality so that it has a positive leading coefficient. For example, to solve $8x - x^2 > 15$, first write the inequality in standard form: $-x^2 + 8x - 15 > 0$. Then make the coefficient of the x^2-term positive by changing the sign of each term to its opposite *and* reversing the direction of the inequality: $x^2 - 8x + 15 < 0$.

General Solution to a Quadratic Inequality

Probably you have already noticed that the solution to a quadratic inequality has a predictable form.

> ### Form of the Solution to a Quadratic Inequality
>
> If r and R are the unequal roots of the quadratic equation $ax^2 + bx + c = 0$ with $r < R$ and $a > 0$, then the corresponding inequalities with their solution intervals are:
>
> - $ax^2 + bx + c < 0; r < x < R$ and $ax^2 + bx + c \leq 0; r \leq x \leq R$.
> - $ax^2 + bx + c > 0; x < r$ or $x > R$ and $ax^2 + bx + c \geq 0; x \leq r$ or $x \geq R$.

Exercise 1   Solving a Quadratic Inequality

Find the solution set of $8 - x^2 \leq 2x$.

Solution:

- Write the inequality in standard form as $-x^2 - 2x + 8 \leq 0$. Then make the coefficient of the x^2-term positive by changing the sign of each term to its opposite *and* reversing the direction of the inequality: $x^2 + 2x - 8 \geq 0$.

- Solve the related quadratic equation. If $x^2 + 2x - 8 = 0$, then $(x + 4)(x - 2) = 0$, so $x = -4$ or $x = 2$.
- Since the solution of $x^2 + 2x - 8 \geq 0$ includes the roots of $x^2 + 2x - 8 = 0$, the solution set has the form $x \leq r$ or $x \geq R$, where $r = -4$ and $R = 2$. Hence, the solution set is $\{x \mid x \leq -4 \ or \ x \geq 2\}$, which is read as "the set of all x such that x is less than or equal to 4, or x is greater than or equal to 2."

CHECKUP EXERCISES

1–18. Multiple Choice

1. If $f(x) = x^2 - 1$, then $f(a + 1) =$

    (1) $a^2 + 1$ (2) a^2 (3) $a^2 + 2a$ (4) $a^2 - 1$

2. Which of the accompanying diagrams describes a function?

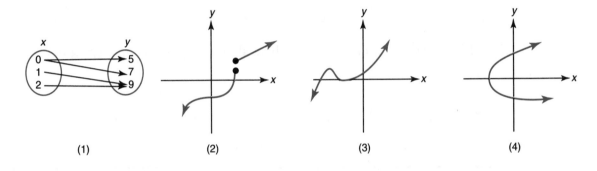

        (1)   (2)   (3)   (4)

3. If $h(x) = \dfrac{\dfrac{1}{x} - 9x}{1 + 3x}$, then $h\left(\dfrac{1}{x}\right) =$

    (1) $\dfrac{3}{x+3}$ (2) $x - 3$ (3) $\dfrac{x}{3}$ (4) $x + 3$

4. For which value(s) of x is the fraction $\dfrac{x+4}{x^2 - 2x - 3}$ *not* defined?

    (1) 1, –3   (2) –1, 3   (3) –3, –1   (4) – 4

5. Let the function g be defined by $g(x) = 1 - 5x$. If the domain of function g is $\{x \mid -3 \le x \le 2\}$, what is the smallest value in the range of the function?

    (1) –17   (2) –9   (3) 9   (4) 17

6. When a current, I, flows through a given electrical circuit, the power, W, of the circuit can be determined by the function $W(I) = 120I - 12I^2$. What is the maximum power that can be supplied to the circuit?

    (1) 5   (2) 240   (3) 300   (4) 600

7. Let the function f be defined by $f(x) = 9 - x^2$. If the domain of function f is $\{x \mid -3 \le x \le 4\}$, what is the largest value in the range of the function?

    (1) 0   (2) 4   (3) 9   (4) 25

8. If the function f is defined by $f(x) = 3x$ and the function g is defined by $g(x) = 7x - 1$, what is the value of $f(g(4))$?

    (1) 12   (2) 81   (3) 83   (4) 324

9. If the function f is defined by

    $$f = \{(0,2), (1,-1), (4,6), (2,5)\}$$

    and the function g is defined by

    $$g = \{(-1,6), (1,0), (2,4), (4,1)\},$$

    then $f(g(4)) =$

    (1) –1   (2) 1   (3) 2   (4) 6

10. Which is an equation of the parabola shown in the accompanying figure?

    (1) $y = -x^2 + 2x + 3$ (3) $y = x^2 + 2x + 3$

    (2) $y = -x^2 - 2x + 3$ (4) $y = x^2 - 2x + 3$

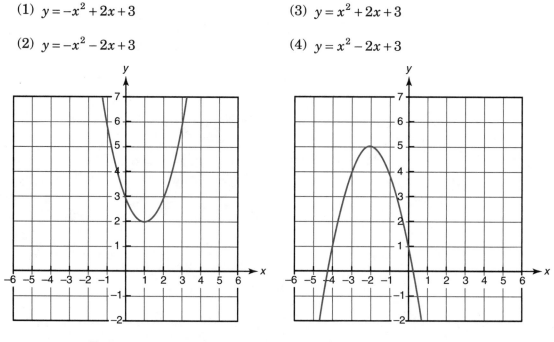

Exercise 10   Exercise 11

11. Which is an equation of the parabola shown in the accompanying figure?

    (1) $y = -x^2 - 4x + 1$ (3) $y = x^2 - 4x + 1$

    (2) $y = -x^2 + 4x + 1$ (4) $y = x^2 + 4x + 1$

12. If the function f is defined by $f(x) = 2x + k$ and the function g is defined by $g(x) = \dfrac{x-5}{2}$, for what value of k is $f(g(x)) = g(f(x))$?

    (1) –15   (2) –5   (3) 5   (4) 15

13. If $f(x) = x^2$, then $\dfrac{f(x+h) - f(x)}{h} =$

    (1) $2x + h$ (2) $x^2 - 2h$ (3) $\dfrac{x^2 - h^2}{h}$ (4) h

14. A ball is thrown into the air in such a way that its height, h, at any time, t, is given by the function

    $$h(t) = -16t^2 + 80t + 10.$$

    What is the maximum height attained by the ball?

    (1) 140   (2) 110   (3) 85   (4) 10

15. An archer shoots an arrow into the air in such a way that its height, h, at any time, t, is given by the function

    $$h(t) = -16t^2 + kt + 3.$$

    If the maximum height of the arrow occurs at time $t = 4$, what is the value of k?

    (1) 128   (2) 64   (3) 8   (4) 4

16. Which graph represents the solution set of $x^2 - x - 12 < 0$?

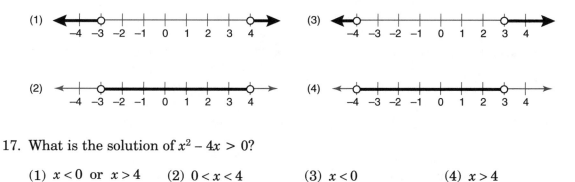

17. What is the solution of $x^2 - 4x > 0$?

    (1) $x < 0$ or $x > 4$ (2) $0 < x < 4$ (3) $x < 0$ (4) $x > 4$

18. What is the solution set of $28 \le x^2 - 3x$?

    (1) $\left\{x \mid -7 \le x \le 4\right\}$ (3) $\left\{x \mid -4 \le x \le 7\right\}$

    (2) $\left\{x \mid x \le 7 \text{ or } x \ge -4\right\}$ (4) $\left\{x \mid x \ge 7 \text{ or } x \le -4\right\}$

19. When a ball is thrown into the air, the height, h, in feet, that the ball will reach after t seconds is given by the function

    $$h(t) = -16t^2 + 30t + 6.$$

    Find, to the *nearest tenth of a foot*, the maximum height that the ball will reach.

20. If $f(x)=x^{\frac{2}{3}}$ and $g(x)=8x^{-\frac{1}{2}}$, find $(f \circ g)(x)$.

21. Let the function f be defined by $f(x) = ax + b$, where a and b are nonzero constants such that $f(5) = 1$ and $f(-3) = 25$. If $f(c) = 0$, what is the value of c?

22. Let the function f be defined by $f(x) = x^2 + 12$. If n is a positive number such that $f(3n) = 3f(n)$, what is the value of n?

23–25. a. Graph each parabola.

   b. From the graph determine the coordinates of the vertex and an equation of the axis of symmetry.

   c. Confirm the equation of the axis of symmetry by obtaining it algebraically using the formula $x = -\dfrac{b}{2a}$.

23. $y = x^2 - 2x - 3$ 24. $y = -x^2 + 6x - 7$ 25. $y = x^2 + x + 4$

26. A ball is thrown into the air so that its height, h, in feet after t seconds is given by the equation $h = 144t - 16t^2$.
   a. Find the number of seconds that the ball is in the air when it reaches a height of 180 feet.
   b. After how many seconds does the ball hit the ground?

27. The price of stock A varied over a 12-month period according to the function $A(x) = 0.75x^2 - 6x + 20$, where $A(x)$ represents the price of stock A after x months. Over the same 12-month period, the price of stock B varied according to the function $B(x) = 2.75x + 1.50$. Using graphical methods, determine the prices, to the *nearest cent*, at which both stocks had the same value.

28. The manufacturing of a metal box starts with a rectangular sheet of metal that is three times as long as it is wide. Four equal squares are cut from the four corners, and the flap at each corner is turned up to form a box 2 inches high that is open at the top. What should be the width of the rectangular sheet of metal in order for the box to have a volume of 102 cubic inches?

29–31. Using a graphing calculator, approximate the roots of the given quadratic equation correct to the *nearest hundredth*.

29. $x^2 - 6x - 1 = 0$ 30. $3x^2 - 9x + 4 = 0$ 31. $2x^2 + 5 = 10x$

32–43. Solve for the variable using algebraic methods.

32. $3x^2 - 18 = 9 - x^2$ 33. $y^2 + 3y + 2 = 0$ 34. $3x = \dfrac{x^2}{2}$

35. $h^3 - 3h^2 = 40h$

38. $5r + 3 = 2r^2$

41. $\dfrac{w-3}{w-2} = \dfrac{w+3}{2w}$

36. $3p^2 + 14p = 5$

39. $6t^2 = 7t + 3$

42. $y^{-4} - 5y^{-2} + 4 = 0$

37. $9x^2 - 12x + 4 = 0$

40. $\dfrac{n+5}{n+1} = \dfrac{n-1}{4}$

43. $2k^{\frac{2}{3}} + 7k^{\frac{1}{3}} - 4 = 0$

44 and 45. The profit that a coat manufacturer makes each day is modeled by the function

$$P(x) = -x^2 + 120x - 2000,$$

where P is the profit and x is the price of each coat sold.

44. For what values of x does the company make a profit?

45. What is the maximum profit that the company can make each day?

46–51. In each case, find the solution set.

46. $x^2 + 4x - 5 \geq 0$

48. $2y^2 < 5y + 3$

50. $6 \leq -t + t^2$

47. $7m > m^2$

49. $2x - 63 \leq -x^2$

51. $p^3 - 14p > 5p^2$

52–54. Solve each system of equations graphically or algebraically.

52.　$y = x^2 - 4x + 9$

$\quad\ y - x = 5$

53.　$y = x^2 - 6x + 6$

$\quad\ y - x = -4$

54.　$y = x^2 - 6x + 5$

$\quad\ y + 7 = 2x$

55. A rocket is launched straight up from ground level. After t seconds its height, h, in feet is given by the function

$$h(t) = -16t^2 + 376t.$$

   a. At what time will the rocket reach its maximum height? What is the maximum height?
   b. During what interval of time will the height of the rocket exceed 2040 feet?

56. A farmer is enclosing a rectangular plot of land that is bounded on one side by a river and will be fenced on the other three sides. What is the maximum area that can be enclosed if 880 yards of fencing material are used?

57. A rectangular sheet of aluminum 25 feet long and 12 inches wide is to be made into a rain gutter by folding up two parallel sides the same number of inches at right angles to the sheet. How many inches on each side should be folded up so that the gutter will have its greatest capacity?

58. Student editors are designing the yearbook so that each rectangular page will have a perimeter of 40 inches. The printer advises the student editors that each page must have a margin of $1\frac{1}{2}$ inches on the bottom and 1 inch on the other three sides. What must be the dimensions of the page if the printed area is to be as great as possible?

59. When 14 apple trees are planted per acre in a certain orchard, the average yield per tree is 480 apples per year. For each additional tree planted per acre in the same orchard, the annual yield per tree decreases by 10 apples. How many additional trees should be planted in the orchard so that the number of apples harvested in this orchard per year is as great as possible?

60. What is the greatest possible perimeter of a rectangle that can be inscribed in the parabola $y = 16 - x^2$ so that two of the vertices of the rectangle are on the x–axis and the other two are on the parabola?

Chapter 5

COMPLEX NUMBERS AND THE QUADRATIC FORMULA

OVERVIEW

The equation $x^2 + 1 = 0$ has no real roots because $x^2 = -1$ and $x = \pm\sqrt{-1}$. To give meaning to solutions of equations that involve square roots of negative numbers, the imaginary unit i is defined as $i = \sqrt{-1}$. If $x^2 + 1 = 0$, then $x = \pm\sqrt{-1} = \pm i$.

A number that has the form $a + bi$, where a and b stand for real numbers, is called a **complex number**. Because any real number a can be written in the form $a + 0 \cdot i$, the set of real numbers is a subset of the set of complex numbers. Thus, 5 is a complex number because $5 = 5 + 0 \cdot i$.

Lesson 5-1	Complex Numbers

KEY IDEAS

If a and b are real numbers, then $a + bi$, where $i = \sqrt{-1}$, is called a **complex number**. When a complex number is written in the **standard form** $a + bi$, a is called the **real part** of the complex number and b is the **imaginary part**. Thus, the real part of the complex number $3 + 2i$ is 3, and the imaginary part is 2. Arithmetic operations can be performed with complex numbers in much the same way that these operations are performed with binomials.

The Imaginary Unit

The **imaginary unit** i is equal to $\sqrt{-1}$, and $i^2 = -1$. Square roots of negative numbers other than -1 can be expressed in terms of i by factoring out $\sqrt{-1}$ and replacing it with i. For example, $\sqrt{-4} = \sqrt{4} \cdot \sqrt{-1} = 2i$. The number $2i$ is called a *pure imaginary number*. A **pure imaginary number** is the product of any nonzero real number and i.

Properties of Complex Numbers

All of the properties of arithmetic that work for real numbers apply also to complex numbers. When performing arithmetic operations with complex numbers, treat terms involving i as if they are monomials.

EXAMPLE: $3i + 5i = 8i$ and $6i - i = 5i$

EXAMPLE: $i \cdot i^2 = i^{1+2} = i^3$ and $\dfrac{i^{13}}{i^4} = i^{13-4} = i^9$

To add or subtract complex numbers of the form $a + bi$, combine first the real parts and then the imaginary parts.

EXAMPLE: $(2 + 3i) + (4 - 5i) = (2 + 4) + (3i - 5i) = 6 - 2i$

EXAMPLE: $(1 - 2i) - (-4 + 6i) = (1 - 2i) + (4 - 6i)$
$$= (1 + 4) + (-2i - 6i) = 5 - 8i$$

EXAMPLE: $\sqrt{-50} + 4\sqrt{-18} = \left(\sqrt{25} \cdot \sqrt{2} \cdot \sqrt{-1}\right) + \left(4\sqrt{9} \cdot \sqrt{2} \cdot \sqrt{-1}\right)$
$$= \quad 5\sqrt{2}\,i \quad + \quad 12\sqrt{2}\,i \quad = 17\sqrt{2}\,i$$

Graphing Complex Numbers

A complex number in $a + bi$ form can be graphed in the *complex* plane by measuring a units along the horizontal, or "real," axis and b units along the vertical, or "imaginary," axis. For example, to graph $Z_1 = 2 + 5i$ in the complex plane, plot $(2,5)$ as shown in the accompanying figure. Graph $Z_2 = -6 - 2i$ by plotting $(-6,-2)$.

The sum $Z = Z_1 + Z_2$ can be obtained graphically by drawing the diagonal of the parallelogram whose adjacent sides correspond to Z_1 and Z_2, as shown in the accompanying figure 5.1. From the graph, the coordinates of Z are $(-4,3)$. Therefore:

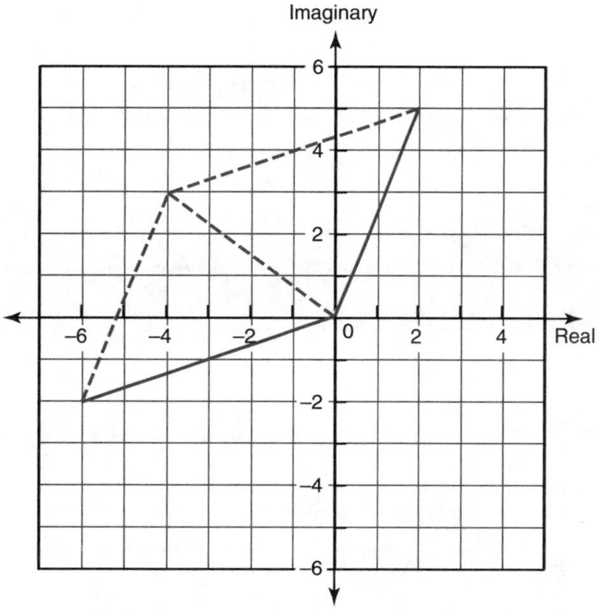

Figure 5.1 Graphing in the complex plane

$$\overbrace{\left(2+5i\right)}^{Z_1} + \overbrace{\left(-6-2i\right)}^{Z_2} = \overbrace{-4+3i}^{Z}.$$

Simplifying Powers of i

The expression i^n, where n is a positive whole number, can be reduced to either ± 1 or $\pm i$, as the following list suggests:

$$i^0 = 1 \qquad\qquad i^4 = i^2 \cdot i^2 = \left(-1\right)\left(-1\right) = 1$$
$$i^1 = i \qquad\qquad i^5 = i^4 \cdot i = \quad 1 \cdot i \quad = i$$
$$i^2 = -1 \qquad\qquad i^6 = i^4 \cdot i^2 = \quad 1 \cdot \left(-1\right) = -1$$
$$i^3 = i^2 \cdot i = -i \qquad\qquad i^7 = i^4 \cdot i^3 = \quad 1 \cdot \left(-i\right) = -i$$

Consecutive positive-integer powers of i follow a cyclic pattern that repeats every four integers. Using the list of powers of i, you should be able to predict that $i^8 = 1$, $i^9 = i$, $i^{10} = -1$, $i^{11} = -i$, ... A large power of i can be simplified by dividing the exponent by 4 and using the remainder as the new power of i. For example, to simplify i^{31}:

- Find the equivalent exponent, which is 3 since $31 \div 4 = 7$ with a remainder of 3.
- Rewrite i^{31} as i^3 and simplify: $i^{31} = i^3 = -i$.

Multiplying and Dividing Complex Numbers

KEY IDEAS

To multiply pure imaginary numbers of the form bi, use the fact that $i^2 = -1$, as in $2i \cdot 4i = 8i^2 = -8$. Complex numbers in $a + bi$ form are multiplied and divided in much the same way that binomials are multiplied and divided.

Multiplying Complex Numbers

Because complex numbers have the form of a binomial, they can be multiplied together using the familiar *FOIL* method:

$$
\begin{aligned}
(2+3i)(4-5i) &= \overbrace{(2)(4)}^{F} + \overbrace{(2)(-5i)}^{O} + \overbrace{(3i)(4)}^{I} + \overbrace{(3i)(-5i)}^{L} \\
&= 8 \quad + \quad [-10i \quad + \quad 12i] \quad + \quad (-15)i^2 \\
&= 8 \quad + \quad\quad\quad 2i \quad\quad\quad + (-15)(-1) \\
&= 23 + 2i
\end{aligned}
$$

Complex Conjugates

Two complex numbers that take the sum and the difference of the same two real and pure imaginary numbers are called **complex conjugates**. Thus, the numbers $3 + 4i$ and $3 - 4i$ are complex conjugates. The product of a pair of complex conjugates is always a positive real number. For example, $(3 + 4i)(3 - 4i) = 3^2 + 4^2 = 25$.

> **Product of Complex Conjugates**
>
> $(a + bi)(a - bi) = a^2 + b^2$

Dividing Complex Numbers

To divide $3 + i$ by $1 - 2i$:

1. Write the division example as a fraction: $\dfrac{3+i}{1-2i}$.

2. Multiply the numerator and the denominator of the fraction by the complex conjugate of the denominator:

$$\frac{3+i}{1-2i} \cdot \frac{1+2i}{1+2i} = \frac{3+7i+2i^2}{1-4i^2}.$$

3. Simplify and write the answer in standard $a + bi$ form:

$$\frac{3+i}{1-2i} \cdot \frac{1+2i}{1+2i} = \frac{5+7i}{1+4}$$

$$= 1 + \frac{7}{5}i$$

Completing the Square

KEY IDEAS

You can easily verify that $(x + k)^2 = x^2 + (2k)x + k^2$. Because $x^2 + (2k)x + k^2$ is the result of squaring a binomial, it is called a **perfect square trinomial**. The last term of the perfect square trinomial, k^2, can be obtained from the coefficient of the middle term by dividing it by 2 and then squaring the result, as in $(2k \div 2)^2 = k^2$. You should now be able to predict what number must be added to $x^2 + 6x + \boxed{?}$ in order to "complete the square" so that a perfect square trinomial is formed. Adding 9 completes the square:

$$(6 \div 2)^2 = 9, \text{ so } x^2 + 6x + \boxed{9} = \left(x+3\right)^2.$$

Completing the square can be used to help solve quadratic equations that cannot be solved by factoring.

Solving Quadratic Equations by Completing the Square

If a quadratic equation in x can be rewritten so that one side is the square of a binomial in x and the other side is a constant, the equation can be solved by taking the square root of each side. To solve $x^2 + 6x - 8 = 0$, rewrite the equation so that only terms involving x are on the same side of the equation, as in $x^2 + 6x = 8$. Then work as follows:

1. Divide the coefficient of the x-term by 2, and then square the result:

$$\frac{6}{2} = 3 \text{ and } 3^2 = \boxed{9}.$$

2. Add 9 to both sides of $x^2 + 6x = 8$:

$$x^2 + 6x + \boxed{9} = 8 + \boxed{9}.$$

3. Rewrite the left side of the equation as the square of a binomial:

$$(x+3)(x+3) = 17, \text{ so } (x+3)^2 = 17.$$

4. Find the square root of each side of the equation:

$$x + 3 = \pm\sqrt{17}, \text{ so } x = -3 \pm \sqrt{17}.$$

Exercise 1   **Solving a Quadratic Equation with Irrational Roots**

Solve for x by completing the square: $3x^2 - 12x + 5 = 0$.

Solution: First rewrite $3x^2 - 12x + 5 = 0$ as $3x^2 - 12x = -5$. Then factor the left side of the equation so that the coefficient of the x^2-term is 1:

$$3(x^2 - 4x) = -5.$$

Now complete the square of the expression inside the parentheses:

- Since $-\dfrac{4}{2} = -2$ and $(-2)^2 = 4$, add 4 to the binomial inside the parentheses:

$$3\left(x^2 - 4x + \boxed{4}\right) = -5 + \boxed{?}.$$

- Because each term inside the parentheses is being multiplied by 3, you have added $3 \times 4 = 12$ to the left side of the equation. To obtain an equivalent equation, add 12 also to the right side of the equation:

$$3\left(x^2 - 4x + \boxed{4}\right) = -5 + \boxed{12},$$

$$3\left(x - 2\right)^2 = 7, \text{ so } \left(x - 2\right)^2 = \frac{7}{3}.$$

- Find the square root of each side of the equation, and then solve for x:

$$x - 3 = \pm\sqrt{\frac{7}{3}}, \text{ so } \boldsymbol{x = 3 \pm \sqrt{\frac{7}{3}}}.$$

Equation of a Circle

Sometimes it is necessary to complete the square in order to put the equation of a circle in the standard form $(x - h)^2 + (y - k)^2 = r^2$, where r is the radius of a circle whose center is at (h,k). For example, if the equation of a circle is $(x - 1)^2 + y^2 + 6x = 16$, then completing the square in y gives

$$\left(x - 1\right)^2 + y^2 + 6x + \boxed{9} = 16 + \boxed{9}$$

$$\left(x - 1\right)^2 + \left(y + 3\right)^2 = 25$$

In the equation $(x - 1)^2 + (y + 3)^2 = 25$, $h = 1$ and $r = 5$. Because $(y + 3)^2 = (y - (-3))^2$, $k = -3$. Hence, $(x - 1)^2 + y^2 + 6x = 16$ describes a circle centered at $(1,-3)$ with radius 5.

Exercise 2   **Writing an Equation of a Circle in Standard Form**

Determine the radius and the coordinates of the center of the circle whose equation is $x^2 + 4x + y^2 - 16y + 19 = 0$.

Solution: Rewrite the given equation in the standard form $(x - h)^2 + (y - k)^2 = r^2$.

- Rewrite $x^2 + 4x + y^2 - 16y + 19 = 0$ as $(x^2 + 4x) + (y^2 - 16y) = -19$.
- Complete the square for both the x- and the y-expression:

$$\left(x^2 + 4x + \boxed{4}\right) + \left(y^2 - 16y + \boxed{64}\right) = -19 + \boxed{4} + \boxed{64}$$

$$\left(x + 2\right)^2 \quad + \quad \left(y - 8\right)^2 \quad = 49$$

- Because $h = -2$, $k = 8$, and $r = \sqrt{49} = 7$, the center of the circle is at **(-2,8)** and the length of the radius is **7**.

The Quadratic Formula

KEY IDEAS

If $ax + b = c$, solving for x gives $x = \dfrac{c-b}{a}$. Any linear equation can be solved by writing it in the form $ax + b = c$, and then substituting the values for a, b, and c in the formula $x = \dfrac{c-b}{a}$. For example, if $4x - 3 = 25$, then $a = 4$, $b = -3$, and $c = 25$, so

$$x = \frac{c-b}{a} = \frac{25 - \left(-3\right)}{4} = \frac{28}{4} = 7.$$

Similarly, if $ax^2 + bx + c = 0$, then x can be solved for in terms of the coefficients a, b, and c. The result is called the **quadratic formula**.

Using the Quadratic Formula

The method of completing the square can be used to obtain a general solution to the quadratic equation $ax^2 + bx + c = 0$, in which the roots of the equation are expressed in terms of the coefficients a, b, and c.

Quadratic Formula

If $ax^2 + bx + c = 0$ and $a \neq 0$, then

$$x = \frac{-b \pm \sqrt{b^2 - 4ac}}{2a}.$$

Any quadratic equation, including equations with rational, irrational, or imaginary roots, can be solved by writing the equation in the form $ax^2 + bx + c = 0$, and then substituting the numerical values of a, b, and c into the quadratic formula.

Exercise 1   Solving a Quadratic Equation by Formula

Find the roots of $2x^2 - 3x - 1 = 0$ in radical form.

Solution: Use the quadratic formula, where $a = 2$, $b = -3$, and $c = -1$:

$$x = \frac{-b \pm \sqrt{b^2 - 4ac}}{2a}$$

$$= \frac{-(-3) \pm \sqrt{(-3)^2 - 4(2)(-1)}}{2(2)}$$

$$= \frac{3 \pm \sqrt{17}}{4}$$

If the two roots of a quadratic equation are denoted by x_1 and x_2, then

$$x_1 = \frac{3 + \sqrt{17}}{4} \quad \text{and} \quad x_2 = \frac{3 - \sqrt{17}}{4}.$$

Although you could use a graphing calculator to estimate the roots of this equation, the quadratic formula allows you to obtain an exact representation of the irrational roots in radical form.

Finding Imaginary Roots of Quadratic Equations

If a quadratic equation has imaginary roots, its graph does not intersect the x-axis. The quadratic formula can be used to find these imaginary roots.

Exercise 2   Finding the Imaginary Roots of a Quadratic Equation

Find the roots of $x^2 + 5 = 2x$ in $a + bi$ form.

Solution: If $x^2 + 5 = 2x$, then $x^2 - 2x + 5 = 0$. Use the quadratic formula, where $a = 1$, $b = -2$, and $c = 5$:

$$x = \frac{-b \pm \sqrt{b^2 - 4ac}}{2a}$$

$$= \frac{-(-2) \pm \sqrt{(-2)^2 - 4(1)(5)}}{2(1)}$$

$$= \frac{2 \pm \sqrt{-16}}{2}$$

$$= \frac{2 \pm 4i}{2} = \frac{2}{2} \pm \frac{4i}{2} = 1 \pm 2i$$

Hence, $x_1 = 1 + 2i$ and $x_2 = 1 - 2i$.

Making Graphical Connections

The expression $b^2 - 4ac$ underneath the radical sign in the quadratic formula is called the **discriminant**. The discriminant determines the nature of the roots of a quadratic equation as well as the number of x-intercepts of its graph. See the accompanying table.

Predicting the Type of Roots of $ax^2 + bx + c = 0$ $(a \neq 0)$

Discriminant	Type of Roots	Graph
$b^2 - 4ac > 0$	Real and unequal roots. If $b^2 - 4ac$ is a perfect square, the roots will be rational.	Two x-intercepts
$b^2 - 4ac = 0$	Two equal rational roots.	One x-intercept at vertex
$b^2 - 4ac < 0$	Two imaginary roots.	No x-intercept

| Exercise 3 | Describing the Nature of the Roots of a Quadratic Equation |

Describe the type of roots of each equation without solving the equation:

   a. $3x^2 - 11x = 4$ *b.* $-2x^2 + 3x - 4 = 0$

Solutions:

a. Find the value of $b^2 - 4ac$ for $3x^2 - 11x - 4 = 0$, where $a = 3$, $b = -11$, and $c = -4$:

$$\begin{aligned} \text{Discriminant} &= (-11)^2 - 4(3)(-4) \\ &= 121 + 48 \\ &= 169 \end{aligned}$$

Because 169 is a perfect square (13×13), the roots of $3x^2 - 11x = 4$ are **rational** and **unequal**.

b. Find the value of $b^2 - 4ac$ for $-2x^2 + 3x - 4 = 0$, where $a = -2$, $b = 3$, and $c = -4$:

$$\begin{aligned} \text{Discriminant} &= (3)^2 - 4(-2)(-4) \\ &= 9 - 32 \\ &= -23 < 0 \end{aligned}$$

The discriminant is less than 0, so $-2x^2 + 3x - 4 = 0$ has **imaginary** roots.

| Exercise 4 | Using the Discriminant |

Find the least integer value of k for which the roots of $x^2 - 5x + k = 0$ are imaginary.

Solution: Express the value of the discriminant of $x^2 - 5x + k = 0$ in terms of k, where $a = 1$, $b = -5$, and $c = k$:

$$\begin{aligned} b^2 - 4ac &= (-5)^2 - 4(1)k \\ &= 25 - 4k \end{aligned}$$

If the roots are imaginary, the discriminant must be less than 0. Hence, $25 - 4k < 0$ and $-4k < -25$, so $k > \dfrac{-25}{-4}$ or, equivalently, $k > 6\dfrac{1}{4}$. Thus, the least integer value of k that makes the inequality a true statement is **7**.

TIPS

- Irrational and imaginary roots of quadratic equations always occur in conjugate pairs. For example, if $3 - \sqrt{2}$ is a root of a quadratic equation, then the other root must be $3 + \sqrt{2}$. If $-4 + 3i$ is a root of a quadratic equation, then the other root must be $-4 - 3i$.

- Finding the discriminant *before* you solve a quadratic equation can help you decide which method of solution to use. For example, if the discriminant is 0 or a perfect square, you know that the quadratic equation is factorable.

Finding the Sum and the Product of the Roots

According to the quadratic formula, the two roots of $ax^2 + bx + c = 0$ are

$$x_1 = \frac{-b - \sqrt{b^2 - 4ac}}{2a} \quad \text{and} \quad x_2 = \frac{-b + \sqrt{b^2 - 4ac}}{2a}.$$

Adding and multiplying these roots together give simple formulas that express the sum and the product of the roots of a quadratic equation in terms of its coefficients.

Sum and Product of the Roots Formulas

If x_1 and x_2 are the roots of $ax^2 + bx + c = 0$ and $a \neq 0$, then:

$$\text{Sum} = x_1 + x_2 = -\frac{b}{a} \quad \text{and} \quad \text{Product} = x_1 \cdot x_2 = \frac{c}{a}.$$

Exercise 5   Applying the Formulas for the Sum and the Product of the Roots

By what amount does the product of the roots of $2x^2 - 7x + 11 = 0$ exceed the sum of the roots?

Solution:

- The sum of the roots is $-\dfrac{b}{a} = -\dfrac{-7}{2} = \dfrac{7}{2}$.

- The product of the roots is $\dfrac{c}{a} = \dfrac{11}{2}$.

- Since $\dfrac{11}{2} - \dfrac{7}{2} = \dfrac{4}{2} = 2$, the product of the roots exceeds the sum of the roots by **2**.

Chapter 5	CHECKUP EXERCISES

1–15. Multiple Choice

1. The expression $i^{50} + i^0$ is equivalent to

    (1) 1   (2) 2   (3) –1   (4) 0

2. Expressed in simplest form, $2\sqrt{-50} - 3\sqrt{-8}$ is equivalent to

    (1) $16i\sqrt{2}$ (2) $3i\sqrt{2}$ (3) $4i\sqrt{2}$ (4) $-i\sqrt{42}$

3. If $Z_1 = 3 - 6i$ and $Z_2 = 2 - 5i$, in which quadrant does $Z_1 - Z_2$ lie?

    (1) I   (2) II   (3) III   (4) IV

4. What is the value of x in the equation $\sqrt{5 - 2x} = 3i$

    (1) 1   (2) 7   (3) –2   (4) 4

5. If $2\sqrt{-2}$ is subtracted from $3\sqrt{-18}$, the difference is

    (1) $7i\sqrt{2}$ (2) $11i\sqrt{2}$ (3) $-7i\sqrt{2}$ (4) $-11i\sqrt{2}$

6. In $a + bi$ form, $(3 + 2i)^2$ is equivalent to

    (1) 5   (2) 13   (3) 5 + 12i (4) 13 + 12i

7. The product of $3\sqrt{-2} + 1$ and $3\sqrt{-2} - 1$ is

    (1) 17   (2) –19   (3) $17 - i\sqrt{2}$ (4) $9\sqrt{2} - 1$

8. The product of $-2 + 6i$ and $3 + 4i$ is

    (1) $-6 + 24i$ (2) $-6 - 24i$ (3) $18 + 10i$ (4) $-30 + 10i$

9. Expressed in $a + bi$ form, $\dfrac{5}{3+i}$ is equivalent to

    (1) $\dfrac{15}{8} - \dfrac{5}{8}i$ (2) $\dfrac{5}{3} - 5i$ (3) $\dfrac{3}{2} - \dfrac{1}{2}i$ (4) $15 - 5i$

10. If –1 and 7 are the roots of the equation $x^2 + kx - 7 = 0$, then k must be equal to

    (1) –7   (2) –6   (3) 6   (4) 8

11. If the roots of the quadratic equation $ax^2 + bx + c = 0$ are real, rational, and equal, what is true about the graph of the function $y = ax^2 + bx + c$?

  (1) It intersects the x-axis in two distinct points.

  (2) It lies entirely below the x-axis.

  (3) It lies entirely above the x-axis.

  (4) It is tangent to the x-axis.

12. The roots of the equation $5x^2 - 2x = -3$ are

  (1) imaginary   (3) real, rational, and unequal

  (2) real, rational, and equal   (4) real, irrational, and unequal

13. The roots of the equation $ax^2 + 4x = -2$ are real, rational, and equal if $a =$

  (1) 1   (2) 2   (3) 3   (4) 4

14. The roots of $3x^2 - 4x + 2 = 0$ are

  (1) $\dfrac{1 \pm \sqrt{2}}{3}$ (2) $\dfrac{2 \pm \sqrt{2}\,i}{3}$ (3) $\dfrac{2 \pm \sqrt{10}}{3}$ (4) $4 \pm \dfrac{\sqrt{2}}{3}\,i$

15. Given the quadratic equation $y = ax^2 + bx + c$, in which a, b, and c are integers. If the graph intersects the x-axis in two distinct points whose abscissas are integers, then $b^2 - 4ac$ may equal

  (1) −16   (2) 0   (3) 8   (4) 9

16. What is the positive value of k for which the graph of $y = x^2 - 2kx + 16$ is tangent to the x-axis?

17. What is the smallest integer value that makes the roots of $2x^2 - 5x + k = 0$ imaginary?

18–26. Write each expression in simplest form in terms of i.

18. $2\sqrt{-9} + 3\sqrt{-25}$ 21. $\left(\dfrac{i^{80}}{i^{30}}\right)^4$ 24. $\sqrt{-121} + 2i^{11}$

19. $3\sqrt{-1} - 4\sqrt{-4}$ 22. $3\sqrt{-12} - 3\sqrt{-75} + \sqrt{-48}$ 25. $i^{37} + i^{99}$

20. $\sqrt{-49} \cdot 2\sqrt{-4}$ 23. $\left(-i^{21}\right)^9$ 26. $\dfrac{2i^{40}}{i^{25} + i^5}$

27. If the product of $(a + bi)$ and $(4 - 3i)$ is $18 - i$, what are the values of a and b?

28–37. Express irrational roots in simplest radical form and imaginary roots in simplest $a + bi$ form.

28. $3x^2 - 5 = x$

33. $\dfrac{2 - 3x}{x} = \dfrac{x + 6}{2}$

29. $2x^2 + 1 = 10x$

34. $5y + \dfrac{5}{y} = 8$

30. $x^2 = 6x - 12$

35. $2 + \dfrac{5}{n^2} = \dfrac{6}{n}$

31. $9x^2 = 2(3x - 1)$

36. $9x + \dfrac{2}{x} = -6$

32. $8x^2 + 29 = 28x$

37. $\dfrac{x + 8}{5} + \dfrac{x + 5}{x} = 1$

38. *a.* Show that any quadratic equation can be expressed in the form
$$x^2 + \left[-\left(\text{sum of roots} \right) \right] x + \left(\text{product of roots} \right) = 0.$$

   b. Using the result from part *a*, write a quadratic equation whose roots are 2 and –5.

39. Write a quadratic equation that has $5 + 2i$ as one of its roots.

40. A rectangle is said to have a golden ratio when its dimensions satisfy the proportion $\dfrac{w}{h} = \dfrac{h}{w - h}$, where w is the width of the rectangle and h is the height. What is the height of a rectangle with a golden ratio whose width is 3?

41. Find the roots of $3x^2 - 6x + 4 = 0$ in simplest $a + bi$ form.

42. *a.* Using an algebraic method, estimate the roots of $2x^2 - 8x + 1 = 0$ to the *nearest tenth*.

   b. Confirm your answer by drawing the graph of the equation $y = 2x^2 - 8x + 1$ and estimating the roots of $2x^2 - 8x + 1 = 0$ from the graph.

43. A homeowner wants to increase the size of a rectangular deck that now measures 15 feet by 20 feet, but the building-code laws state that a homeowner cannot have a deck larger than 900 square feet. If the length and the width are to be increased by the same amount, find, to the *nearest tenth*, the maximum number of feet by which the length of the deck may legally be increased.

44. A manufacturer is designing a cardboard box from a rectangular sheet of cardboard that is twice as long as it is wide. From each of the four corners of the rectangular sheet of cardboard, a square 2 centimeters on a side is cut off and the flaps are turned up to form an uncovered box. Find, to the *nearest tenth*, the smallest dimensions of the rectangular sheet of cardboard so that the volume of the box formed will be at least 300 cubic centimeters?

45. A rocket is launched straight up from ground level. After t seconds, its height, h, in feet is given by the function $h(t) = -16t^2 + 376t$.

    a. How many seconds after the rocket is launched will it strike the ground?

    b. During what interval of time, to the *nearest tenth of a second*, will the height of the rocket exceed 1800 feet?

STUDY UNIT II: FUNCTIONS AND THEIR GRAPHS

Chapter 6

SPECIAL FUNCTIONS AND EQUATIONS

OVERVIEW

Although only linear and quadratic functions have been discussed so far, there are other important types of functions, including absolute value, square root, cube, and reciprocal. Special methods are needed to solve absolute-value equations and inequalities, rational equations and inequalities, and radical equations.

Lessons in Chapter 6
Lesson 6-1: Absolute-Value Equations and Inequalities
Lesson 6-2: Transformations of Graphs
Lesson 6-3: Special Functions and Their Graphs
Lesson 6-4: Radical Equations
Lesson 6-5: Rational Equations and Inequalities

Lesson 6-1   Absolute-Value Equations and Inequalities

KEY IDEAS

Because the absolute value of a nonnegative number is that same number and the absolute value of a negative number is its opposite, the graph of $y = |x|$ consists of the union of the rays $y = x$ and $y = -x$ above the x–axis, with the origin as their common endpoint. Absolute-value equations and inequalities can be solved either algebraically or graphically.

Algebraic Definition of Absolute Value

The **absolute value** of a number x is written as $|x|$ and is defined so that $|x| = x$ when $x \geq 0$ and $|x| = -x$ when $x < 0$. For example, $|3| = 3$ and $|-3| = -(-3) = 3$.

The Absolute-Value Function

To graph $y = |x|$, open the $Y =$ editor and press

Right cursor key

$\boxed{\text{MATH}}$ $\boxed{\triangleright}$ $\boxed{\text{ENTER}}$ $\boxed{X,T,\theta,n}$ $\boxed{\ )\ }$.

Figure 6.1 shows the graph in a Decimal window. From the graph you know that $y = |x|$ is a function since the graph passes the vertical-line test. The domain is the set of real numbers, the range is the set of all nonnegative real numbers, and the graph has the y-axis as a line of symmetry.

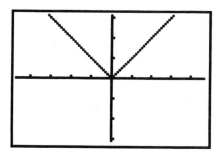

Figure 6.1 Graph of $y = |x|$

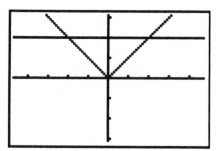

Figure 6.2 Graphs of $y = |x|$ and $y = 2$

Solving Absolute-Value Equations and Inequalities

The strategy for solving an absolute-value equation or inequality is to remove the absolute-value sign and then solve the resulting expression using familiar methods. Figure 6.2 suggests how the absolute-value sign can be removed.

- The graphs of $y = |x|$ and $y = 2$ intersect at $x = -2$ and $x = 2$, so the solution to $|x| = 2$ is $x = -2$ *or* $x = 2$. In general, if $|x| = k$ and k is nonnegative, $x = -k$ *or* $x = +k$.

- From $x = -2$ to $x = 2$, the graph of $y = |x|$ is *below* the graph of $y = 2$. Thus, the solution to $|x| < 2$ is $-2 < x < 2$. Thus, if $|x| < k$ and k is nonnegative, $-k < x < +k$.

- For $x < -2$ and $x > 2$, the graph of $y = |x|$ is *above* the graph of $y = 2$. Thus, the solution to $|x| > 2$ is $x < -2$ or $x > 2$. Thus, if $|x| > k$ and k is nonnegative, $x < -k$ *or* $x > +k$.

The solutions for $|x| = k$, $|x| < k$, and $|x| > k$ can be generalized further by replacing x with $ax - b$.

Rules for Removing the Absolute-Value Sign

If $k > 0$, then:

- $|ax - b| < k$ is equivalent to $-k < ax - b < k$.

- $|ax - b| = k$ is equivalent to $ax - b = -k$ or $ax - b = k$.

- $|ax - b| > k$ is equivalent to $ax - b < -k$ or $ax - b > k$.

These relationships also hold when $<$ is replaced by $\leq$ and when $>$ is replaced by $\geq$.

Exercise 1   Solving an Absolute-Value Equation

Solve $\left|2x+3\right|=1$ for x.

Solution: If $\left|2x+3\right|=1$, then $2x + 3 = -1$ *or* $2x + 3 = 1$.

- If $2x + 3 = -1$, then $2x = -4$, so $x = \dfrac{-4}{2} = \mathbf{-2}$.

- If $2x + 3 = 1$, then $2x = -2$, so $x = \dfrac{-2}{2} = \mathbf{-1}$.

Example 2 illustrates the importance of checking roots in the original absolute-value equation.

Exercise 2   Checking the Roots of an Absolute-Value Equation

Solve and check: $\left|x-3\right|=2x$.

Solution: If $\left|x-3\right|=2x$, then $x - 3 = -2x$ *or* $x - 3 = 2x$.

- If $x - 3 = -2x$, then $3x = 3$, so $x = 1$.
- If $x - 3 = 2x$, then $x = -3$.

Check:

- If $x = 1$, then $\left|1-3\right|=2\left(1\right)$ and $\left|-2\right| = 2$, which is true. Hence, **1** is a root.

- If $x = -3$, then $\left|-6\right|=2\left(-3\right)$, which is false since the absolute value of a number must be nonnegative. Hence, -3 is not a root.

Exercise 3   Absolute-Value Inequality: "Less Than" Case

Solve and graph the solution set: $\left|2x-1\right|\leq 7$.

Solution: If $\left|2x-1\right|\leq 7$, then $-7 \leq 2x - 1 \leq 7$.

- Add 1 to each member of the combined inequality:

$$-7 \leq 2x-1 \leq 7$$
$$\underline{+1 \qquad +1 +1}$$
$$-6 \leq 2x \quad \leq 8$$

- Divide each member of the combined inequality by 2:

$$\frac{-6}{2} \le \frac{2x}{2} \le \frac{8}{2}$$

$$-3 \le x \le 4$$

- Graph the solution set:

Absolute-Value Inequality: "Greater Than" Case

Solve and graph the solution set of $|3x-1|>5$.

Solution: If $|3x-1|>5$, there are two possibilities:

- $3x + 1 < -5$, so $3x < -6$ and $x < -2$;

    or

- $3x + 1 > 5$, so $3x > 4$ and $x > \dfrac{4}{3}$.

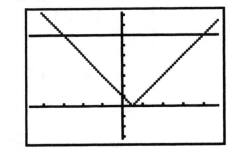

Exercise 5   **Solving an Absolute-Value Inequality Graphically**

Using a graphing calculator, solve:

a. $|2x-1| \le 7$ *b.* $|2x-1| > 7$

Solutions: Graph $Y_1 = \text{abs}(2X - 1)$ and $Y_2 = 7$ in a friendly window in which the graph fits, such as $[-4.7, 4.7] \times [-3.1, 9.3]$, as shown in the accompanying figure.

Use the INTERSECT or TRACE feature to find that the x-coordinates of the points of intersection are $x = -3$ and $x = 4$.

a. The solution of $|2x-1| \le 7$ is **$-3 \le x \le 4$** since, from $x = -3$ to $x = 4$, the graph of Y_1 is *below* ($<$) the graph of Y_2.

b. The solution of $|2x-1| > 7$ is **$x < -3$ *or* $x > 4$** since at these points the graph of Y_1 is *above* ($>$) the graph of Y_2.

Lesson 6-2 — Transformations of Graphs

KEY IDEAS

A **transformation** of a graph "moves" each point of the graph according to a given rule. Flipping a graph over a line is a **reflection** over that line. Shifting a graph vertically, horizontally, or both vertically and horizontally is a **translation**. Reflections and translations do not affect the size or shape of a graph. If two functions are related to each other by a simple transformation, then knowing what the graph of one function looks like enables you to easily determine what the graph of the other function looks like.

Reflecting Over a Coordinate Axis

Reflecting a graph over a line can be thought of as flipping the graph over the line so that, if the page that contains the two graphs is "folded" along the line, the two graphs will coincide.

- **Reflecting $y = f(x)$ over the x-axis.** Replacing $f(x)$ with $-f(x)$ reflects the graph of $y = f(x)$ over the x-axis. For example, the graph of $y = -|x|$ is the reflection of the graph of $y = |x|$ over the x-axis, as shown in Figure 6.3.

- **Reflecting $y = f(x)$ over the y-axis.** Replacing x with $-x$ reflects the graph of $y = f(x)$ over the y-axis. For example, the graph of $y = (-x)^2 - x$ is the reflection of the graph of $y = x^2 + x$ over the y-axis, as shown in Figure 6.4.

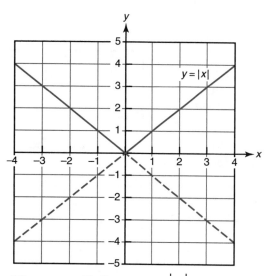

Figure 6.3 Reflecting $y = |x|$ over the x-axis

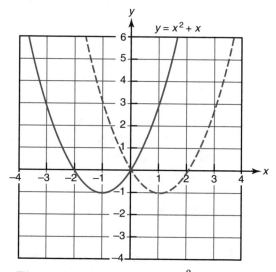

Figure 6.4 Reflecting $y = x^2 + x$ over the y-axis

Shifting Graphs Up or Down

Adding a number k to a function shifts it *up* if k is positive and shifts it *down* if k is negative:

- If $y = x^2$, adding $(+3)$ to y shifts the graph 3 units *up*, as shown in Figure 6.5. The equation of the new graph is $y = x^2 + 3$.
- If $y = x^2$, adding (-3) to y shifts the graph 3 units *down*, as shown in Figure 6.6. The equation of the new graph is $y = x^2 - 3$.

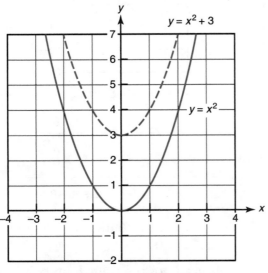

Figure 6.5 Shifting $y = x^2$ up 3 units

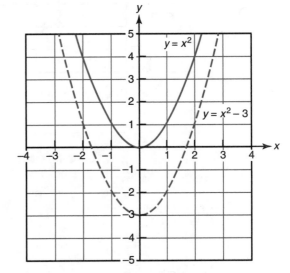

Figure 6.6 Shifting $y = x^2$ down 3 units

Shifting Graphs Right or Left

Subtracting a number h from x in a function shifts the graph to the *right* if h is positive and shifts it to the *left* if h is negative:

- If $y = x^2$, subtracting $(+3)$ from x shifts the graph to the *right* 3 units, as shown in Figure 6.7. The equation of the new graph is $y = (x - 3)^2$.
- If $y = x^2$, subtracting (-3) from x shifts the graph to the *left* 3 units, as shown in Figure 6.8. The equation of the new graph is $y = (x - (-3))^2 = (x + 3)^2$.

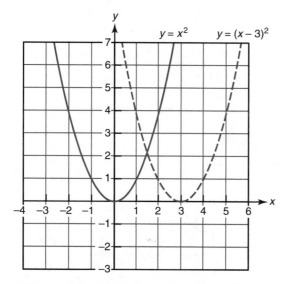

Figure 6.7 Shifting $y = x^2$ to the right

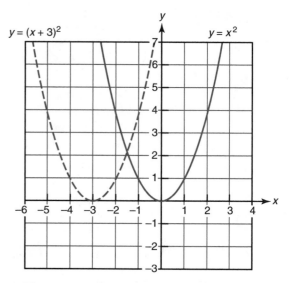

Figure 6.8 Shifting $y = x^2$ to the left

Rules for Shifting the Graph of $y = f(x)$	
Do this . . .	**To Get this result**
Add k to y: $y = f(x) + k$	Shift graph up k units if $k > 0$, or down $\left| k \right|$ units if $k < 0$.
Subtract h from x: $y = f(x - (h))$	Shift graph to the right h units if $h > 0$, or to the left $\left| h \right|$ units if $h < 0$.

EXAMPLE: The graph of $y = (x - 1)^3$ is the graph of $y = x^3$ shifted 1 unit to the *right*.

EXAMPLE: The graph of $y = \left| x + 2 \right|$ is the graph of $y = \left| x \right|$ shifted 2 units to the *left* because $y = \left| x + 2 \right| = \left| x - \left(-2 \right) \right|$.

EXAMPLE: The graph of $y = \left| x - 2 \right|$ is the graph of $y = \left| x \right|$ shifted 2 units *down*.

Combining Horizontal and Vertical Shifts

A graph may be shifted both horizontally and vertically. The graph of $y = (x - 2)^2 + 3$ can be obtained by shifting the graph of $y = x^2$ to the right 2 units and up 3 units, as shown in Figure 6.9. Notice that $(0,0)$, the vertex of $y = x^2$, moves to $(2,3)$, which is, therefore, the vertex of $y = (x - 2)^2 + 3$.

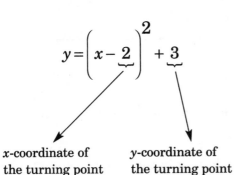

$$y = \left(x - \underbrace{2} \right)^{2} + \underbrace{3}$$

x-coordinate of
the turning point   y-coordinate of
the turning point

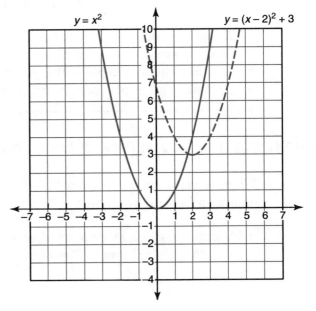

Figure 6.9 Shifting $y = x^2$ to the right 2 units and up 3 units

It can be generalized that $y = a(x - h)^2 + k$ is a parabola with vertex at (h,k).

Exercise 1   Locating the Vertex of a Parabola

If $g(x) = (x + 1)^2 - 4$, what are the coordinates of the vertex of the graph?

Solution: The graph of a function of the form $g(x) = a(x - h)^2 + k$ is a parabola with vertex at (h,k). Because

$$g\left(x\right) = \left(x + 1\right)^{2} - 4 = \left(x - \left(-1\right)\right)^{2} - 4,$$

$h = -1$ and $k = -4$, so the vertex of the graph is at **(−1,−4)**.

Lesson 6-3   Special Functions and Their Graphs

KEY IDEAS

When studying the properties of a function, you should ask questions such as these:

- What is the domain? The range?
- Does the graph of the function have line or origin symmetry?
- What are the intercepts of the graph, if any?
- Over what intervals does the function increase? Decrease?

Line and Origin Symmetry

When working with graphs of functions, it may be helpful to know whether the function enjoys any special symmetry.

- A graph has **line symmetry** if, after a reflection in the line, the new graph coincides with the original graph. Figure 6.10 shows a graph with vertical line symmetry.

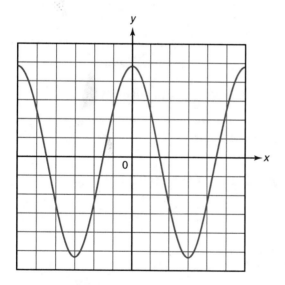

Figure 6.10 Vertical line symmetry

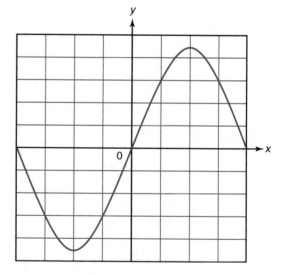

Figure 6.11 Origin symmetry

- A graph is **symmetric to the origin** if, after it is reflected over the x- and y-axes, in either order, the new graph coincides with the original graph. The graph in Figure 6.11 is symmetric to the origin.

Even and Odd Functions

Functions that have y-axis symmetry or origin symmetry are given special names.

- A function f is an **even function** if its equation remains unchanged when x is replaced with $-x$, that is, when $f(-x) = f(x)$ for all x. Even functions are symmetric to the y-axis.

EXAMPLE: The graph in Figure 6.10 describes an even function since it has y-axis symmetry.

EXAMPLE: The function $f(x) = x^2$ is an even function since replacing x with $-x$ gives $f(-x) = (-x)^2 = x^2 = f(x)$.

- A function f is an **odd function** if replacing x with $-x$ changes the sign of each term of the equation to its opposite, that is, when $f(-x) = -f(x)$ for all x. The graph of an odd function is symmetric to the origin.

EXAMPLE: The graph in Figure 6.11 describes an odd function since it has origin symmetry.

EXAMPLE: The function $f(x) = x^3 - x$ is an odd function since replacing x with $-x$ gives $f(-x) = (-x)^3 - (-x) = -x^3 + x = -f(x)$.

- Typically, a function is neither even nor odd.
  If $g(x) = x^2 + x$, then

$$g(-x) = (-x)^2 + (-x) = x^2 - x, \text{ so } g(-x) \neq g(x)$$

and function g is *not* even.
Also, because $-g(x) = -(x^2 + x) = -x^2 - x \neq g(-x)$, function g is *not* odd.

Cube Function: $f(x) = x^3$

If $y = x^3$, then x and y must have the same sign so that the graph is confined to the first and third quadrants. Furthermore, there is no restriction on the domain or range. The graph of $f(x) = x^3$ in Figure 6.12 confirms that the cube function:

- has the set of real numbers as its domain and range;
- has origin symmetry, so it is an odd function;
- includes point (0,0);
- increases throughout its domain.

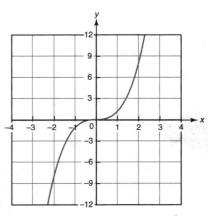

Figure 6.12 Graph of $y = x^3$

Square Root Function: $f(x) = \sqrt{x}$

If $y = \sqrt{x}$, then x must be nonnegative, so y is also nonnegative. Therefore, the graph of the square root function is confined to Quadrant I. The graph of $f(x) = \sqrt{x}$ in Figure 6.13 confirms that:

- the domain and the range of the square-root function are limited to the sets of all nonnegative real numbers;
- the square-root function is neither odd nor even;
- the graph of the square-root function includes point (0,0);
- the square-root function increases throughout its domain.

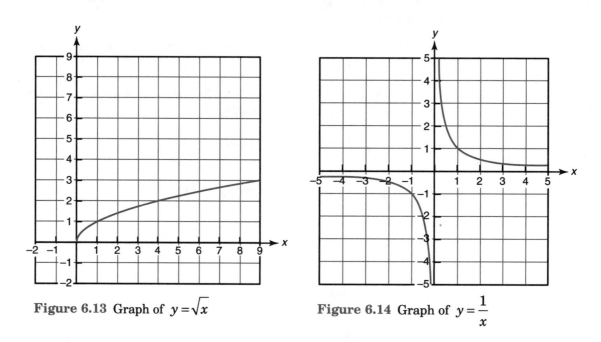

Figure 6.13 Graph of $y = \sqrt{x}$ Figure 6.14 Graph of $y = \dfrac{1}{x}$

Reciprocal Function: $f(x) = \dfrac{1}{x}$

If $y = \dfrac{1}{x}$, then $x \neq 0$ and x and y must have the same sign so that the graph is confined to the first and third quadrants. When x approaches 0 from the positive side of 0, y increases without bound. Similarly, when x approaches 0 from the negative side of 0, y decreases without bound. The graph of $f(x) = \dfrac{1}{x}$ in Figure 6.14 confirms that:

- the domain of the reciprocal function is the set of all real numbers except 0, and the range is the set of all real numbers;
- the reciprocal function is odd;
- the graph of the reciprocal function has no intercepts, and the x- and y-axes are asymptotes of the graph;
- the reciprocal function decreases on the intervals (–∞,0) and (0,∞).

Piecewise-Defined Functions

A function may change its definition over its domain. For example, the absolute-value function $f(x) = |x|$ is a piecewise-defined function; when $x \geq 0$, it is defined as x, but when $x < 0$, it is defined as $-x$. Consider this piecewise-defined function:

$$f(x) = \begin{cases} x, & x > 2 \\ x^2 - 2, & x \leq 2. \end{cases}$$

Function f is defined by two different equations. When $x > 2$, function $f(x) = x$. The definition of the function changes to $f(x) = x^2 - 2$ when $x \leq 2$. For example,

$$f(-3) = (-3)^2 - 2 = 7 \text{ and } f(3) = 3.$$

The graph of function f, shown in Figure 6.15, consists only of the set of points of the parabola $y = x^2 - 2$ for which $x \leq 2$, and only the set of points of the line $y = x$ for which $x > 2$.

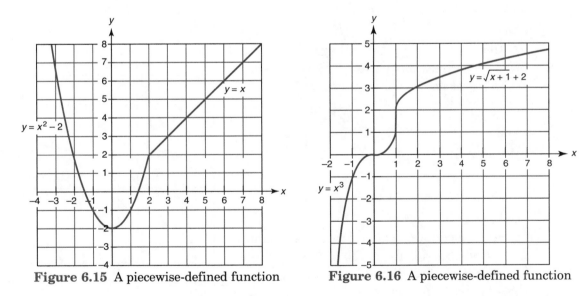

Figure 6.15 A piecewise-defined function   **Figure 6.16** A piecewise-defined function

For a further illustration, study the graph of

$$f(x) = \begin{cases} x^3, & x < 1 \\ \sqrt{x-1} + 2, & x \geq 1 \end{cases}$$

in Figure 6.16.

Types of Discontinuities

The graph of a function may have a break or may be disconnected in some other way so that it is not possible to trace it using a pencil without lifting the pencil off the page. Such a function is said to be **discontinuous**. There are three different types of discontinuities:

- **Infinite discontinuity**. An infinite discontinuity occurs when a function increases or decreases without bound as x gets closer and closer to a particular value. For example, the reciprocal function in Figure 6.14 is not continuous; it consists of two disconnected branches. The function has an infinite discontinuity because, when x gets closer and closer to 0, approaching 0 from the negative side, y decreases without bound. Because the graph comes increasingly close to the negative y-axis without ever touching it, the negative y-axis is said to be an **asymptote** of the graph; when x gets closer and closer to 0, approaching 0 from the positive side, y increases without bound. It's easy to see from the graph that the x-axis is also an asymptote of the graph.
- **Jump discontinuity**. A jump discontinuity occurs when the graph of a function has a gap, as illustrated in Exercise 1.
- **Point discontinuity**. A point discontinuity is a "hole" in the graph of a function that occurs at a particular point if the function is not defined for the x-value at that point. This situation occurs in Exercise 2.

| **Exercise 1** | **Graphing a Function with a Jump Discontinuity** |

Graph the function $f(x) = \begin{cases} -x^2 + 5, & x < 0 \\ \sqrt{x}, & x \geq 0 \end{cases}$.

Solution: When x takes on negative values, f is defined by the equation $f(x) = -x^2 + 5$. When x takes on nonnegative values, f is defined by the equation $f(x) = \sqrt{x}$. At the crossover point, $x = 0$, the function changes definition from

$$f(x) = -x^2 + 5 \text{ to } f(x) = \sqrt{x}.$$

Since these equations do not return the same y-value when $x = 0$, the graph of this function has a vertical *jump discontinuity*, as shown in the accompanying figure.

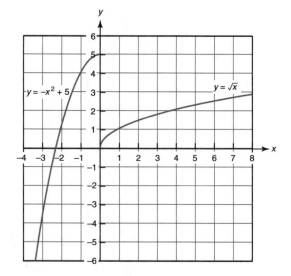

Exercise 2   **Graphing a Function with a Point Discontinuity**

Graph function f, where $f(x) = \begin{cases} \dfrac{x^2 - 9}{x - 3}, & x \neq 3 \\ 1, & x = 3 \end{cases}$.

Solution:

- For all x-values except 3, the function is defined by the equation

$$y = \frac{x^2 - 9}{x - 3} = \frac{(x + 3)(x \cancel{- 3})}{\cancel{(x - 3)}} = x + 3.$$

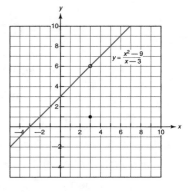

- Graph $y = x + 3$. At $x = 3$, the line has a "hole" or *point discontinuity* because, at this single point, $y = \dfrac{x^2 - 9}{x - 3}$ has no y-value.

- When $x = 3$, the function is defined to have a value of 1. Indicate this by graphing the single point $(3,1)$, as shown in the accompanying figure.

Lesson 6-4 — Radical Equations

KEY IDEAS

An equation in which the variable appears underneath a radical sign, such as $\sqrt{4 - x} - 8 = x$, is called a **radical equation**. To solve a radical equation, isolate the radical and then raise both sides of the equation to the power that eliminates the radical. The transformed equation, however, may include extraneous roots that do not satisfy the original equation.

Solving a Radical Equation Algebraically

To solve $\sqrt{4 - x} - 8 = x$:

1. Isolate the radical: $\qquad\qquad\qquad\qquad \sqrt{4 - x} = x + 8$

2. Eliminate the square-root radical: $\qquad \left(\sqrt{4 - x}\right)^2 = (x + 8)^2$

$$4 - x = x^2 + 16x + 64$$
$$0 = x^2 + 17x + 60$$

3. Solve the transformed equation: $0 = (x+5)(x+12)$

$$x = -5 \; or \; x = -12$$

Because extraneous roots are possible, confirm each possible root algebraically by testing it in the original equation. Since $x = -5$ works in $\sqrt{4-x} - 8 = x$, it is a root. If $x = -12$, then $\sqrt{4-(-12)} - 8 = -12$ or, equivalently, $\sqrt{16} - 8 = -12$, which is not a true statement. Reject $x = -12$; $x = -5$ is the only root.

Solving a Radical Equation Graphically

You can also solve $\sqrt{4-x} - 8 = x$ by graphing $Y_1 = \sqrt{(4-x)} - 8$ and $Y_2 = x$ in a window with a friendly x-readout, such as $[-9.4, 9.4] \times [-6.2, 6.2]$. Then use the TRACE feature to determine that the x-coordinate of the point of intersection of the two graphs is $x = -5$, as shown in Figure 6.17.

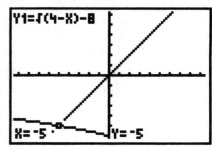

Figure 6.17 Graph of
$$Y_1 = \sqrt{(4-x)} - 8 \text{ and } Y_2 = x$$

Solving a Radical Equation with Two Radicals

If an equation contains two radicals, as in $\sqrt{x+4} + \sqrt{1-x} = 3$, solve for one radical and eliminate it. Then solve for the other radical and eliminate it. For example:

$$\sqrt{x+4} + \sqrt{1-x} = 3$$
$$\sqrt{x+4} = 3 - \sqrt{1-x}$$
$$\left(\sqrt{x+4}\right)^2 = \left(3 - \sqrt{1-x}\right)^2$$
$$x+4 = 9 - 6\sqrt{1-x} + (1-x)$$

Next, isolate the remaining radical term and then square both sides of the equation to eliminate it:

$$\left(6\sqrt{1-x}\right)^2 = \left(6-2x\right)^2$$
$$36\left(1-x\right) = 66-24x+4x^2$$
$$4x^2+12x = 0$$
$$4x\left(x+3\right) = 0$$
$$x = 0 \quad or \quad x = -3$$

You should verify that both roots satisfy the original equation.

Solving an Equation with a Rational Exponent

To solve an equation that can be put into the form $x^{\frac{p}{r}} = k$, raise both sides of the equation to the power that makes the exponent of x equal to 1. For example, if $x^{\frac{3}{5}}+9=1$, then $x^{\frac{3}{5}} = -8$. Since the reciprocal of the exponent of x is $\frac{5}{3}$, raise both sides to the $\frac{5}{3}$ power:

$$\left(x^{\frac{3}{5}}\right)^{\frac{5}{3}} = \left(-8\right)^{\frac{5}{3}}$$
$$x^1 = \left(\sqrt[3]{-8}\right)^5$$
$$x = \left(-2\right)^5 = -32$$

Rational Equations and Inequalities

KEY IDEAS

A **rational equation** is an equation in which a variable appears in the denominator of one or more fractions. To solve a rational equation, eliminate the denominators by multiplying each member of the equation by the lowest common denominator (LCD) of all the denominators. To solve a *rational inequality*, rewrite it as a single fraction on one side of the inequality sign and 0 on the other side. Then use a number line and test values to help locate these intervals over which the fraction takes on values that satisfy the inequality.

Solving a Rational Equation

When clearing an equation of any fractional terms, make sure that you multiply *both* sides of the equation by the LCD of all the denominators. Then, since the equation produced by clearing an equation of its variable denominators may include *extraneous roots*, carefully check the roots of the transformed equation in the original equation.

Exercise 1   **Solving a Rational Equation**

Solve for y: $\dfrac{2}{y} - \dfrac{9}{10} = \dfrac{1}{5y}$.

Solution: Since the LCD of all the denominators is $10y$, clear the equation of its fractions by multiplying each member of the equation by $10y$. Then solve the resulting equation:

$$10y\left(\frac{2}{y}\right) - 10y\left(\frac{9}{10}\right) = 10y\left(\frac{1}{5y}\right)$$

$$10\cancel{y}\left(\frac{2}{\cancel{y}}\right) - \overset{1}{\cancel{10}}y\left(\frac{9}{\underset{1}{\cancel{10}}}\right) = \overset{2}{\cancel{10}}\cancel{y}\left(\frac{1}{\cancel{5}\cancel{y}}\right)$$

$$20 \quad - \quad 9y \quad = 2$$

$$-9y = -18$$

$$y = \frac{-18}{-9} = \mathbf{2}$$

Exercise 2   **Identifying Extraneous Roots**

Solve for x: $\dfrac{x}{x+2} - \dfrac{1}{x-2} = \dfrac{8}{x^2 - 4}$.

Solution:

• Rewrite the equation with each denominator in factored form:

$$\frac{x}{x+2} - \frac{1}{x-2} = \frac{8}{\left(x+2\right)\left(x-2\right)}.$$

The LCD is $(x + 2)(x - 2)$.

- Clear the fractions by multiplying each term of the equation by the LCD:

$$
\begin{aligned}
(x-2)x \;-\; (x+2)\cdot 1 \;&= 8 \\
x^2 - 2x \;-\; x - 2 \;&= 8 \\
x^2 \;-\; 3x - 10 \;&= 0 \\
(x+2)(x-5) \;&= 0 \\
x+2 = 0 \quad or \quad x - 5 &= 0 \\
x = -2 \qquad\qquad x &= 5
\end{aligned}
$$

- Check for extraneous roots. Look back at the original equation. Any value of x that makes a denominator evaluate to 0 is excluded from the domain of x. Hence, $x = -2$ is not in the domain of x; it represents an extraneous root and must be rejected. You should also verify that $x = 5$ checks and is the only root of the equation.

Solving a Motion Problem

Solving a motion problem depends on the relationship

$$R \times T = D,$$

where R represents the average rate of speed and D represents the distance traveled in time T.

Exercise 3   **Using Rate × Time = Distance**

On a 75-mile trip, Frank's average rate of driving speed for the first 15 miles was 10 miles per hour less than his average rate of driving speed for the rest of the trip. If the total driving time for the trip was 2 hours, find the average rate of speed for the first 15 miles.

Solution: Frank drove 15 miles at one rate and 75 – 15 = 60 miles at the faster rate. Hence, if x represents the average rate of speed for the last 60 miles, then $x - 10$ represents the average rate of speed for the first 15 miles.

- Organize the information in a table, as shown below. Since $R \times T = D$, $T = \dfrac{D}{R}$. Complete the column headed "Time" by dividing the distance traveled by the average rate of speed over that distance.

	Rate	×	Time	=	Distance
First 15 miles of trip	$x - 10$		$T = \dfrac{D}{R} = \dfrac{15}{x-10}$		15
Last 60 miles of trip	x		$T = \dfrac{D}{R} = \dfrac{60}{x}$		60

- Since the total time for the trip was 2 hours:

$$\frac{15}{x-10} + \frac{60}{x} = 2.$$

Clear the fractions by multiplying each member of the equation by $x(x - 10)$:

$$x\left(x - 10\right)\left[\frac{15}{x - 10}\right] + x\left(x - 10\right)\left[\frac{60}{x}\right] = 2x\left(x - 10\right)$$

$$15x \quad + \quad 60x - 600 \quad\quad = 2x^2 - 20x$$

- Simplify and rearrange terms. Then $2x^2 - 95x + 600 = 0$, so

$$(2x - 15)\,(x - 40) = 0, \text{ making } x = 7.5 \quad or \quad x = 40.$$

If $x = 7.5$, then $x - 10 < 0$, which is not possible, so discard 7.5. If $x = 40$, then $x - 10 = 40 - 10 = 30$. Frank drove at an average rate of **30 miles per hour** for the first 15 miles of the trip.

Solving a Work Problem

The solution of a work problem depends on the relationship

$$\left(\begin{array}{c}\text{Rate of}\\\text{work}\end{array}\right) \times \left(\begin{array}{c}\text{Time}\\\text{worked}\end{array}\right) = \left(\begin{array}{c}\text{Part of job}\\\text{completed}\end{array}\right),$$

where the rate of work is the reciprocal of the total time required to do the entire job when working alone. For example, if it takes 4 hours to complete a job, the rate of work is $\dfrac{1}{4}$ of the job per hour.

Exercise 4   Solving a Work Problem

John takes 2 days longer than Bill to build the scenery for a school play. After John worked 1 day alone, he was joined by Bill. Working together, they completed the whole job in 4 additional days. How long would Bill, working alone, have taken to do the entire job?

Solution:

- If x represents the number of days Bill takes to do the whole job working alone, then $x + 2$ represents the number of days John takes to do the entire job working alone. Since John works alone for 1 day, he works a total of 5 days while Bill works 4 days. Organize the information in a table, as shown below.

	Rate of Work	× Days Worked	= Part of Job Completed
John	$\dfrac{1}{x+2}$	5	$\dfrac{5}{x+2}$
Bill	$\dfrac{1}{x}$	4	$\dfrac{4}{x}$

- Since the sum of the fractional parts of the job completed by John and by Bill must equal one whole job:
$$\frac{5}{x+2} + \frac{4}{x} = 1.$$

- Clear the fractions by multiplying each member by $x(x + 2)$:
$$x\left(x+2\right)\left[\frac{5}{x+2}\right] + x\left(x+2\right)\left[\frac{4}{x}\right] = 1\left(x\right)\left(x+2\right).$$

- After simplifying and rearranging terms, $x^2 - 7x - 8 = 0$. Then
$$(x - 8)\,(x + 1) = 0, \text{ so } x = 8 \text{ or } x = -1.$$

Since the number of days cannot be negative, $x = 8$ is the only solution. Hence, Bill, working alone, would have taken **8 days** to finish the whole job.

Solving a Rational Inequality

To solve a rational inequality such as

$$\frac{3x}{x+2} \geq 2,$$

rewrite the inequality in the form $\dfrac{f(x)}{g(x)} \geq 0$. Then determine over what intervals the fraction takes on positive values or is equal to 0:

1. Subtract 2 from both sides of the inequality. Then rewrite the left side as a single fraction:

    $$\frac{3x}{x+2} - 2 \geq 0$$
    $$\frac{3x - 2(x+2)}{x+2} \geq 0$$
    $$\frac{x-4}{x+2} \geq 0$$

2. Determine the critical numbers for which the fraction is 0 or undefined. When $x = 4$, the numerator is 0, so the fraction is 0; when $x = -2$, the denominator is 0, and the fraction is undefined.

3. Locate the critical numbers on a number line. Use a closed circle to indicate that the critical number is included in the solution set, as shown in Figure 6.18. Use an open circle if the number is not included in the solution set, as shown in Figure 6.18.

4. Determine the sign of the fraction on each of the intervals determined by the critical numbers. To discover the sign of $\dfrac{x-4}{x+2} \geq 0$ on an interval, pick any convenient test value for x in that interval. Using the test value, determine the signs of the numerator, the denominator, and then their quotient, as shown in Figure 6.18. For example, consider the leftmost interval. Pick $x = -3$ as a test value. It is not necessary to determine the value of the fraction; you need only its sign:

    $$\frac{x-4}{x+2} = \frac{-3-4}{-3+1} = \frac{(-)}{(-)} = (+) \geq 0.$$

5. Summarize the solution intervals: $(-\infty, -2)$ and $[4, \infty)$, which may be written also as $x < -2$ and $x \geq 4$.

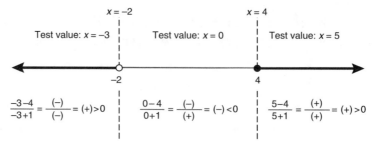

Figure 6.18 Finding solution intervals for $\dfrac{x-4}{x+2} \geq 0$

Exercise 5   Solving a Rational Inequality

Solve: $\dfrac{5}{x-3} \le \dfrac{2}{x}$.

Solution: Rewrite the inequality in the form $\dfrac{f(x)}{g(x)} \le 0$:

$$\frac{5}{x-3} - \frac{2}{x} \le 0$$

$$\frac{5x - 2(x-3)}{x(x-3)} \le 0$$

$$\frac{3x+6}{x(x-3)} \le 0$$

$$\frac{3(x+2)}{x(x-3)} \le 0$$

- The fraction is 0 when $x = -2$ and is undefined when $x = 0$ or $x = 3$. Hence, the critical numbers are –2, 0, and 3.
- Locate the critical numbers on a number line. Place a closed circle around –2 since the fraction can be equal to 0. Use open circles around 0 and 3 since, for these values of x, the fraction is undefined.

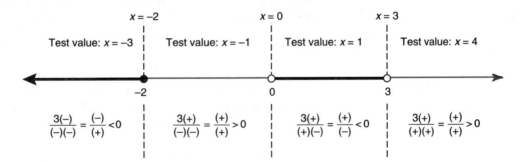

- Determine the sign of the fraction on each of the four intervals determined by the three critical numbers, as shown in the accompanying figure.
- Summarize the solution intervals: $(-\infty, -2]$, or $(0,3)$, which may be written also as $x \le -2$, or $0 < x < 3$.

TIPS

- When the numerator and the denominator consist of distinct linear factors in x or odd powers of x, the critical numbers divide the number line so that every other interval belongs to the solution. Thus, in Exercise 5, once you determine that the solution includes the leftmost interval, $x \leq -2$, you do not need to do any additional work because the solution intervals alternate thereafter:

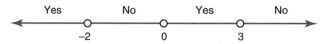

If the leftmost interval did not belong to the solution, the solution would alternate beginning with the next consecutive interval.

- If the original inequality is $\leq$ or $\geq$, *include* in the final solution critical values that make the numerator 0; otherwise, do not include critical values. Always *exclude* from the solution critical values for which the denominator is 0.

Exercise 6   Analyzing a Rational Inequality

Solve for x: $\dfrac{(x+5)(x-3)^2(x-4)(x^2+9)}{(x-1)^3} \geq 0$.

Solution: The factor $(x^2 + 9)$ never changes sign, so it can be ignored. The critical numbers are -5, 1, 3, and 4. The denominator contains an *odd* power of $(x - 1)$, so there is a sign change on either side of 1. However, since the numerator contains an *even* power of $(x - 3)$, there is *no* sign change as x increases through 3.

- Locate the critical numbers on a number line with closed circles around -5, 4, and 3, and an open circle around 1, as shown in the accompanying figure.
- Pick a test value in the leftmost interval, say $x = -6$:

$$\frac{(-6+5)(-6-3)^2(-6-4)}{(-6-1)^3} = \frac{(-)(+)(-)}{(-)} = \frac{(+)}{(-)} < 0.$$

Since $x = -6$ does not satisfy the inequality, the solution intervals alternate starting with the next interval, except that there is no sign change as x increases through 3:

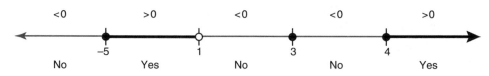

- The solution is **$-5 \leq x < 1$ or $x = 3$ or $x \geq 4$.**

Chapter 6	**CHECKUP EXERCISES**

1—15. Multiple Choice

1. The solution set of $\left|x-3\right| > 5$ is

   (1) $\{x \mid x < 8 \text{ and } x < -2\}$ (3) $\{x \mid x < 8 \text{ and } x > -2\}$

   (2) $\{x \mid x < 8 \text{ or } x < -2\}$ (4) $\{x \mid x > 8 \text{ or } x < -2\}$

2. What is the solution set of the inequality $\left|3x+6\right| \le 30$?

   (1) $\left\{x \mid -12 \le x \le 8\right\}$ (3) $\left\{x \mid x \le -12 \;\; or \;\; x \ge 8\right\}$

   (2) $\left\{x \mid -8 \le x \le 12\right\}$ (4) $\left\{x \mid x \le -8 \;\; or \;\; x \ge 12\right\}$

3. Which graph represents the solution set of $\left|2x-1\right| < 7$?

4. For the equation $\sqrt{x+21} = x+1$, the solution set for x is

   (1) $\{ \ \}$ (2) $\{-5\}$ (3) $\{-5, 4\}$ (4) $\{4\}$

5. The inequality $-3 < x < 7$ is the solution of

   (1) $\left|x-2\right| > 5$ (2) $\left|x-2\right| < 5$ (3) $\left|x+2\right| > 5$ (4) $\left|x+2\right| < 5$

6. Which inequality states that the speed of a car, s, is less than 5 miles per hour from the speed limit, L?

   (1) $\left|s-5\right| < L$ (2) $\left|s-L\right| < 5$ (3) $\left|L-5\right| < s$ (4) $\left|s\right| < L-5$

7. The graph below represents $f(x)$.

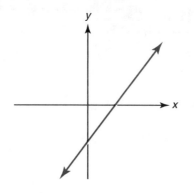

Which graph best represents $\left|f(x)\right|$?

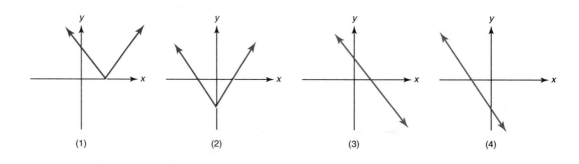

(1)   (2)   (3)   (4)

8. Which equation is represented by the accompanying graph?

(1) $y = \left|x\right| - 3$ (3) $y = \left|x + 3\right| - 1$

(2) $y = \left(x - 3\right)^2 + 1$ (4) $y = \left|x - 3\right| + 1$

9. What is the solution of the inequality $\left|3 - 2x\right| \geq 4$?

(1) $-\dfrac{7}{2} \leq x \leq \dfrac{1}{2}$ (3) $x \leq -\dfrac{1}{2}$ or $x \geq \dfrac{7}{2}$

(2) $-\dfrac{1}{2} \leq x \leq \dfrac{7}{2}$ (4) $x \leq \dfrac{7}{2}$ or $x \geq -\dfrac{1}{2}$

10. The graph below represents the graph of $y = f(x)$.

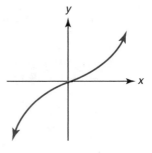

Which is the graph of $y = f(-x)$?

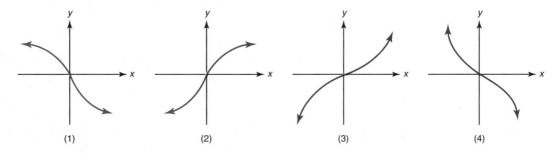

(1)   (2)   (3)   (4)

11. What is an equation of the function f whose graph can be obtained by shifting the graph of $h(x) = 2\sqrt{x}$ to the left 3 units and down 1 unit?

(1) $f(x) = 2\sqrt{x-3} - 1$

(3) $f(x) = 2\sqrt{x+1} - 3$

(2) $f(x) = 2\sqrt{x-1} - 3$

(4) $f(x) = 2\sqrt{x+3} - 1$

12. What is the solution set for the equation $|3x - 1| = x + 5$?

(1) {–1}   (2) {–1, 3}   (3) {3}   (4) {1, –3}

13. What is the solution set for $\sqrt{5-x} - 3 = x$?

(1) {1}   (2) {4, 1}   (3) { }   (4) {4}

14. The solution set of the equation $\sqrt{2x + 15} = x$ is

(1) {5, –3}   (2) {5}   (3) {–3}   (4) { }

15. Which statement is true about the equation $\sqrt{x^2 - 5x + 5} - 1 = 0$?

(1) The only real root is 1.   (3) Both 1 and 4 are roots.

(2) The only real root is 4.   (4) Neither 1 nor 4 is a root.

16–27. Solve for the variable using algebraic methods.

16. $\sqrt{2n} = 4\sqrt{5}$

20. $\dfrac{y-1}{\sqrt{13-y^2}} = 1$

24. $2\sqrt{1-3p} = 3-p$

17. $\sqrt[3]{3x-4} = 2$

21. $n - \sqrt{3n+4} = 2$

25. $\sqrt{2x-1} - 1 = 2\sqrt{x-4}$

18. $\sqrt{5x+1} + 3x = 27$

22. $8x^{-\frac{3}{4}} = 27$

26. $\sqrt{9-x} + \sqrt{x+11} = 6$

19. $26 - x^{\frac{2}{3}} = 1$

23. $\sqrt{1-2x} = \sqrt{x^2-7}$

27. $\sqrt{11-5x} + \sqrt{3x+4} = 5$

28. The equation
$$V = 20\sqrt{C + 273}$$

relates the speed of sound, V, in meters per second, to air temperature, C, in degrees Celsius. What is the temperature, in degrees Celsius, when the speed of sound is 320 meters per second?

29. The number of seconds in the period of a pendulum is the length of time that the pendulum takes to make one complete swing back and forth. The formula
$$T = 2\pi\sqrt{\dfrac{L}{32}}$$

gives the period, T, for a pendulum of length L, in feet. If Todd wants to build a grandfather clock with a pendulum that swings back and forth every 2 seconds, how long, to the *nearest tenth of a foot*, should he make the pendulum?

30–45. Solve for the variable.

30. $\left|4x-1\right| \geq 5$

35. $\dfrac{1}{b-3} - \dfrac{3}{2b+6} = \dfrac{b}{b^2-9}$

31. $\dfrac{\left|2-x\right|}{5} - 1 \leq 0$

36. $\dfrac{1}{y} - \dfrac{y+1}{8} = \dfrac{y-1}{4y}$

32. $11 < \left|4 - 3x\right|$

37. $1 - 3x < 7$

33. $\left|1-x\right| \leq x + 2$

38. $\dfrac{1}{\left|3x-1\right|} < \dfrac{2}{x}$

34. $\dfrac{1}{x} + \dfrac{3}{2x} = \dfrac{x-4}{2}$

39. $\dfrac{t}{t-3} - \dfrac{t-2}{2} = \dfrac{5t-3}{4t-12}$

40. $\dfrac{4x}{x+2} = 1 + \dfrac{12}{x}$

43. $\dfrac{2}{r-6} = \dfrac{10}{r^2-7r+6} - \dfrac{r}{r-1}$

41. $\dfrac{1}{x-2} + \dfrac{x+2}{x+5} = \dfrac{3}{x^2+3x-10}$

44. $|2x+3| = x-1$

42. $\dfrac{4}{m^2-16} + \dfrac{m-3}{m+4} = \dfrac{1}{m-4}$

45. $|x+6| = |2x-3|$

46. An archer shoots an arrow into the air with an initial velocity of 128 feet per second. Because the speed is the absolute value of the velocity, the speed of the arrow, s, in feet per second after t seconds is $|-32t + 128|$. Find the values of t for which s is less than 48 feet per second.

47. Solve the equation $\sqrt{x+2} + \sqrt{2x-1} = 5$ algebraically, and confirm your answer graphically.

48. Electrical circuits can be connected in series, one after another, or in parallel circuits that branch off a main line. If circuits are hooked up in parallel, the reciprocal of the total resistance in the circuit is found by adding the reciprocals of the individual resistances, as shown in the accompanying diagram. The second resistance in a parallel circuit is 3 ohms greater than the first resistance, and the total resistance of the circuit is 2.25 ohms. Determine the first resistance, correct to the *nearest tenth of an ohm*.

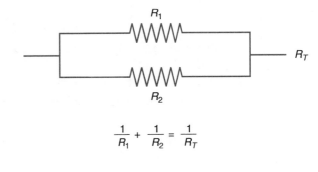

$$\dfrac{1}{R_1} + \dfrac{1}{R_2} = \dfrac{1}{R_T}$$

49. Solve for y: $2(y+1)^{\frac{2}{3}} - 5(y+1)^{\frac{1}{3}} + 2 = 0$.

50. Working by herself, Mary requires 16 minutes more than Antoine to solve a mathematics problem. Working together, Mary and Antoine can solve the problem in 6 minutes. Determine algebraically the number of minutes Antoine would take to solve the problem if he works by himself.

51. Dave can mow his lawn in 20 minutes less time with his power mower than with his hand mower. One day his power mower broke down 15 minutes after he started mowing, and he needed 25 minutes more time to complete the job with his hand mower. How many minutes does Dave take to mow the lawn with the power mower?

52. A teacher drove 280 miles to attend a mathematics conference and arrived an hour late. The teacher figured out that, had she increased her average speed by 5 miles per hour, she would have arrived at the time for which the conference was scheduled. What was her average rate of speed?

53. Solve the equation $\sqrt{10-3x} + \sqrt{x^2+5} = 7$ graphically, and confirm your answer algebraically.

54. A mechanic's helper required 4 hours longer than the mechanic to repair a car. The mechanic began the repair job alone and worked on it for 3 hours before he was called away. His helper needed 5 hours to finish the job. How many hours would the mechanic, working alone, have taken to repair the car?

55. At 9:00 A.M. Mike started from home on a hike to a mountain 12 miles away. After arriving, he took 1 hour for lunch and then returned home over the same route, reaching home at 5:00 P.M. If his average rate returning was 1 mile per hour less than his rate going, find his rate on the return trip.

56. On a business trip a salesman traveled 40 miles per hour for the first third of the distance and 50 miles per hour for the rest of the distance. If the entire trip took 4 hours and 20 minutes, how many miles did he travel?

57. Sean started on a trip across a lake by motorboat. After he had traveled 15 miles, the motor failed and he had to row the remaining 6 miles to shore. His average speed by motor was 4 miles per hour faster than his average speed while rowing. If the entire trip took $5\frac{1}{2}$ hours, what was Sean's average speed while rowing?

58. A building contractor knows that a small truck would take 4 days longer to haul an order of gravel to a construction site than a large truck would take to haul the same order. By using 8 small and 6 large trucks working together, the contractor delivered the gravel in 1 day. Find the number of days that one small truck, working alone, would have taken to deliver the gravel.

59. The members of a running club planned to contribute equally to raise $896 to pay for refreshments and prizes at a race. When it was discovered that $1080 would be needed, instead of $896, four members withdrew from the club. Each of the remaining members had to increase his or her contribution by $13 to raise the necessary amount. How many members were there in the club originally?

60. The cost, C, of selling x calculators in a store is modeled by the equation

$$C = \frac{3,200,000}{x} + 60,000.$$

The store profit, P, on these sales is modeled by the equation $P = 500x$. What is the *least* number of calculators that must be sold for the profit to be greater than the cost?

Chapter 7

POLYNOMIAL AND RATIONAL FUNCTIONS

OVERVIEW

Let's look at polynomial functions and equations again, but from a more advanced point of view. You have seen how to solve quadratic equations by factoring, extracting roots, or using the quadratic formula. Unfortunately, these methods are usually not options when you need to solve a polynomial equation whose degree is greater than 2. If some of the roots of a polynomial equation of degree greater than 2 are discovered, it may be possible to divide out linear factors corresponding to those roots, thereby reducing the degree of the polynomial equation so that a quadratic equation eventually results.

Some topics needed for the study of calculus are also introduced in this chapter, including breaking an algebraic fraction down into the sum of two or more simpler fractions.

Lessons in Chapter 7
Lesson 7-1: Division of Polynomials
Lesson 7-2: Zeros of Polynomial Functions
Lesson 7-3: Solving Polynomial Equations
Lesson 7-4: Graphing Rational Functions
Lesson 7-5: Decomposing Rational Expressions

Lesson 7-1 — Division of Polynomials

KEY IDEAS

Polynomials can be divided using a long-division process in much the same way that whole numbers were divided before calculators became available. Division of polynomials is important since it is related to finding factors of polynomials. If 6 is divided by 2, the remainder is 0, so 2 is a factor of 6. Similarly, you can tell whether a polynomial is a factor of another polynomial by determining whether their division produces a remainder of 0. Solving a polynomial equation $p(x) = 0$ often depends on finding factors of $p(x)$.

Polynomial Functions and Related Terms

The function $s(x) = x^3 - \sqrt{x}$ is *not* a polynomial function since it contains $\sqrt{x} \left(= x^{\frac{1}{2}} \right)$.

The function $h(x) = x^2 - \dfrac{1}{x}$ is *not* a polynomial function since it contains $\dfrac{1}{x} \left(= x^{-1} \right)$.

Since each of the exponents of x in $p(x) = 7x^4 - 5x^3 + \dfrac{1}{2}x + 9$ is a whole number, $p(x)$ is a *polynomial function*.

Definition of Polynomial Function

A **polynomial function**, p, is a function that can be written in the form

$$p(x) = a_n x^n + a_{n-1} x^{n-1} + a_{n-2} x^{n-2} + \cdots + a_1 x + a_0,$$

where n is a positive integer called the **degree** of the polynomial, provided that $a_n \neq 0$. Each member of the sum is a **term** of the polynomial. The number constants $a_n, a_{n-1}, a_{n-2}, \ldots, a_1, a_0$ are **coefficients**. The **leading coefficient** is a_n, and the **constant term** is a_0.

If $p(x) = 7x^4 - 5x^3 + \dfrac{1}{2}x + 9$:

- The leading coefficient is 7, and the constant term is 9.

- The values of the coefficients are $a_3 = -5$, $a_4 = 7$, $a_2 = 0$, $a_1 = \dfrac{1}{2}$, and $a_0 = 9$.

- The corresponding polynomial *equation* is $x^3 - 7x^2 + 4x + 9 = 0$.

Naming the Parts of a Division Example

In the division example $8 \div 5$,

$$8 \div 5 = \frac{8}{5} = 1 + \frac{3}{5},$$

8 is the **dividend**, 5 is the **divisor**, 1 is the **quotient**, and 3 is the **remainder**.
In general,

$$\frac{\text{dividend}}{\text{divisor}} = \text{quotient} + \frac{\text{remainder}}{\text{divisor}}.$$

Division of Polynomial Functions

The division relationship for numbers is true also for polynomials. If $D(x)$ and $d(x)$ are polynomial functions with the degree of the dividend, D, greater than or equal to the degree of the divisor, d, there are unique polynomials $q(x)$ and $r(x)$, called the quotient and the remainder, such that the following algorithm can be written:

Division Algorithm for Polynomials

$$\frac{D(x)}{d(x)} = q(x) + \frac{r(x)}{d(x)},$$

where $d(x) \neq 0$ and either $r(x) = 0$ or the degree of $r(x)$ is less than or equal to the degree of $d(x)$.

Long Division of Polynomials

The long division of polynomials in standard form follows a pattern similar to that for the long division of whole numbers. To divide $2x^3 + 5x^2 - 3x + 6$ by $x^2 + 1$, set up the long division example in the usual way, where $x^2 + 1$ is the divisor and $2x^3 + 5x^2 - 3x + 6$ is the dividend. Then work as follows:

1. Divide the first term of the dividend, $2x^3$, by the first term of the divisor, x^2. Since $2x^3 \div x^2 = 2x$, write $2x$ above the leading term of the dividend:

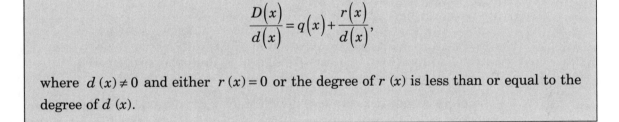

$$x^2 + 1 \,\overline{\smash{)}\, 2x^3 + 5x^2 - 3x + 6}.$$

2. Multiply $2x$ by the divisor, x^2+1, and write the result on the line below the dividend, aligning like terms in the same vertical column:

$$
\begin{array}{r}
2x \\
x^2+1\overline{\smash{)}2x^3+5x^2-3x+6} \\
\underline{2x^3 +2x}
\end{array}
\qquad \leftarrow \ 2x\left(x^2+1\right)=2x^3+2\jmath
$$

3. Subtract and bring down the remaining terms:

$$
\begin{array}{r}
2x \\
x^2+1\overline{\smash{)}2x^3+5x^2-3x+6} \\
\underline{2x^3 +2x}
\end{array}
\qquad
\begin{array}{l}
\\
\\
\leftarrow \ \text{Subtract.}
\end{array}
$$

$$5x^2-5x+6 \qquad \leftarrow \ \text{Bring down } 5x^2 \text{ and } 6.$$

4. Repeat the first three steps, using $5x^2-5x+6$ as the dividend.

$$
\begin{array}{r}
2x\ +5 \\
x^2+1\overline{\smash{)}2x^3+5x^2-3x+6} \\
\underline{2x^3 +2x} \\
5x^2-5x+6 \\
\underline{5x^2 +5} \\
-5x+1
\end{array}
$$

$\leftarrow 5x^2 \div x^2 = 5$

$\leftarrow 5\left(x^2+1\right)=5x^2+5$

$\leftarrow$ Result after subtraction

The process ends here because dividing $-5x$ by x^2 does not result in a monomial. The quotient is $2x+5$, and the remainder is $-5x+1$. Using the division algorithm of polynomials, write the answer in "mixed number" form:

$$\frac{2x^3+5x^2-3x+6}{x^2+1} = 2x+5 \ + \frac{-5x+1}{x^2+1}.$$

Synthetic Division

When a polynomial $p(x)$ is divided by a binomial of the form $x-k$, a condensed form of long division, called **synthetic division**, can be used. Synthetic division uses only the detached numerical coefficients of $p(x)$ as the dividend. The original binomial divisor, $x-k$, is shortened to k. For example, to divide $p(x)=4x^3-7x^2-16x+5$ by $x-3$ using synthetic division, follow these steps:

1. Set up the synthetic division table, using 3 as the synthetic divisor. On the same line write the detached coefficients of $p(x)$. Then draw a horizontal line under the dividend, leaving room for a row of numbers above it:

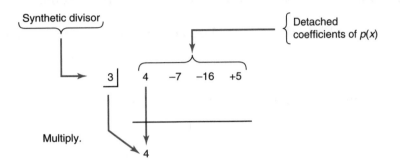

2. Bring down the leading coefficient, 4.
3. Multiply 4 by the synthetic divisor, 3. Enter the product, $+12$, above the horizontal line in the next column under -7. Add the figures in the second column to get $+5$:

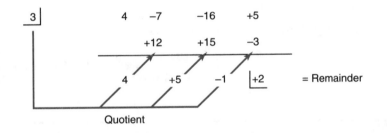

4. Repeat the third step for the remaining columns.
5. Determine the degree of the quotient. Since the divisor is a first-degree binomial, the degree of the quotient will be 1 less than the degree of the dividend. In this example, the dividend is a third-degree polynomial, so the quotient will be a second-degree polynomial of the form $ax^2 + bx + c$, where $a = 4$, $b = 5$, and $c = -1$.
6. Write the quotient and the remainder:

$$\frac{4x^3 - 7x^2 - 16x + 5}{x - 3} = 4x^2 + 5x - 1 + \frac{2}{x - 3}.$$

TIPS

- If necessary, rewrite the divisor in the form $x - k$. Then use k as the synthetic divisor. For example, if the divisor is $x - 3$, then $k = 3$; if the divisor is $x + 2$, rewrite it as $x - (-2)$, so $k = -2$.

- Write the detached coefficients of $p(x)$, where p represents a polynomial function in standard form. If a power of x is missing from $p(x)$, include it with 0 as its coefficient.

- On the last line of the synthetic division table, working from left to right, match the numbers to the coefficients of a polynomial in standard form whose degree is 1 less than the degree of the dividend. The last number on this line is the remainder.

Exercise 1   Dividing Polynomials Using Synthetic Division

Divide $x^4 - 8x^2 + 13x + 1$ by $x + 4$ using synthetic division.

Solution: Since the x^3-term does not appear in the dividend, rewrite the dividend as

$$x^4 + 0 \cdot x^3 - 8x^2 + 13x + 1.$$

Put the divisor $x + 4$ in the form $x - k$ by rewriting it as $x - (-4)$, where $k = -4$ is the synthetic divisor:

$$
\begin{array}{r|rrrrr}
-4 & 1 & 0 & -8 & +13 & +1 \\
   &   & -4 & +16 & -32 & +76 \\
\hline
   & 1 & -4 & +8 & -19 & +\boxed{77} \quad \Leftarrow \text{Remainder} \\
   & \downarrow & \downarrow & \downarrow & \downarrow \\
   & 1 \cdot x^3 & -4x^2 & +8x & -19 & \Leftarrow \text{Quotient}
\end{array}
$$

The answer in "mixed number" form is

$$\frac{x^4 - 8x^2 + 13x + 1}{x + 4} = x^3 - 4x^2 + 8x - 19 + \frac{77}{x+4}.$$

Exercise 2   Synthetic Division with a Divisor of the Form $ax - b$

Divide $4x^3 - 8x^2 + x + 11$ by $2x - 3$ using synthetic division.

Solution: For synthetic division to be used, the coefficient of x in the linear divisor must be 1. Rewrite $2x - 3$ as $2\left(x - \frac{3}{2}\right)$, and perform the synthetic division using $x - \frac{3}{2}$ as the divisor. Since this divisor does not include the factor of 2, you must remember to divide the result by 2 to get the final answer.

$$
\begin{array}{r|rrrr}
\frac{3}{2} & 4 & -8 & +1 & +11 \\
   &   & +6 & -3 & -3 \\
\hline
   & 4 & -2 & -2 & +\boxed{8}
\end{array}
$$

The quotient is $4x^2 - 2x - 2$, and the remainder is 8. Remember that you must adjust the answer to reflect the fact that the actual divisor is $2x - 3$:

- Write the answer: $\dfrac{4x^3 - 8x^2 + x + 11}{x - \dfrac{3}{2}} = 4x^2 - 2x - 2 + \dfrac{8}{x - \dfrac{3}{2}}$

- Divide by 2: $\dfrac{4x^3 - 8x^2 + x + 11}{2\left(x - \dfrac{3}{2}\right)} = \dfrac{4x^2 - 2x - 2}{2} + \dfrac{8}{2\left(x - \dfrac{3}{2}\right)}$

- Simplify: $\dfrac{4x^3 - 8x^2 + x + 11}{2x - 3} = \mathbf{2x^2 - x - 1} + \dfrac{\mathbf{8}}{\mathbf{2x - 3}}$

Figuring Out the Remainder

You saw previously on page 171 that, when $p(x) = 4x^3 - 7x^2 - 16x + 5$ is divided synthetically by $x - 3$, the remainder is 2. Evaluate $p(x)$ at $x = 3$:

$$
\begin{aligned}
p(3) &= 4(3)^3 - 7(3)^2 - 16(3) + 5 \\
&= 108 \quad - 63 \quad - 48 \quad + 5 \\
&= 113 - 111 \\
&= 2
\end{aligned}
$$

Notice that $p(3)$ has the same value as the remainder in the synthetic division. The generalization of this result is called the **remainder theorem**. The remainder theorem tells you that, if a polynomial $p(x)$ is divided by a binomial of the form $x - k$, you can figure out the remainder simply by evaluating $p(k)$. If you also want to know the quotient, use long or synthetic division.

| Exercise 3 | Applying the Remainder Theorem |

Using the remainder theorem, find the remainder when $x^3 + 3x^2 + x - 1$ is divided by $x + 2$.

Solution: Let $p(x) = x^3 + 3x^2 + x - 1$. For the remainder theorem to be used, the divisor must be in the form $x - k$. Rewrite $x + 2$ as $x - (-2)$, so $k = -2$. Now evaluate $p(-2)$:

$$
\begin{aligned}
p(x) &= \quad x^3 \quad + 3x^2 \quad + x \quad - 1 \\
p(-2) &= (-2)^3 + 3(-2)^2 + (-2) - 1 \\
&= -8 \quad + \quad 12 \quad - 2 \quad - 1 \\
&= 1
\end{aligned}
$$

The remainder is **1**.

The Factor Theorem

When a number is divided by one of its factors, as in $6 \div 3$, the remainder is 0. Conversely, if a zero remainder results from the division of two quantities, the divisor must be a factor of the dividend. The generalization of this result is called the **factor theorem**.

The Remainder and Factor Theorems

- Remainder theorem: If $p(x)$ is divided by a binomial of the form $x - k$, the remainder is $p(k)$.
- Factor theorem: If $x - k$ is a factor of $p(x)$, then $p(k) = 0$.
  Also, if $p(k) = 0$, then $x - k$ is a factor of $p(x)$.

Zeros of Functions

A **zero of function** f is a root of $f(x) = 0$. To find a zero of $f(x) = 3x - 6$, set $f(x)$ equal to 0 and solve for x: $x = 2$. Because $f(2) = 0$, the graph of function f has an x-intercept of 2 at its zero. The graph of a function has an x-intercept at each of its real zeros, if any.

Making Mathematical Connections

The function $f(x) = x^3 - 3x^2 - 6x + 8$ can be looked at in different ways.

- **Consider the function numerically.** The table feature of a graphing calculator can be used to quickly display the function values of f, as shown in Figure 7.1. Because $y = 0$ when $x = -2$, $x = 1$, and $x = 4$, these three values of x are the zeros of function f and the roots of the equation $x^3 - 3x^2 - 6x + 8 = 0$.

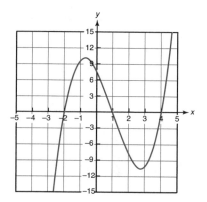

Figure 7.1 Table for $y = x^3 - 3x^2 - 6x + 8$ **Figure 7.2** Graph of $f(x) = x^3 - 3x^2 - 6x + 8$

- **Consider the function graphically.** Because the graph of
  $y = f(x) = x^3 - 3x^2 - 6x + 8$ has x-intercepts at $(-2,0)$, $(1,0)$, and $(4,0)$, as shown in Figure 7.2, -2, 1, and 4 are the zeros of function f and the roots of the equation $x^3 - 3x^2 - 6x + 8 = 0$.
- **Consider the function algebraically.** Function f can be written in factored form as

$$\begin{aligned} f(x) &= x^3 - 3x^2 - 6x + 8 \\ &= (x-1)(x-4)(x+2) \\ &= (x-1)(x-4)(x-(-2)) \end{aligned}$$

Each linear factor of $f(x)$ is the difference between x and a zero of function f. Setting each linear factor equal to 0 gives a root of the equation $x^3 - 3x^2 - 6x + 8 = 0$.

TIPS

The following statements are equivalent:

- k is a zero of function f.
- $x = k$ is a root of the equation $f(x) = 0$.
- $x - k$ is a factor of $f(x)$.
- $(k,0)$ is an x-intercept of the graph of $y = f(x)$, provided that k is a real number.

Exercise 4   Factoring a Polynomial, Given One of Its Factors

Verify, using algebraic methods, that $x - 2$ is a factor of $f(x) = 2x^3 + x^2 - 13x + 6$. Then factor $f(x)$ completely. Confirm your answer using a graphing calculator.

Solution: To verify that $x - 2$ is a factor of $f(x)$, divide $f(x)$ by $x - 2$ synthetically. Check that the remainder is 0. Then factor the quotient. The factors of the quotient must also be factors of $f(x)$.

- Divide $2x^3 + x^2 - 13x + 6$ by $x - 2$:

$$\begin{array}{r|rrrr} 2| & 2 & +1 & -13 & +6 \\ & & +4 & +10 & -6 \\ \hline & 2 & +5 & -3 & \boxed{0} \leftarrow \text{Remainder} \end{array}$$

$$\underbrace{2 \quad +5 \quad -3}$$
Quotient $= 2x^2 + 5x - 3$

Since the remainder is 0, $x - 2$ is a factor of $2x^3 + x^2 - 13x + 6$.

- Factor the quotient:

$$2x^2 + 5x - 3 = (2x - 1)(x + 3).$$

- Write $f(x)$ as the product of its three linear factors:

$$2x^3 + x^2 - 13x + 6 = (x - 2)(2x - 1)(x + 3).$$

Confirm graphically.

Using your graphing calculator, set $Y_1 = 2x^3 + x^2 - 13x + 6$. Set the window size as shown in the figure below at the right. Confirm that the x-intercepts of the graph are at $(2,0)$, $(0.5,0)$, and $(-3,0)$:

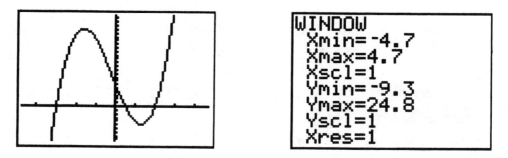

From the graph you see that the x-intercepts or zeros of function f appear to be at $x = -3$, at $x = 2$, and at another value of x between $x = 0$ and $x = 1$. You can use the ZERO or ROOT feature of your calculator to confirm that these three x-values are -3, 0.5, and 2. Alternatively, you could use the table-building feature of your calculator, with a step value of 0.5, to verify that $Y_1 = 0$ for each of the suspected zeros of f:

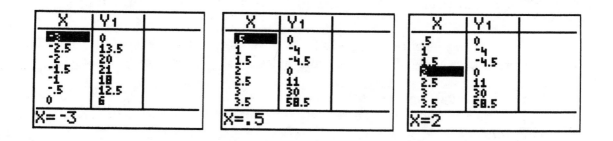

Lesson 7-2

Zeros of Polynomial Functions

KEY IDEAS

This lesson looks at several theorems that can be used to help locate zeros of polynomial functions.

Locating Zeros of Polynomial Functions

Locate $(1,-2)$ and $(2,3)$ on the graph of $f(x) = -x^3 + 4x^2 - 5$ in Figure 7.3.

As x varies from 1 to 2, function f will assume every real number from $f(1) = -2$ to $f(2) = 3$. The generalization of this observation is called the **intermediate value theorem**. Because $f(1)$ and $f(2)$ have opposite signs, the graph crosses the x-axis at some point between $x = 1$ and $x = 2$. Therefore, whenever $f(a)$ and $f(b)$ have opposite signs, there is *at least* one real zero between $x = a$ and $x = b$.

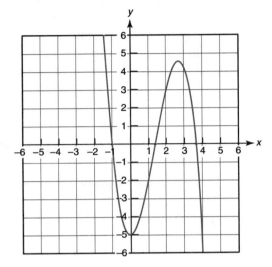

Figure 7.3 Graph of $f(x) = -x^3 + 4x^2 - 5$

Intermediate Value Theorem

Let f represent a polynomial function such that $f(a) \neq f(b)$, where a and b stand for real numbers with $a < b$.

- In the interval from $x = a$ to $x = b$, f takes on every number between $f(a)$ and $f(b)$.
- If $f(a)$ and $f(b)$ have opposite signs, at least one real zero is located in the interval from $x = a$ to $x = b$, as shown in the accompanying figure.

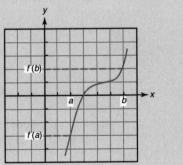

Exercise 1   **Locating Real Zeros Between Consecutive Integers**

In the interval $[-2,4]$, find any pairs of consecutive integers between which the zeros of $p(x) = 6x^4 - 13x^3 + 7x^2 - 26x - 10$ are located.

Solution: Use the table-building feature of your graphing calculator with $Y_1 = 6x^4 - 13x^3 + 7x^2 - 26x - 10$, as shown in the accompanying figure.

- The sign of Y_1 changes from positive to negative as x increases from $x = -1$ to $x = 0$, so there is at least one real zero between –1 and 0.
- The sign of Y_1 changes from negative to positive as x increases from $x = 2$ to $x = 3$, so there is at least one real zero between 2 and 3.

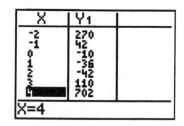

Revisiting the Intermediate Value Theorem

If $f(x)$ changes sign between two different x-values, say $x = 1$ and $x = 2$, is there exactly one zero between 1 and 2? If $f(x)$ does not change sign between two different x-values, are there no zeros between those x-values? In Figure 7.4:

- $f(0)$ and $f(1)$ have the *same* sign, and f has two zeros between 0 an 1;
- $f(1)$ and $f(2)$ have *opposite* signs, and f has three zeros between 1 and 2;
- $f(2)$ and $f(3)$ have the *same* sign, and f has no zeros between 2 and 3.

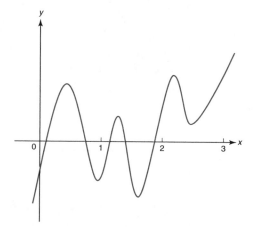

Figure 7.4 Locating zeros of a function

Extending the Intermediate Value Theorem

Let f represent a polynomial function such that $f(a) \neq f(b)$, where a and b stand for real numbers with $a < b$.

- If $f(a)$ and $f(b)$ have opposite signs, there is an odd number of real zeros of f between a and b.
- If $f(a)$ and $f(b)$ have the same sign, either there are no real zeros of f between a and b or there is an even number of real zeros.

Looking for Possible Rational Zeros

The rational zeros theorem relates the coefficients of a polynomial function to the real zeros that the function may have. Consider the function

$$p(x) = 3x^4 - 7x^3 + 8x^2 - 14x + 4.$$

According to the **rational zeros theorem**, if this function has rational zeros, then each of these zeros, when expressed as a fraction in lowest terms, must have a numerator that is a factor of 4, the constant term of $3x^4 - 7x^3 + 8x^2 - 14x + \boxed{4}$, and a denominator that is a factor of 3, the leading coefficient of $\boxed{3}x^4 - 7x^3 + 8x^2 - 14x + 4$.

Rational Zeros Theorem

If $\dfrac{m}{n}$ is a rational zero in lowest terms of a polynomial function

$$p(x) = a_n x^n + a_{n-1}x^{n-1} + \cdots + a_1 x + a_0 \text{ with integer coefficients,}$$

then m is a factor of a_0, and n is a factor of a_n, provided that $a_n \neq 0$.

Exercise 2   Identifying Possible Rational Zeros

List all possible rational zeros of $p(x) = 3x^4 - 7x^3 + 8x^2 - 14x + 4$ using the rational zeros theorem. Determine whether any of these possible zeros is a zero of the function. Confirm your answer graphically.

Solutions: The factors of 4, the constant term of the polynomial, are ± 1, ± 2, and ± 4. The factors of 3, the leading coefficient of the polynomial, are ± 1 and ± 3. Hence, the possible rational zeros are:

$$\pm\frac{1}{3}, \ \pm\frac{2}{3}, \ \pm 1, \ \pm\frac{4}{3}, \ \pm 2 \left(= \pm\frac{2}{1} \right), \text{ and } \pm 4 \left(= \pm\frac{4}{1} \right).$$

When you test the potential rational zeros, you will find that only $p\left(\dfrac{1}{3}\right) = 0$ and $p(2) = 0$.

Hence, $x = \dfrac{1}{3}$ and $x = 2$ are the only rational zeros of the function.

Confirm graphically: Here are the graph and the window variables used to obtain it:

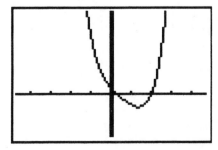 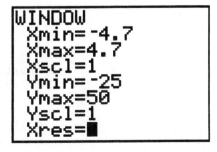

Use the ZERO or ROOT feature of your calculator to confirm that the first x-intercept is at $x \approx .33333333 \approx \dfrac{1}{3}$ and the other x-intercept is at $x = 2$.

- Use the intermediate value theorem, together with the table-building feature of your graphing calculator to help locate any real zeros of a function. This procedure was illustrated in Exercise 1.
- After using the rational zeros theorem to compile a list of potential rational zeros of a function, look at the graph of the function. You may be able to shorten the list of potential rational zeros by eliminating any possible zeros that do not match the x-intercepts of the graph.

How Many Zeros Can a Function Have?

A first-degree polynomial such as $f(x) = 3x - 6$ has one zero, namely, $x = 2$. A second-degree polynomial such as $f(x) = x^2 - 4x$ has two zeros, $x = 4$ and $x = 0$. Although difficult to prove, every polynomial function of degree 1 or greater has at least one zero. This fact is called the **fundamental theorem of algebra.**

> ### The Fundamental Theorem of Algebra
>
> If p is a polynomial function of degree n with $n > 0$, then $p(x)$ has at least one zero in the set of complex numbers.

How Many Linear Factors Can a Function Have?

- A second-degree polynomial, such as $x^2 - x - 12$, has two linear factors since $x^2 - x - 12 = (x - 4)(x + 3)$.
- The third-degree polynomial $p(x) = x^3 + 4x$ can be written as the product of *three* linear factors:

$$p(x) = x^3 + 4x = x(x^2 + 4) = \overbrace{(x - 0)(x + 2i)(x - 2i)}^{\text{Three linear factors}}.$$

- The fourth-degree polynomial $p(x) = x^4 - 2x^2 - 3$ has *four* linear factors:

$$p(x) = x^4 - 2x^2 - 3 = \overbrace{(x + \sqrt{3})(x - \sqrt{3})(x - i)(x + i)}^{\text{Four linear factors}}.$$

The generalization of this factoring pattern is called the **linear factorization theorem**.

The Linear Factorization Theorem

If p is a polynomial function of degree n with $n > 0$, then $p(x)$ can be factored as the product of n linear binomial factors of the form $x - r$, where r belongs to the set of complex numbers.

According to the linear factorization theorem, a third-degree polynomial equation has exactly three roots, a fourth-degree polynomial equation has exactly four roots, and an nth-degree polynomial equation has exactly n roots. These roots may be real or imaginary, rational or irrational, equal or unequal.

Exercise 3   Using Factoring to Find Complex Zeros of Functions

Find the zeros of each function: $a.\ f(x) = x^4 - 9$ $b.\ f(x) = x^3 - 2x^2 + 5x$

Solutions: Factor each polynomial completely.

$a.$
$$f(x) = x^4 - 9 = \left(x^2 - 3\right)\left(x^2 + 3\right)$$
$$= \left(x - \sqrt{3}\right)\left(x + \sqrt{3}\right)\left(x^2 + 3\right)$$
$$= \left(x - \sqrt{3}\right)\left(x + \sqrt{3}\right)\left(x - \sqrt{3}\,i\right)\left(x + \sqrt{3}\,i\right)$$

The four zeros are $\sqrt{3}$, $-\sqrt{3}$, $\sqrt{3}\,i$, and $-\sqrt{3}\,i$. The irrational zeros occur in conjugate pairs, as do the nonreal zeros.

$b.\ f(x) = x^3 - 2x^2 + 5x = x(x^2 - 2x + 5)$. Hence, $x = 0$ is a zero. To find the other two zeros, set $x^2 - 2x + 5 = 0$ and solve for x, using the quadratic formula. You should verify that $x = 1 \pm 2i$.

The three zeros are $\mathbf{0}$, $\mathbf{1 - 2i}$, and $\mathbf{1 + 2i}$. The nonreal zeros occur in conjugate pairs.

Exercise 4   Finding a Polynomial Function, Given Its Zeros

Find a polynomial function f with real coefficients of degree 4 that has the zeros –1, 2, and $-1 + i$.

Solution: Since $-1 + i$ is a zero of f, then so is its conjugate, $-1 - i$.

- If k is a zero of f, then $x - k$ is a factor of $f(x)$. The factors that correspond to the four zeros are $x + 1, x - 2, x - (-1 + i)$, and $x - (-1 - i)$.

- If c is some nonzero constant, then

$$f(x) = c(x+1)(x-2)\left[x-(-1+i)\right]\left[x-(-1-i)\right]$$

describes the family of functions that have the given numbers as zeros.
- To find the simplest function with the given numbers as zeros, set $c = 1$. Then multiply the factors together.

$$
\begin{aligned}
f(x) &= 1(x+1)(x-2)\left[x-(-1+i)\right]\left[x-(-1-i)\right] \\
&= \left(x^2-x-2\right)\left[x^2-x(-1+i)-x(-1-i)+(-1+i)(-1-i)\right] \\
&= \left(x^2-x-2\right)\left[x^2+x-ix+x+ix+\left(1-i^2\right)\right] \\
&= \left(x^2-x-2\right)\left(x^2+2x+2\right) \\
f(x) &= x^4+x^3-2x^2-6x-4
\end{aligned}
$$

Counting Repeated Zeros

The linear factorization theorem does not guarantee that the linear factors are all different. For example, $p(x) = x^2 - 6x + 9 = (x-3)(x-3)$. Hence, $p(x)$ has two equal zeros, namely, 3. Since $x - 3$ is repeated two times in the linear factorization of $p(x)$, we say that "3 is a *zero of multiplicity two*." If all of the factors of the polynomial function

$$p(x) = \left(x^2+1\right)(x-6)^3$$

are multiplied together, the term with the greatest power will be $x^2 \cdot x^3 = x^5$. Because $p(x)$ is a fifth-degree polynomial function, it has a total of *five* complex zeros. It is easy to see from the factored form of the polynomial that, if $x^2 + 1 = 0$, then $x = \pm i$, so two of the five zeros are imaginary. The factor of $x - 6$ is repeated three times, so the zero of 6 must be counted three times.

TIPS

When counting the zeros of a function, use these facts:

- A polynomial function of degree n with $n > 0$ has exactly n complex zeros, provided that repeated zeros are counted the number of times they occur as factors in the complete linear factorization of the function.
- If a polynomial function has real coefficients, then any nonreal zeros occur in conjugate pairs. For example, if $1 + 3i$ is a zero of such a function, then so is $1 - 3i$.
- If a polynomial function has rational coefficients and the irrational number $a + \sqrt{b}$ is a zero of the function, then its conjugate, $a - \sqrt{b}$, is also a zero of the function. For example, if $2 + \sqrt{5}$ is a zero of the function, then so is $2 - \sqrt{5}$.

Exercise 5   Writing a Polynomial Function with Repeated Zeros

Find a polynomial function f for which 2 is a zero of multiplicity 2 and $\sqrt{3}$ is also a zero.

Solution:

- Since 2 is a zero of multiplicity 2 for function f, $(x-2)^2$ is a factor of the function.

- If $\sqrt{3}$ is a zero of function f, then so is $-\sqrt{3}$. Therefore, $\left(x-\sqrt{3}\right)$ and $\left(x+\sqrt{3}\right)$ are also factors of function f.

- Hence:

$$f(x) = (x-2)^2\left(x-\sqrt{3}\right)\left(x+\sqrt{3}\right)$$
$$= \left(x^2 - 4x + 4\right)\left(x^2 - 3\right)$$
$$f(x) = x^4 - 4x^3 + x^2 + 12x - 12$$

Recognizing Duplicate Zeros

The function $f(x) = (1-x)^3(x+2)^2$ has five zeros. Since $(1-x)$ occurs as a factor three times, the zero of 1 is repeated three times. The factor $(x+2)$ occurs two times, so the zero of –2 is repeated two times. The graph of function f has only two distinct x-intercepts, as can be seen in Figure 7.5.

- At $(-2,0)$, the graph touches the x-axis and remains on the same side of it. As the function passes through this point, the graph of f does *not* change sign since f contains an *even* power of $(x+2)$.

- At $(1,0)$, the graph crosses over the x-axis. As the function passes through this point, the graph of f changes sign since f contains an odd power of $(x-1)$.

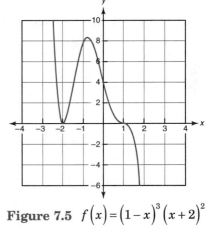

Figure 7.5 $f(x) = (1-x)^3(x+2)^2$

TIPS

Let f represent a polynomial function with a real zero, k.

- If the zero k is repeated an *even* number of times, the graph of f touches the x-axis at $(k,0)$ but does not cross over it.
- If the zero k is repeated an *odd* number of times or is not repeated, the graph of f crosses over the x-axis at $(k,0)$.

Exercise 6   Finding a Root with a Multiplicity Greater Than 1

Determine whether any of the real zeros of the function $f(x) = 2x^3 - 9x^2 + 27$ is (are) repeated. Confirm your answer algebraically.

Solution: Using your graphing calculator, display the graph of the given function:

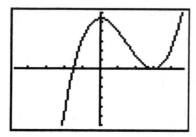 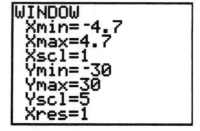

- Using the zero or root feature, confirm that the zeros of the function occur at $x = -\dfrac{3}{2}$ and $x = 3$.
- Since the graph crosses the x-axis at $x = -\dfrac{3}{2}$, if $-\dfrac{3}{2}$ is a repeated zero, it must be repeated at least three times. This is not possible since the function has a maximum of three zeros and you know that there is also a zero at $x = 3$.
- Because function f touches the x-axis at $(3,0)$, but does not cross over it, $x = 3$ is repeated an even number of times. Therefore, the zero of 3 can be repeated two times, but only two times.

Confirm algebraically. To confirm that 3 is a zero of multiplicity 2, divide function f by $x - 3$ two times:

$$
\begin{array}{r|rrrr}
3 & 2 & -9 & 0 & 27 \\
  &   & +6 & -9 & -27 \\
\hline
  & 2 & -3 & -9 & \underline{0} \quad \leftarrow \text{Thus, } (x-3) \text{ is a factor of } f(x).
\end{array}
$$

Use as the new dividend.

Divide again, using 3 as the synthetic divisor:

$$
\begin{array}{r|rrr}
3 & 2 & -3 & -9 \\
  &   & +6 & +9 \\
\hline
  & 2 & +3 & \underline{0} \quad \leftarrow \text{Thus, } (x-3)^2 \text{ is a factor of } f(x).
\end{array}
$$

Lesson 7-3

Solving Polynomial Equations

KEY IDEAS

The degree of a polynomial function, p, is reduced by 1 each time the function is divided by a linear factor of $p(x)$. If the degree of function p is greater than 2, then it may be possible to find enough linear factors of $p(x)$ so that through division the polynomial equation $p(x) = 0$ can be reduced to a quadratic equation.

Decreasing the Degree of a Polynomial Equation

To solve the equation $p(x) = x^3 - 2x^2 + 3x - 6 = 0$, you could try to get the roots by looking at the graph of $y = p(x)$ in Figure 7.6.

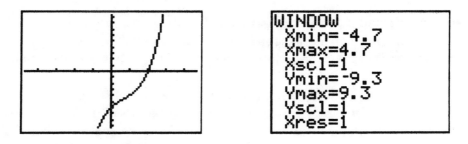

Figure 7.6 Graph of $p(x) = x^3 - 2x^2 + 3x - 6$ with the window settings

Although the graph suggests that $x = 2$ is a root of $p(x) = 0$, it cannot help you find the remaining two roots, which are imaginary. You can, however, reason as follows:

- Because $x = 2$ is a root of $p(x) = 0$, $x - 2$ is a factor of $p(x)$.
- "Dividing out" the factor of $x - 2$ from the third-degree equation $x^3 - 2x^2 + 3x - 6 = 0$ produces a second-degree equation.
- The second-degree equation can be solved either by factoring or by using the quadratic formula. The roots of this equation are also roots of the original third-degree equation.

Here is the mathematical procedure:

1. Divide $p(x) = x^3 - 2x^2 + 3x - 6$ by $x - 2$ synthetically:

$$
\begin{array}{r|rrrr}
2\!\!\!\! & 1 & -2 & +3 & -6 \\
& & +2 & 0 & +6 \\
\hline
& 1 & 0 & +3 & \big|\,0 \;\leftarrow \text{Remainder} \\
\end{array}
$$

$$\text{Quotient} = x^2 + 0 \cdot x + 3$$

2. Since the quotient $x^2 + 3$ is a factor of $p(x)$, $p(x) = (x - 2)(x^2 + 3)$. If $p(x) = 0$, then any factor of $p(x)$ may be equal to 0. Setting $x^2 + 3$ equal to 0 forms what is called the *depressed equation*.
3. Solve the depressed equation. If $x^2 + 3 = 0$, $x^2 = -3$ and $x = \pm\sqrt{-3} = \pm\sqrt{3}\,i$.

Hence, the three roots of the equation $x^3 - 2x^2 + 3x - 6 = 0$ are $x = 2, -\sqrt{3}\,i, \sqrt{3}\,i$.

Some Facts About Depressed Equations

-   A **depressed equation** arises from "dividing out" a *linear* factor of a polynomial function and then setting the quotient equal to 0. Thus, the degree of the depressed equation is always 1 less than the degree of the polynomial function from which it was derived.
-   Each root of the depressed equation is also a root of the original polynomial equation.
-   If you know, or can find, one or more linear factors of a higher degree polynomial equation, you may be able to solve it completely by successively dividing out each of the known linear factors from the polynomial and its quotients until arriving at a quadratic equation. Solving the quadratic equation by standard methods gives the remaining two roots.

Exercise 1   Solving a Polynomial Equation by Dividing Out a Linear Factor

Find an integer root of the equation $p(x) = x^3 - 13x + 12 = 0$. Then, using that integer root, find the remaining roots of the given equation. Confirm your answer graphically.

Solution:

-   For certain types of polynomial equations, it may be possible to arrive at a root simply by inspecting the coefficients and then using a "guess and check" process. For instance, it is easy to see by looking at the coefficients of $x^3 - 13x + 12 = 0$ that you should first try as possible roots small integer values of x that are also factors of 12, such as 1 and –1. Since $p(1) = 1^3 - 13 \cdot 1 + 12 = 0$, $x = 1$ is a root of the given equation.

- Because $x = 1$ is a root, $x - 1$ is a factor of $p(x)$. Divide $p(x) = x^3 - 13x + 12$ by $x - 1$ synthetically, and then set the quotient equal to 0 to get the depressed equation: $x^2 + x - 12 = 0$. Then $(x - 3)(x + 4) = 0$ and, as a result, $x = 3$ or $x = -4$.

- The three roots of the original polynomial equation are **1**, **3**, and **–4**.

Confirm graphically. Display the graph in an appropriate viewing window, and then use the ZERO or ROOT feature of your calculator to verify that the x-intercepts fall at $x = 1, x = 3$, and $x = -4$:

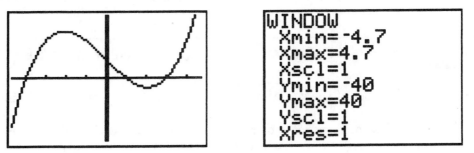

```
WINDOW
 Xmin=-4.7
 Xmax=4.7
 Xscl=1
 Ymin=-40
 Ymax=40
 Yscl=1
 Xres=1
```

Exercise 2   Solving an Equation by Repeated Division of Linear Factors

If two roots of the equation $p(x) = x^4 + x^3 - 15x^2 - 22x + 8 = 0$ are –2 and 4, find the exact values of the other two roots.

Solution: Since –2 and 4 are roots, $x - (-2)$ and $x - 4$ are factors of $p(x)$.

- Divide $p(x)$ by $x - (-2)$ synthetically:

$$
\begin{array}{r|rrrrr}
-2 & 1 & 1 & -15 & -22 & +8 \\
   &   & -2 & +2 & +26 & -8 \\
\hline
   & 1 & -1 & -13 & +4 & \boxed{0}
\end{array}
$$

The last line on the synthetic division table represents a third-degree polynomial quotient that is used as the dividend in the next division.

- Divide the preceding quotient by $x - 4$:

$$
\begin{array}{r|rrrr}
4 & 1 & -1 & -13 & +4 \\
  &   & +4 & +12 & -4 \\
\hline
  & 1 & +3 & -1 & \boxed{0}
\end{array}
$$

$$\text{Quotient} = x^2 + 3x - 1$$

- Solve the depressed equation $x^2 + 3x - 1 = 0$ by using the quadratic formula, where $a = 1$, $b = 3$, and $c = -1$:

$$x = \frac{-3 \pm \sqrt{9 - 4(1)(-1)}}{2} = \frac{-3 \pm \sqrt{13}}{2}.$$

- The exact values of the two irrational roots are $\dfrac{-3 + \sqrt{13}}{2}$ and $\dfrac{-3 - \sqrt{13}}{2}$.

Exercise 3   Solving a Polynomial Equation, Given an Imaginary Root

If $1 + 2i$ is a root of the equation $p(x) = x^4 + 4x^3 - 4x^2 + 24x + 15 = 0$, find the other roots.

Solution: Imaginary roots of polynomial functions with real coefficients occur in conjugate pairs. Hence, if $1 + 2i$ is a root, then so is $1 - 2i$. The corresponding binomial factors of these imaginary roots are $x - (1 + 2i)$ and $x - (1 - 2i)$. Divide $p(x)$ successively by the two binomial factors so that a second-degree depressed equation results. Although the synthetic divisors are complex numbers, the synthetic division is performed in exactly the same way as when the synthetic divisors are real numbers.

- Divide $p(x)$ by $x - (1 + 2i)$ synthetically:

$$
\begin{array}{r|ccccc}
(1+2i) & 1 & +4 & -4 & +24 & +15 \\
 & & 1+2i & 1+12i & -27+6i & -15 \\
\hline
 & 1 & 5+2i & -3+12i & -3+6i & \boxed{0}
\end{array}
$$

- Divide the quotient by $x - (1 - 2i)$ synthetically:

$$
\begin{array}{r|cccc}
(1-2i) & 1 & 5+2i & -3+12i & -3+6i \\
 & & 1-2i & 6-12i & -3+6i \\
\hline
 & 1 & 6 & +3 & \boxed{0}
\end{array}
$$

$$\text{Quotient} = x^2 + 6x + 3$$

- The depressed equation is $x^2 + 6x + 3 = 0$. Solve the equation by using the quadratic formula, where $a = 1$, $b = 6$, and $c = 3$:

$$x = \frac{-6 \pm \sqrt{6^2 - 4(1)(3)}}{2(1)} = -3 \pm \sqrt{6}.$$

- In addition to the given root, $1 + 2i$, the other roots of the equation are $\mathbf{1 - 2i, 3 + \sqrt{6}}$, and $\mathbf{3 - \sqrt{6}}$.

Exercise 4   Applying a Four-Step Strategy

Solve: $p(x) = 3x^4 - 16x^3 + 24x^2 + 8x - 35 = 0$.

Solution: Use the following four-step strategy:

STEP 1. List all possible rational zeros of $p(x) = 3x^4 - 16x^3 + 24x^2 + 8x - 35$:

- The factors of the constant term are ±1, ±5, ±7, and ±35.
- The factors of the leading coefficient are ±1 and ±3.
- The possible rational zeros of $p(x)$ are $\pm\dfrac{1}{3}, \pm 1, \pm\dfrac{5}{3}, \pm\dfrac{7}{3}, \pm 5, \pm 7, \pm\dfrac{35}{3}$, and ±35 .

STEP 2. Use a graphing calculator to help eliminate any of the possible zeros identified in STEP 1. You can quickly determine by using the table feature of your calculator that, since Y is not 0 when $x = \pm 5, \pm 7$, or ±35 , none of these values can be a zero of the function. The graph is shown in the accompanying figure. Since there is an x-intercept at $(-1,0)$ and another x-intercept between $x = 2$ and $x = 3$, you can eliminate from the list of possible rational zeros all values except $x = -1$ and $x = \dfrac{7}{3}$.

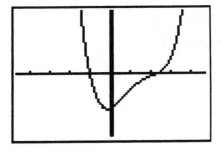

$[-4.7, 4.7] \times [-62, 62]$

STEP 3. Use the remainder theorem to confirm that $x = -1$ and $x = \dfrac{7}{3}$ are zeros of the function. At the same time, work toward obtaining a quadratic depressed equation:

- Test $x = -1$:

$$
\begin{array}{r|rrrrr}
-1 & 3 & -16 & +24 & +8 & -35 \\
   &   & -3 & +19 & -43 & +35 \\
\hline
   & 3 & -19 & +43 & -35 & \underline{|\,0} \leftarrow \text{This confirms that } x = -1 \text{ is a zero.}
\end{array}
$$

Use as the next dividend.

- Test $x = \dfrac{7}{3}$:

$$
\begin{array}{r|rrrr}
\dfrac{7}{3} & 3 & -19 & +43 & -35 \\
             &   & +7 & -28 & \cdot +35 \\
\hline
             & 3 & -12 & +15 & \underline{|\,0} \leftarrow \text{This confirms that } x = \dfrac{7}{3} \text{ is a zero.}
\end{array}
$$

Quotient = $3x^2 - 12x + 15$

STEP 4.   Form and then solve the depressed equation:

$$3x^2 - 12x + 15 = 0$$

$$\frac{3x^2}{3} - \frac{12x}{3} + \frac{15}{3} = \frac{0}{3}$$

$$x^2 - 4x + 5 = 0$$

Use the quadratic formula, where $a = 1$, $b = -4$, and $c = 5$:

$$x = \frac{-(-4) \pm \sqrt{(-4)^2 - 4(1)(5)}}{2(1)} = \frac{4 \pm \sqrt{-4}}{2} = 2 \pm i.$$

The roots of the given equation are -1, $\dfrac{7}{3}$, $2 + i$, and $2 - i$.

Exercise 5   Finding and Approximating an Irrational Root

Find all real roots of the equation $p(x) = x^3 + x - 7 = 0$.

Solution:

- The possible rational zeros of $p(x)$ are ± 1 and ± 7.
- Use a graphing calculator to help determine whether any of the possible zeros should be tested. Here is the graph and the window settings used to obtain it.

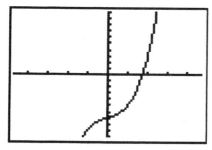

Since the only x-intercept is between $x = 2$ and $x = 3$, ± 1 and ± 7 cannot be zeros of the function. The only real root of the given equation is, therefore, irrational.

- Use the ZERO or ROOT feature of your calculator to find that the irrational root shown in the graph is approximately **1.7392039**.

Lesson 7-4 — Graphing Rational Functions

KEY IDEAS

A rational function, f, has the form $f(x) = \dfrac{P(x)}{Q(x)}$, where $P(x)$ and $Q(x)$ are polynomial functions. The graph of a rational function may have asymptotes as x approaches the values for which the function is undefined. When discussing the behavior of a rational function, it is helpful to know any intercepts and asymptotes.

Intercepts of $f(x) = P(x)/Q(x)$

To determine the points at which the graph of $f(x) = \dfrac{P(x)}{Q(x)}$ intersects the coordinate axes:

- Set $P(x) = 0$. Any solution of this equation that does not also make $Q(x)$ equal to 0 is an x-intercept.
- Set $x = 0$ and evaluate $f(0)$. The y-intercept is $(0, f(0))$, provided that it is defined.

EXAMPLE: If $f(x) = \dfrac{x^2 - 16}{x + 8}$, then the x-intercepts are the roots of the equation

$x^2 - 16 = 0$, provided that these roots do not make the denominator evaluate to 0. The roots are $x = \pm 4$. Since the denominator is not 0 when $x = \pm 4$, the x-intercepts are $(-4, 0)$ and $(4, 0)$.

EXAMPLE: If $f(x) = \dfrac{x^2 - 16}{x + 8}$, then $f(0) = \dfrac{0^2 - 16}{0 + 8} = -2$. Therefore, the y-intercept is $(0, -2)$.

Vertical Asymptotes

The graph of a rational function has a **vertical asymptote** at each value of x, if any, for which the denominator evaluates to 0 but the numerator does not evaluate to 0.

Locating Vertical Asymptotes

Let $f(x) = \dfrac{P(x)}{Q(x)}$. The line $x = a$ is a vertical asymptote of the graph of function f

when $Q(a) = 0$ and $P(a) \neq 0$.

EXAMPLE: To find the vertical asymptote of $f(x) = \dfrac{x^2 - 16}{x + 8}$, set $x + 8 = 0$, so $x = -8$. Since

the numerator of the function does not equal 0 when $x = -8$, the line $x = -8$ is a vertical asymptote of the graph of the function. The graph is shown in Figure 7.7.

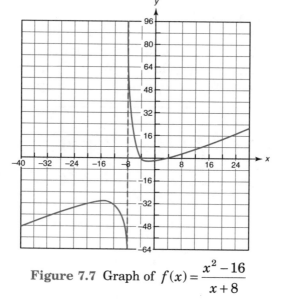

Figure 7.7 Graph of $f(x) = \dfrac{x^2 - 16}{x + 8}$

Horizontal Asymptotes

The graph of $f(x) = \dfrac{P(x)}{Q(x)}$ has a **horizontal asymptote** when the degree of the numerator is less than or equal to the degree of the denominator.

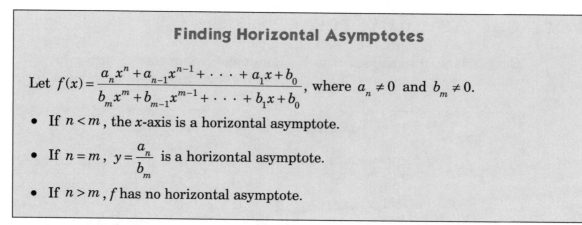

Finding Horizontal Asymptotes

Let $f(x) = \dfrac{a_n x^n + a_{n-1} x^{n-1} + \cdots + a_1 x + b_0}{b_m x^m + b_{m-1} x^{m-1} + \cdots + b_1 x + b_0}$, where $a_n \neq 0$ and $b_m \neq 0$.

- If $n < m$, the x-axis is a horizontal asymptote.

- If $n = m$, $y = \dfrac{a_n}{b_m}$ is a horizontal asymptote.

- If $n > m$, f has no horizontal asymptote.

EXAMPLE: The degree of the numerator of $\dfrac{3x}{x^2-9}$ is 1 less than the degree of the denominator, so the x-axis is a horizontal asymptote graph of $f(x)=\dfrac{3x}{x^2-9}$, as shown in Figure 7.8. When checking for vertical asymptotes, you will find that the graph has vertical asymptotes at $x=-3$ and $x=3$ since for these x-values the denominator of the function is not defined.

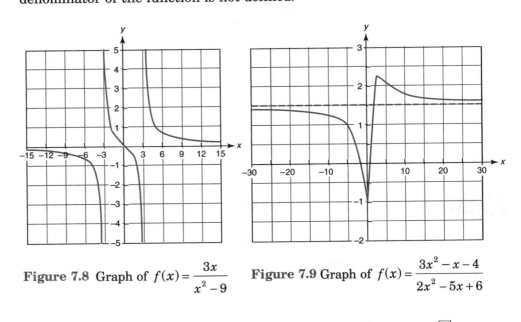

Figure 7.8 Graph of $f(x)=\dfrac{3x}{x^2-9}$ **Figure 7.9** Graph of $f(x)=\dfrac{3x^2-x-4}{2x^2-5x+6}$

EXAMPLE: Since the degrees of the numerator and the denominator of $f(x)=\dfrac{\boxed{3}x^2-x-4}{\boxed{2}x^2-5x+6}$ are equal, $y=\dfrac{3}{2}$ is a horizontal asymptote of the graph, as shown in Figure 7.9.

Slant Asymptotes

If the degree of the numerator of a rational function is 1 more than the degree of the denominator, its graph has a **slant asymptote**. The graph of

$$f(x)=\frac{3x^2-12x+2}{x-4}$$

has a slant asymptote. To find its equation, do the long division:

1. Divide the numerator by the denominator:

$$
\begin{array}{r}
3x \\
x-4\overline{)3x^2-12x+2} \\
-\underline{3x^2-12x} \\
+2
\end{array}
$$

The answer is $3x+\dfrac{2}{x-4}$.

2. Delete the remainder from the answer:

$$3x + \frac{\cancel{2}}{\cancel{x} - 4}.$$

3. Write the equation of the slant asymptote by setting y equal to the remaining part of the quotient: $y = 3x$. Figure 7.10 shows the graph of the function with its asymptote.

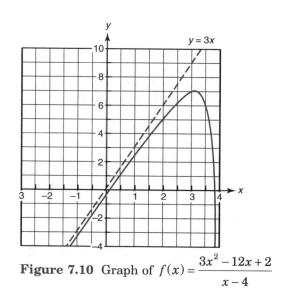

Figure 7.10 Graph of $f(x) = \dfrac{3x^2 - 12x + 2}{x - 4}$

Exercise 1   Finding Key Features of the Graph of a Rational Function

Find the intercepts and asymptotes of the graph of $f(x) = \dfrac{x^2 - 36}{x^2 - 9}$. Discuss any symmetry that the graph displays.

Solution:

- **x-intercepts:** (**–6,0**) and (**6,0**) because $f(x) = 0$ when $x^2 - 36 = 0$, so $x = \pm\sqrt{36} = \pm 6$.

- **y-intercepts:** (**0,4**) because, if $x = 0$, $y = f(0) = \dfrac{0^2 - 36}{0^2 - 9} = 4$.

- **asymptotes**: $x = +3$ and $x = -3$ because, when $x^2 - 9 = 0$, $x = \pm\sqrt{9} = \pm 3$ and these values of x do not make the denominator evaluate to 0.

- **symmetry:** The given function is an even function since $f(-x) = f(x)$, so that its graph is symmetric with respect to the y-axis, as shown in the accompanying figure, which was obtained using a graphing calculator set to a Decimal window.

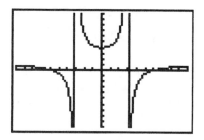

Exercise 2   Finding a Slant Asymptote

Find the equation of the slant asymptote of the graph of $f(x) = \dfrac{x^2 - 16}{x + 8}$.

Solution: Divide the numerator by the denominator using long division:

$$\begin{array}{r} x - 8 \\ x+8\overline{\smash{\big)}\,x^2 - 16} \\ -\underline{x^2 + 8x} \\ -8x - 16 \\ -\underline{8x - 64} \\ 48 \end{array}$$

The answer is $x - 8 + \dfrac{48}{x+8}$. Deleting the remainder leaves $x - 8$. The equation of the slant asymptote is $y = x - 8$, as shown in the accompanying figure.

Lesson 7-5   Decomposing Rational Expressions

KEY IDEAS

You already know how to find the sum of two fractions:

$$\frac{2}{x} + \frac{5}{x-3} = \frac{7x-6}{x^2-3x}.$$

Sometimes, as in the study of calculus, it is necessary to reverse this process by starting with the fraction on the right side of the equation above and "decomposing it" by writing it as the sum or difference of two or more simpler fractions, called **partial fractions**.

The rational expression $\dfrac{7x-6}{x^2-3x}$ is *proper*. A rational expression is **proper** if the degree of the polynomial in the numerator is lower than the degree of the polynomial in the denominator. Otherwise, the rational expression is **improper**. Only proper fractions can be decomposed into the sum of two or more partial fractions.

Revisiting the Linear Factorization Theorem

Every polynomial of degree n greater than 1 can be factored over the set of complex numbers into the product of n linear factors. If the coefficients of the polynomial are real, however, any nonreal linear factors must occur in conjugate pairs. The product of each such conjugate pair is an *irreducible* quadratic. The quadratic expression $ax^2 + bx + c$ is **irreducible** if it cannot be factored into the product of linear factors with real coefficients. For example, if $x - 2i$ is a factor of a polynomial with real coefficients, then so is $x + 2i$.

The product $(x - 2i)(x + 2i)$ is the irreducible quadratic $x^2 + 4$.

> ## Linear-Quadratic Factorization Theorem for Polynomials
>
> Every polynomial function with real coefficients can be uniquely factored over the real numbers into a product of linear or irreducible quadratic factors.

The Decomposition Principle: Four Cases to Consider

The method used to decompose a proper rational expression into partial fractions depends largely on the nature of the factors in its denominator. Because of the linear-quadratic factorization theorem, there are four cases to consider:

- The denominator contains only linear factors, all of which are different.
- The denominator contains at least one linear factor that is repeated.
- The denominator includes an irreducible quadratic factor.
- The denominator contains at least one irreducible quadratic factor that is repeated.

When studying the four cases, assume that the rational expression is proper and written in lowest terms.

Case I: Nonrepeated linear case

If the denominator of a rational expression can be written as the product of different non-repeated linear factors of the form $ax+b$, then, for each of these factors, the decomposition of the rational expression includes a partial fraction of the form $\dfrac{A}{ax+b}$, where A is some real number.

EXAMPLE: To decompose $\dfrac{7x-6}{x^2-3x}$, factor the denominator and then provide a partial fraction for each linear factor:

$$\frac{7x-6}{x\left(x-3\right)}=\frac{A}{x}+\frac{B}{x-3}.$$

Next, clear this equation of its fractions by multiplying each term by $x\left(x-3\right)$:

$$7x-6=A\left(x-3\right)+Bx$$
$$=Ax-3A+Bx$$
$$=\left(A+B\right)x-3A$$

Compare the coefficients of the x-terms on both sides of the equation, and compare the constant terms on both sides of the equation. It must be the case that $3A = 6$, so $A = 2$, and also that $A + B = 7$, so $B = 7 - A = 7 - 2 = 5$. Hence:

$$\frac{7x - 6}{x^2 - 3x} = \frac{2}{x} + \frac{5}{x - 3}.$$

Case II: Repeated linear case

If the factored form of the denominator of a rational expression includes a factor of the form $(ax + b)^m$, where m is a positive integer greater than 1, then the decomposition of the rational expression includes the sum of m partial fractions of the form

$$\frac{A_1}{ax + b} + \frac{A_2}{(ax + b)^2} + \cdots + \frac{A_m}{(ax + b)^m},$$

where A_1, A_2, ..., A_m are m real-valued constants.

EXAMPLE: To decompose $\dfrac{8x + 7}{(2x - 1)^3}$, provide three partial fractions because the exponent of $2x - 1$ is 3:

$$\frac{8x + 7}{(2x - 1)^3} = \frac{A}{2x - 1} + \frac{B}{(2x - 1)^2} + \frac{C}{(2x - 1)^3}.$$

Applying the Decomposition Principle

To find the partial-fraction decomposition of $\dfrac{3x^2 - 29x + 80}{x(x - 4)^2}$, work as follows:

1. Determine the partial fraction that corresponds to each factor of the denominator. The partial fraction $\dfrac{A}{x}$ corresponds to the factor x. For the factor $(x - 4)^2$, there corresponds the partial fraction sum

$$\frac{B}{x - 4} + \frac{C}{(x - 4)^2}.$$

2. Rewrite the original fraction in terms of its partial-fraction decomposition:

$$\frac{3x^2 - 29x + 80}{x(x - 4)^2} = \frac{A}{x} + \frac{B}{x - 4} + \frac{C}{(x - 4)^2}.$$

3.   Determine the values of the constants.

*   Clear the equation of its fractional terms by multiplying each term by $x(x-4)^2$, the lowest common multiple of the denominators:

$$3x^2 - 29x + 80 = A(x-4)^2 + Bx(x-4) + Cx.$$

*   On the right side of the equation, perform the indicated operations and then collect like terms:

$$3x^2 - 29x + 80 = A(x^2 - 8x + 16) + B(x^2 - 4x) + Cx$$
$$= (A+B)x^2 + (-8A - 4B + C)x + 16A$$

*   Write a system of equations. The numerical coefficients of like variable terms on the two sides of the equation, as well as the constant terms, must be equal. Thus:

x^2-coefficients:	$3 = A + B$
x-coefficients:	$-29 = -8A - 4B + C$
constant terms:	$80 = 16A$

*   Solve the system of three equations in three unknowns by starting with the simplest of the three equations. Since $80 = 16A$, $A = \dfrac{80}{16} = 5$. Substituting $A = 5$ in the first equation makes $B = 3 - A = 3 - 5 = -2$. Substituting 5 for A and -2 for B in the second equation yields $C = 3$.

4.   Use the values of the constants to write the complete partial-fraction decomposition:

$$\frac{3x^2 - 29x + 80}{x(x-4)^2} = \frac{5}{x} + \frac{-2}{x-4} + \frac{3}{(x-4)^2}.$$

Case III: Nonrepeated irreducible quadratic case

For each nonrepeated irreducible quadratic factor of the form $ax^2 + bx + c$ that the denominator of a rational expression contains, the decomposition of that fraction requires a partial fraction of the form $\dfrac{Ax+B}{ax^2+bx+c}$.

EXAMPLE: $\dfrac{3x^2 + 7}{(x-1)(x^2 + 2x + 5)} = \dfrac{A}{x-1} + \dfrac{Bx+C}{x^2+2x+5}$

Case IV: Repeated irreducible quadratic case

For each repeated irreducible quadratic factor of the form $\left(ax^2 + bx + c\right)^m$ that the denominator of a proper fraction contains, where m is a positive integer greater than 1, the decomposition of that fraction requires a partial fraction of the form

$$\frac{A_1 x + B_1}{ax^2 + bx + c} + \frac{A_2 x + B_2}{\left(ax^2 + bx + c\right)^2} + \cdots + \frac{A_m x + B_m}{\left(ax^2 + bx + c\right)^m},$$

where A_1, A_2, ..., A_m and B_1, B_2, ..., B_m are real-valued constants.

EXAMPLE: $\dfrac{x}{\left(x^2 + 1\right)^3} = \dfrac{A_1 x + B_1}{x^2 + 1} + \dfrac{A_2 x + B_2}{\left(x^2 + 1\right)^2} + \dfrac{A_3 x + B_3}{\left(x^2 + 1\right)^3}$

Decomposing Improper Rational Expressions

If the original rational expression is improper, use long division to rewrite it as the sum of a polynomial (the quotient) and a proper fraction (the remainder). Then decompose the proper fraction. Before you can apply the decomposition principle to $\dfrac{2x^2 + x - 6}{x^2 - 3x}$, you must rewrite it as the sum of a polynomial and a proper fraction.

1. Divide $2x^2 + x - 6$ by $x^2 - 3x$ using long division:

$$\begin{array}{r} 2 \\ x^2 - 3x \overline{)\, 2x^2 + x - 6} \\ \underline{-\ 2x^2 - 6x } \\ 7x - 6 \end{array}$$

   Hence, $\dfrac{2x^2 + x - 6}{x^2 - 3x} = 2 + \dfrac{7x - 6}{x^2 - 3x}$.

2. Decompose the proper fraction (the remainder). From page 197, you already know that

$$\frac{7x - 6}{x^2 - 3x} = \frac{2}{x} + \frac{5}{x - 3}.$$

3. Rewrite the original rational expression as the sum of a polynomial and partial fractions:

$$\frac{2x^2 + x - 6}{x^2 - 3x} = 2 + \frac{2}{x} + \frac{5}{x - 3}.$$

Chapter 7	**CHECKUP EXERCISES**

1—5. Multiple Choice

1. Three of the roots of a polynomial equation with rational coefficients are $3, 4i,$ and $1 - 2\sqrt{3}$. What is the lowest possible degree of the equation?

   (1) 5   (2) 6   (3) 3   (4) 4

2. Let $f(x)$ represent a third-degree polynomial function with real coefficients. When $y = f(x)$ is graphed, it is found that $f(2) = -3$ and $f(3) = 1$. Which statement *must* be true?

   (1) The function $f(x)$ has exactly one real zero in the interval $\left[2, 3\right]$.

   (2) The function $f(x)$ has exactly one real zero in the interval $\left[-3, 1\right]$.

   (3) The function $f(x)$ has one or three real zeros in the interval $\left[2, 3\right]$.

   (4) The function $f(x)$ has one or two real zeros in the interval $\left[-3, 1\right]$.

3. Let $f(x)$ represent a fourth-degree polynomial function with real coefficients. When $y = f(x)$ is graphed, it is found that $f(4) = 2$ and $f(7) = 3$. Which statement *must* be true?

   (1) The function $f(x)$ has an odd number of real zeros in the interval $[4, 7]$.

   (2) The function $f(x)$ has an even number of real zeros in the interval $[4, 7]$.

   (3) The function $f(x)$ has at least one real zero in the interval $[2, 3]$.

   (4) No conclusion is possible.

4. According to the rational zeros theorem, which of the following could *not* be a zero of the function $f(x) = 3x^4 - 5x^3 + 8x + 6$?

   (1) $-\dfrac{1}{3}$ (2) $-\dfrac{2}{3}$ (3) $\dfrac{3}{2}$ (4) 2

5. Based on the accompanying graph of the polynomial function $y = f(x)$, which statement *must* be true?

   (1) The degree of function f is even.

   (2) The function f is divisible by $x^2 - 2x - 3$.

   (3) The function f has no imaginary zeros.

   (4) $x = -1$ is a zero of multiplicity 1 and $x = 3$ is a zero of multiplicity 2.

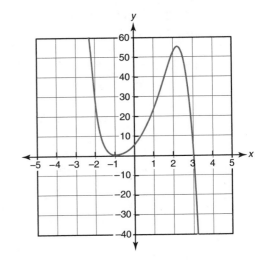

6. Find the remainder when $p(x) = 2x^3 - 5x^2 + 7x + 4$ is divided by $x - 2$.

7. Find the remainder when $p(x) = x^4 + 4x^3 - 2x^2 + x - 4$ is divided by $x + 3$.

8. For what value of k is $(x + 3)$ a factor of $x^4 + x^3 - x^2 + kx - 12$?

9–12. Use factoring to find each of the zeros of the given function.

9. $f(x) = x^3 + 2x^2 - 3x$

10. $f(x) = x^3 + 5x$

11. $f(x) = 3x^4 - 48$

12. $f(x) = 4x^5 - x^3 - 32x^2 + 8$

13–15. Divide $p(x)$ by the given binomial.

13. $p(x) = 4x^3 - x^2 + 8x - 1; x - 2$

14. $p(x) = x^4 - 2x^3 + 4x + 11; x + 1$

15. $p(x) = 6x^3 - x^2 + 3x - 6; 2x - 1$

16. Divide $x^4 + 6x^3 - 5x^2 + 13x + 1$ by $x^2 + 2$ using long division.

17. Given $p(x) = 2x^3 - 7x^2 + 2x + 3$.

   a. Verify that $x - 3$ is a factor of $p(x)$.

   b. Factor $p(x)$ as the product of three linear binomial factors.

18. Find a polynomial function f with real coefficients of degree 4 that has the zeros 2, –1, and $1 + 2i$.

19. Find a polynomial function f for which –3 is a zero of multiplicity 2 and $1 - \sqrt{2}$ is also a zero.

20. Given the equation $2x^3 - 5x + 6 = 0$.

    a. Find an integer root.

    b. Use your answer to part *a* to find the remaining two roots.

21. If two roots of the equation $x^4 + 2x^3 - 9x^2 - 16x - 6 = 0$ are –1 and 3, find the exact values of the other two roots.

22. If $2 - i$ is a root of the equation $2x^4 - 23x^2 + 44x - 5 = 0$, find the other three roots.

23. Use synthetic division to show that all of the real zeros of $f(x) = x^3 - 3x^2 - 4x + 5$ lie in the interval [–2, 4].

24. Approximate the irrational root of $-x^3 - 3x^2 + 4x - 5 = 0$ to five decimal places.

25–27. In each case, determine any intercepts and an equation for the slant asymptote.

25. $f(x) = \dfrac{x^3 + 1}{x^2 - 1}$ 26. $f(x) = \dfrac{x^3 + 8}{x^2 - 2x}$ 27. $f(x) = \dfrac{x^3 - 8}{x^2 - 3x - 4}$

28. Given $p(x) = 6x^4 - 7x^3 - 8x^2 + 13x - 4$.

    a. Show that $x = 1$ is a zero of multiplicity 2.

    b. Factor $p(x)$ as the product of linear binomials.

29–32. Solve each polynomial equation.

29. $x^3 - 4x^2 - 13x - 2 = 0$ 31. $2x^4 - 3x^3 - 13x^2 + 2x + 12 = 0$

30. $2x^3 - 9x^2 + 18x - 20 = 0$ 32. $2x^4 - 3x^3 + 3x^2 + 77x - 39 = 0$

33—38. For each rational function, discuss its intercepts and asymptotes and sketch the graph.

33. $f(x) = \dfrac{3x}{16 - x^2}$

36. $f(x) = \dfrac{x^2 + 2x}{x - 1}$

34. $f(x) = \dfrac{3x^2 - 11x - 4}{2x^2 - 5x - 12}$

37. $f(x) = \dfrac{2x^2 + 5x - 3}{x - 3}$

35. $f(x) = \dfrac{x^2 - 25}{x^2 - 4}$

38. $f(x) = \dfrac{x^4 - 3x^3 - 4}{1 - x^2}$

39—47. Write the partial-fraction decomposition of each rational expression.

39. $\dfrac{6x + 5}{x^2 + 2x}$

42. $\dfrac{3x^2 + 7}{(x - 1)(x^2 + 2x + 5)}$

45. $\dfrac{2x + 9}{x^3 - 1}$

40. $\dfrac{2x + 1}{x^2 - 5x - 6}$

43. $\dfrac{6x}{(x - 2)(x^2 - 4)}$

46. $\dfrac{x^2 + 2x + 8}{x^3 - 4x^2 + 4x}$

41. $\dfrac{x - 1}{x(x + 3)^2}$

44. $\dfrac{x^2}{x^3 - 8}$

47. $\dfrac{x^2 + 1}{2x^3 + 5x^2 - 3x}$

48—50. Write the improper rational expression as the sum of a polynomial and a proper rational expression.

48. $\dfrac{3x^2 + 1}{x^2 + 4x}$

49. $\dfrac{2x^2 - 5x + 1}{x^2 - 3x + 2}$

50. $\dfrac{x^3 + x + 2}{x^2 - 3x}$

Chapter 8

EXPONENTIAL AND LOGARITHMIC FUNCTIONS

OVERVIEW

Much of our work so far has been with algebraic functions. **Algebraic functions** are functions that can be defined in terms involving sums, differences, products, quotients, powers, or roots of polynomial functions. Nonalgebraic functions often arise when trying to represent real-life processes mathematically. The **exponential function** and its inverse, the **logarithmic function**, are important types of nonalgebraic functions that can be used to model certain types of growth processes.

Lessons in Chapter 8
Lesson 8-1: Inverse Functions
Lesson 8-2: The Exponential Function
Lesson 8-3: The Logarithmic Function
Lesson 8-4: Logarithm Laws and Equations
Lesson 8-5: Exponential and Logarithmic Models

Inverse Functions

KEY IDEAS

The area, A, of a square is a function of the length of a side, s. Using the function $A = s^2$, you can calculate the area of the square, given s. Suppose you now want to undo that process and find s when the area is given. The function $s = \sqrt{A}$ does that job nicely. Functions that undo each other's effect are called **inverse functions**. The functions $A = s^2$ and $s = \sqrt{A}$ are inverse functions.

Consider the function $A = s^2$. When $s = 3$, $A = 3^2 = 9$, so (3,9) belongs to this function.

Now consider its inverse, $s = \sqrt{A}$. When $A = 9$, $s = \sqrt{9} = 3$, so (9,3) belongs to this function. When functions are *inverses* of each other, the domain and the range of one function are interchanged in the inverse of that function.

One-to-One Functions

A function is a **one-to-one function** if the same value of y is never paired with two different values of x. A "marriage function" that pairs husbands to their wives consists of ordered pairs of the form (husband,wife). This is a one-to-one function since a wife cannot be paired with two different husbands. Not all functions, however, are one to one. For example, the function $y = x^2$ includes $\left(2,\underline{4}\right)$ and $\left(-2,\underline{4}\right)$, so it is not a one-to-one function.

Functions That Have Inverse Functions

Let f = {(–1,5), (1,2), (3,5)}. The inverse of function f will have the domain and range interchanged:

$$g = \{(\underline{5},-1),\ (2,1),\ (\underline{5},3)\}.$$

Notice that g is not a function since the same x-value is paired with two different y-values. How did it happen that f is a function but g is *not* a function? Because f is not a one-to-one function, at least two different ordered pairs have the same y-values. Therefore, when x and y are interchanged, there will be at least two different ordered pairs with the same x-value. Only one-to-one functions have inverses that are functions.

Horizontal-Line Test

Because only one-to-one functions have inverses that are functions, it is important to be able to tell at a glance whether a function is one to one.

Horizontal-Line Test

A function is a **one-to-one** function if no horizontal line intersects the graph of the function in more than one point.

- The graph of the function in Figure 8.1 passes the horizontal-line test because any horizontal line that is drawn will intersect the graph in at most one point. Because no two points on the graph have the same y-values but different x-values, the graph represents a one-to-one function. If the graph of a function passes the horizontal-line test, then its inverse is a function.

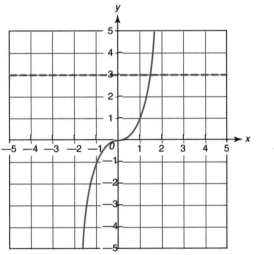 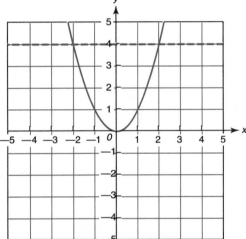

Figure 8.1 Graph *passes* the horizontal-line test.

Figure 8.2 Graph *fails* the horizontal-line test.

- The graph in Figure 8.2 fails the horizontal-line test, so the function it represents does not have an inverse function.

Finding an Equation of an Inverse Function

To find an equation for the inverse of a function, solve for x in terms of y. Since functions in the form "y = function of x" are preferable, interchange x and y. To find the inverse of $f(x) = 3x + 6$:

1. Let $y = f(x)$: $\qquad\qquad\qquad\qquad y = 3x + 6$

2. Solve for x: $\qquad\qquad\qquad\qquad x = \dfrac{y-6}{3} = \dfrac{1}{3}y - 2$

3. Interchange x and y: $\qquad\qquad\qquad y = \dfrac{1}{3}x - 2$

Composing Inverse Functions

If $f(x) = 3x + 6$ and $g(x) = \dfrac{1}{3}x - 2$, then

$$g\big(f(3)\big) = g(15) = 3 \text{ and } f\big(g(3)\big) = f(-1) = 3.$$

Thus, $g(f(3)) = f(g(3)) = 3$. Composing a pair of inverse functions in either order always produces the starting x-value since one function undoes the effect of the other, as illustrated in Figure 8.3.

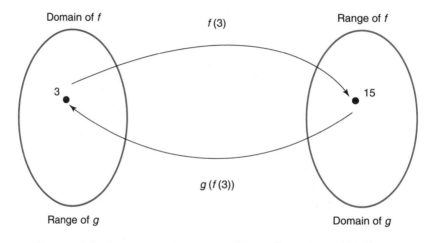

Figure 8.3 Composing inverse functions f and g: $g(f(x)) = x$

Definition of an Inverse Function

Let f represent a one-to-one function. The **inverse** of function f, denoted as f^{-1}, is the function for which

$$f\big(f^{-1}(x)\big) = f^{-1}\big(f(x)\big) = x$$

for all values of x for which the compositions are defined. The notation f^{-1} is read as "f-inverse."

Testing Whether Two Functions Are Inverses

You can determine whether f and g are inverse functions, even if you did not derive one equation from the other, by verifying that $f(g(x)) = x$ for all x in the domain of g *and* that $g(f(x)) = x$ for all x in the domain of f.

Exercise 1   Testing Whether Two Functions Are Inverses

Decide whether $f(x) = x^3 - 1$ and $g(x) = (x+1)^{\frac{1}{3}}$ are inverse functions.

Solution: Determine whether the statements $f(g(x)) = x$ and $g(f(x)) = x$ are both true:

$$f\big(g(x)\big) = f\left((x+1)^{\frac{1}{3}}\right)$$

$$= \left((x+1)^{\frac{1}{3}}\right)^3 - 1 \qquad \text{and}$$

$$= (x+1) - 1$$

$$= x$$

$$g\big(f(x)\big) = g\big(x^3 - 1\big)$$

$$= \big((x^3 - 1) + 1\big)^{\frac{1}{3}}$$

$$= \big(x^3\big)^{\frac{1}{3}}$$

$$= x$$

Since both statement are true, functions f and g **are** inverse functions.

Exercise 2   Testing Whether Two Functions Are Inverses

Decide whether $f(x) = 2x - 1$ and $g(x) = \dfrac{1}{2}x + 1$ are inverse functions.

Solution: Determine whether the statements $f(g(x)) = x$ and $g(f(x)) = x$ are both true:

$$f\big(g(x)\big) = f\left(\frac{1}{2}x + 1\right) = 2\left(\frac{1}{2}x + 1\right) - 1$$

$$= x + 2 - 1$$

$$= x + 1$$

You need not do any more work. Because $f(g(x)) = x + 1 \ne x$, f and g **are *not*** inverse functions.

TIPS

- The inverse of a function, f, may or may not be a function. If the inverse is a function, it is denoted by f^{-1}, where -1 is *not* an exponent. Do not confuse inverse-function notation with exponent notation by thinking of f^{-1} as $\dfrac{1}{f}$.

- Because the inverse of a function involves switching the roles of x and y, the domain and range of the inverse function are found by interchanging the domain and range of the original function:

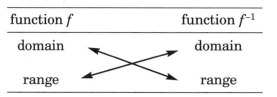

Be careful to exclude any value of x for which it is not possible to calculate either $f(f^{-1}(x))$ or $f^{-1}(f(x))$.

- Functions such as $y = x^3$ and $y = \sqrt{x}$ that always increase or always decrease over their entire domains are one-to-one functions, so their inverses are functions.

Exercise 3   Finding the Inverse of a Function

If $f(x) = \dfrac{2x+1}{x-3}$, find $f^{-1}(x)$.

Solution: Find $f^{-1}(x)$ by interchanging x and y.

- Replace $f(x)$ with y:

$$y = \frac{2x+1}{x-3}$$

- Solve for x:

$$y(x-3) = 2x+1$$
$$xy - 3y = 2x + 1$$
$$xy - 2x = 3y + 1$$
$$x(y-2) = 3y + 1$$
$$x = \frac{3y+1}{y-2}$$

- Interchange x and y:

$$y = f^{-1}(x) = \frac{3x+1}{x-2}$$

The denominator of a fraction cannot be equal to 0, so $x \neq 3$ for function f and $x \neq 2$ for function f^{-1}. Hence, f and f^{-1} are inverse functions for all real numbers x, provided that $x \neq 2$ and $x \neq 3$.

Symmetry and Inverse Functions

The graphs of $f(x) = 3x + 6$ and its inverse, $f^{-1}(x) = \dfrac{1}{3}x - 2$, are shown in Figure 8.4 with their line of symmetry, $y = x$. Figure 8.5 shows that the graphs of $f(x) = x^3 - 1$ and its inverse, $f^{-1}(x) = (x+1)^{\frac{1}{3}}$, are also symmetric to the line $y = x$.

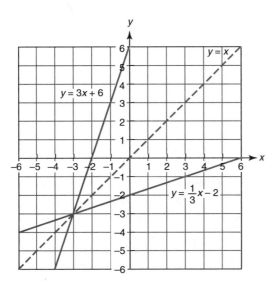

Figure 8.4 Symmetry of inverse functions

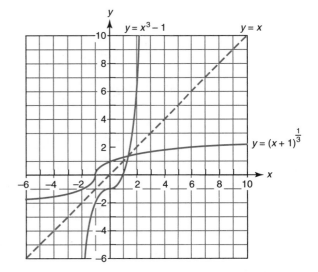

Figure 8.5 Symmetry of inverse functions

Symmetry Property for Inverse Functions

- The graphs of f and f^{-1} are symmetric to the line $y = x$.
- If you know what the graph of a one-to-one function looks like, you can obtain the graph of its inverse by reflecting the original graph in the line $y = x$.
- If (a,b) is on the graph of f, then (b,a) is on the graph of f^{-1}.

Exercise 4   Finding and Graphing the Inverse of a Function

If $f(x) = \sqrt{2x + 3}$, find $f^{-1}(x)$. Sketch the graphs of f and f^{-1}. Determine the domain and the range of each function.

Solution: To find f^{-1}, let $y = \sqrt{2x+3}$. When the variables are switched, $x = \sqrt{2y+3}$.

Solving for y gives $y = f^{-1}(x) = \dfrac{x^2-3}{2}$.

The graphs of f and f^1 are shown in the accompanying figure. The *domain* of f is the set of all real numbers x such that $2x+3 \geq 0$, which, after solving for x, is the interval $\left[-\dfrac{3}{2}, \infty\right)$. You now know that the *range* of f^{-1} is the interval $\left[-\dfrac{3}{2}, \infty\right)$, which is consistent with the graph of f^{-1}, shown as the broken curve in the accompanying figure. Since

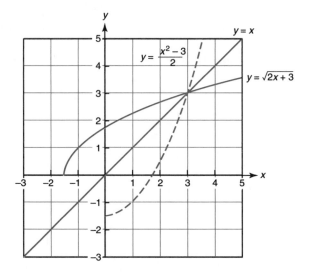

$y = \sqrt{2x+3}$, y must be either 0 or a positive number. Thus, the *range* of f is the interval $[0,\infty)$, which implies that the *domain* of f^{-1} is $[0,\infty)$, as can be seen from its graph.

Here is a summary of what was learned about the functions:

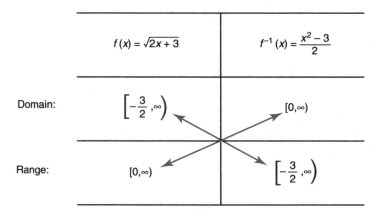

Lesson 8-2

The Exponential Function

KEY IDEAS

The equations $y = 2x, y = x^2$, and $y = 2^x$ are fundamentally different. The first equation is a linear equation, and its graph is a straight line. The second equation is a quadratic equation, and its graph is a parabola. The third equation is an *exponential equation*. Unlike the first two equations, this equation has an exponent that is a variable. An *exponential function* is a nonalgebraic function in which the independent variable is in the exponent.

The Graphs of $y = 2^x$ and $y = \left(\dfrac{1}{2}\right)^x$

An **exponential function** is a function of the form $f(x) = b^x$, where b is a positive number other than 1 and x is any real number. The graphs of $y = 2^x$ and $y = \left(\dfrac{1}{2}\right)^x$ can be used to make generalizations about how exponential functions behave. As x increases, the graph of $y = 2^x$ rises while the graph of $y = \left(\dfrac{1}{2}\right)^x$ falls, as shown in Figures 8.6 and 8.7. From the graphs, it is evident that, for an exponential function:

$$\begin{aligned}
\textbf{domain:} &\quad \{\text{real numbers}\} \\
\textbf{range:} &\quad \{\text{positive real numbers}\} \\
\textbf{\textit{y}-intercept:} &\quad (0,1)
\end{aligned}$$

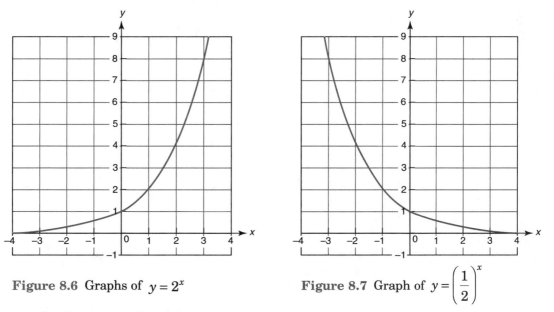

Figure 8.6 Graphs of $y = 2^x$ Figure 8.7 Graph of $y = \left(\dfrac{1}{2}\right)^x$

You can further generalize that:

- The graph of $y = b^x$ has the same shape as the graph of $y = 2^x$ when $b > 1$. The graph rises from left to right and, as x decreases, the negative x-axis is a horizontal asymptote.

- The graph of $y = b^x$ has the same shape as the graph of $y = \left(\dfrac{1}{2}\right)^x$ when $0 < b < 1$. The graph falls from left to right and, as x increases, the positive x-axis is a horizontal asymptote.

- The graphs of $y = b^x$ and $y = \left(\dfrac{1}{b}\right)^x$ are reflections of each other in the y-axis.

- An exponential function is a one-to-one function since its graph passes the horizontal-line test and, as a result, has an inverse function.

Exponential Equations with Like Bases

An **exponential equation** is an equation such as $3^{x-4} = 9$ in which the variable is in an exponent. If an exponential equation can be written so that each side is a power of the same base, the equation can be solved by setting the exponents equal to each other.

Exercise 1   **Solving an Exponential Equation**

Solve for x: *a.* $9^{x+1} = 27^x$ *b.* $64^{1-x} = \dfrac{1}{16^{2x}}$

Solution: Write each side as a power of the same base. Then equate the exponents and solve for x.

a. Since $3^2 = 9$ and $3^3 = 27$:

$$\left(3^2\right)^{x+1} = \left(3^3\right)^x$$
$$3^{2(x+1)} = 3^{3x}$$

Setting the exponents equal makes $2(x+1) = 3x$, so $2x + 2 = 3x$ and $\boldsymbol{x = 2}$.

b. Since $4^3 = 64$ and $4^{-2} = \dfrac{1}{4^2} = \dfrac{1}{16}$:

$$\left(4^3\right)^{1-x} = \left(4^2\right)^{-2x}$$
$$4^{3(1-x)} = 4^{-4x}$$

Setting the exponents equal makes $3(1-x) = -4x$, so $3 - 3x = -4x$ and $\boldsymbol{x = -3}$.

Lesson 8-3

The Logarithmic Function

KEY IDEAS

The inverse of $y = b^x$ is $x = b^y$. There is no algebraic method for solving $x = b^y$ for y in terms of x. This algebraic limitation can be overcome, however, by defining the *logarithmic function* as $y = \log_b x$, where $\log_b x$ is the *exponent* needed for base b so that the result is x. For example, to evaluate $\log_2 8$, find the exponent, say y, of the base 2 such that $2^y = 8$ and $y = 3$. Thus, $2^3 = 8$ and $\log_2 8 = 3$ are equivalent equations.

The Logarithmic Function

A logarithm is an *exponent*. A logarithm is defined so that $\log_b x = y$ means $b^y = x$.

Because $b^y = x$ is the inverse of $y = b^x$, $\log_b x = y$ is the inverse of $y = b^x$.

> ## Definition of the Logarithmic Function
>
> The **logarithmic function with base b**, denoted as $y = \log_b x$, is the inverse of the exponential function $y = b^x$ with base b, where x and b are positive and $b \neq 1$. Thus:
>
> $$y = \log_b x \text{ means } b^y = x.$$

The equation $y = \log_b x$ is read as "y is the logarithm of x with base b." In the equation $y = \log_b x$, y is the *exponent* of base b that produces x.

EXAMPLE: If $\log_8 64 = 2$, then $8^2 = 64$, where 2 is the power to which base 8 must be raised to produce 64.

EXAMPLE: The expression $\log_3 81$ refers to an exponent: the exponent of base 3 needed to produce 81. Hence, $\log_3 81 = 4$ because $3^4 = 81$.

The Graph of $y = \log_b x$

The graph of $y = \log_2 x$ is shown in Figure 8.8, and the graph of $y = \log_{\frac{1}{2}} x$ appears in Figure 8.9.

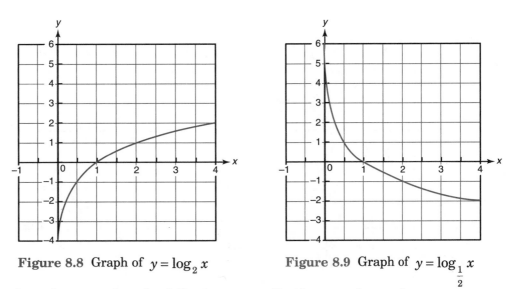

Figure 8.8 Graph of $y = \log_2 x$ **Figure 8.9** Graph of $y = \log_{\frac{1}{2}} x$

Based on these graphs, the following generalizations can be made about all logarithmic functions:

- The domain is limited to the set of positive real numbers, while the range is the set of all real numbers.
- The x-intercept of the graph is (0,1).

- If the base is greater than 1, as in Figure 8.8, the graph rises as x increases. As x decreases, the graph is asymptotic to the negative y-axis.
- If the base is between 0 and 1, as in Figure 8.9, the graph falls as x increases. As x decreases, the graph is asymptotic to the positive y-axis.

These properties should come as no surprise, since exponential and logarithmic functions are inverse functions. The domain and range of the two functions are interchanged. Because point $(0,1)$ is on the graph of an exponential function, point $(1,0)$ is on the graph of the logarithmic function. Table 8.1 provides a more complete comparison of the graphs of the two functions.

Table 8.1 Comparing the Graphs of an Exponential and a Logarithmic Function

Property	Graph of $y = b^x$	Graph of $y = \log_b x$
Quadrants	I and II	I and IV
Domain	{real numbers}	{positive real numbers}
Range	{positive real numbers}	{real numbers}
x-Intercept	None	$(1,0)$
y-Intercept	$(0,1)$	None
Asymptote	x-axis	y-axis
Rises as x increases	$b > 1$	$b > 1$
Falls as x increases	$0 < b < 1$	$0 < b < 1$
Line symmetry	Symmetric to the line $y = x$.	See Figures 8.10 and 8.11.

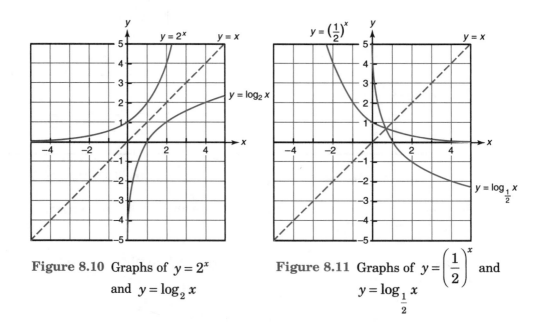

Figure 8.10 Graphs of $y = 2^x$ and $y = \log_2 x$

Figure 8.11 Graphs of $y = \left(\dfrac{1}{2}\right)^x$ and $y = \log_{\frac{1}{2}} x$

Common Logarithm

The **common logarithm** is the logarithm whose base is 10. Common logarithms are indicated when the base is omitted from the logarithm expression. Thus, $\log x$ is understood to mean $\log_{10} x$.

1. The common logarithm of a power of 10 is simply the exponent of 10. For example, $\log 100 = 2$ because $100 = 10^2$. Also, $\log\left(\dfrac{1}{10}\right) = -1$ since $\dfrac{1}{10} = 10^{-1}$.

2. To find the value of a common logarithm such as log 45, quit to the home screen of your graphing calculator and press $\boxed{\text{LOG}}$. Enter 45 and close the parentheses. Pressing $\boxed{\text{ENTER}}$ displays 1.653212514. Thus, $\log(45) = 1.653212514$, indicating that $10^{1.653212514} \approx 45$.

Change-of-Base Formula

You can rewrite a logarithm with base b as an equivalent base-10 logarithm by using a simple formula.

Change-of-Base Formula

$$\log_b x = \frac{\log x}{\log b}$$

For example, $\log_3 x = \dfrac{\log x}{\log 3}$. To graph $y = \log_3 x$ using your calculator, set $Y_1 = \dfrac{\log(x)}{\log(3)}$.

Writing Equivalent Equations

To change the exponential to the logarithmic form of an equation, or vice versa, use the fact that $y = \log_b x$ and $x = b^y$ are equivalent equations.

Exponential Form	Logarithmic Form
$3^2 = 9$	$\log_3 9 = 2$
$10^3 = 1000$	$\log_{10} 1000 = 3$
$4^{-2} = \dfrac{1}{16}$	$\log_4\left(\dfrac{1}{16}\right) = -2$

| Exercise 1 | Solving Logarithm Equations |

For each equation, find the value of x:

a. $\log_2 x = -5$ *b.* $\log_x 8 = \dfrac{3}{4}$ *c.* $\log\left(\dfrac{x}{4}\right) = 1.6$

Solutions:

a. If $\log_2 x = -5$, then, in exponential form, $2^{-5} = x$. Hence, $x = \dfrac{1}{2^5} = \dfrac{1}{\mathbf{32}}$.

b. If $\log_x 8 = \dfrac{3}{4}$, then, in exponential form, $x^{\frac{3}{4}} = 8$. Thus:

$$\left(x^{\frac{3}{4}}\right)^{\frac{4}{3}} = (8)^{\frac{4}{3}}, \text{ so } x = \left(\sqrt[3]{8}\right)^4 = 2^4 = \mathbf{16}.$$

c. If $\log\left(\dfrac{x}{4}\right) = 1.6$, then $\dfrac{x}{4} = 10^{1.6}$. Use your graphing calculator to find an approximate value for $10^{1.6}$: $x = 4 \times 10^{1.6} \approx \mathbf{159.24}$.

| Lesson 8-4 | **Logarithm Laws and Equations** |

KEY IDEAS

Because a logarithm is an exponent, the laws for finding the logarithms of a product, quotient, and power follow the laws of exponents for those operations. For example, to multiply powers of the same base, *add* the exponents. The logarithm of a *product* is the *sum* of the logarithms of the factors of that product. Because powers of the same base are divided by *subtracting* their exponents, the logarithm of a fraction is the logarithm of the numerator *minus* the logarithm of the denominator.

Decomposing a Logarithm Expression

Logarithms of products, quotients, powers, and roots can be broken down into their component parts by using one or more of the three logarithm laws in Table 8.2.

Table 8.2 Logarithm Laws

Law	General Rule	Examples
Product	$\log_b\left(xy\right)=\log_b x+\log_b y$	• $\log 21 = \log\left(7\times3\right)=\log 7+\log 3$ • $\log 100a = \log 100+\log a = 2+\log a$
Quotient	$\log_b\left(\dfrac{x}{y}\right)=\log_b x-\log_b y$	• $\log\left(\dfrac{pq}{2}\right)=\log p+\log q-\log 2$ • $\log\left(\dfrac{k}{10}\right)=\log k-\log 10=\log k-1$
Power	$\log_b\left(x^n\right)=n\log_b x$	• $\log 25 = \log 5^2 = 2\log 5$ • $\log\sqrt{x}=\log x^{\frac{1}{2}}=\dfrac{1}{2}\log x$

Exercise 1   Decomposing a Logarithm

Express $\log\left(\sqrt[3]{\dfrac{m}{n}}\right)$ in terms of $\log m$ and $\log n$.

Solution: Use the fact that the cube root of a number can be written in exponent form with an exponent of $\dfrac{1}{3}$:

$$\log\left(\sqrt[3]{\frac{m}{n}}\right)=\log\left(\frac{m}{n}\right)^{\frac{1}{3}}=\frac{1}{3}\log\left(\frac{m}{n}\right)=\frac{1}{3}\left(\log m-\log n\right).$$

Exercise 2   Applying Logarithm Laws

If $\log 2 = x$ and $\log 3 = y$, express each of the following in terms of x and y:

a. $\log\sqrt{6}$ b. $\log 24$

Solution:

a. Factor 6 as 2×3. Then use the power and product laws of logarithms:

$$\log \sqrt{6} = \log \left(2 \times 3\right)^{\frac{1}{2}} = \frac{1}{2} \log \left(2 \times 3\right)$$

$$= \frac{1}{2} \left(\log 2 + \log 3\right)$$

$$= \frac{1}{2} \left(x + y\right)$$

b. Rewrite 24 as $8 \times 3 = 2^3 \times 3$. Then use the product and power laws of logarithms:

$$\log 24 = \log \left(2^3 \times 3\right) = \log 2^3 + \log 3$$

$$= 3 \log 2 + \log 3$$

$$= 3x + y$$

Solving Equations Containing Logarithms

When each term of an equation is a logarithm with the same base, rewrite each side of the equation as a single logarithm. Then write an equation without logarithms, using the one-to-one function property of logarithms: if $\log_b A = \log_b B$, then $A = B$.

| Exercise 3 | Solving a Logarithm Equation |

If $\log N = 2 \log x + \log y$, solve for N in terms of x and y.

Solution:

Write the right side of $\log N = 2 \log x + \log y$ as a single logarithm:

- Undo the power law: $\log N = \log x^2 + \log y$

- Undo the product law: $\log N = \log x^2 y$

- Set the numbers equal: $N = x^2 y$

| Exercise 4 | Solving a Logarithm Equation |

Solve for x: $\log x - \frac{1}{3} \log 8 = \log 7$.

Solution: Solve for $\log x$:

- Isolate $\log x$: $\qquad\qquad\qquad\qquad$ $\log x = \dfrac{1}{3}\log 8 + \log 7$

- Undo the power law: $\qquad\qquad\qquad$ $= \log 8^{\frac{1}{3}} + \log 7$
$$= \log 2 + \log 7$$

- Undo the product law: $\qquad\qquad\quad$ $= \log\left(7 \cdot 2\right)$
$$= \log 14$$

Because $\log x = \log 14$, $\boldsymbol{x = 14}$.

Solving Logarithm Equations by Exponentiating

To solve an equation in which only some of the terms are logarithms:

1. Collect and then consolidate the logarithms with the same base on the same side of the equation.
2. Write the equation in the form $\log_b N = \text{number}$.
3. Change to exponential form, and solve that equation.
4. Check each root in the original equation. Reject any root that leads to taking the logarithm of a negative number.

Exercise 5   **Solving a Logarithm Equation by Changing to Exponential Form**

Solve for x: $\log_3 80x = 2 + \log_3\left(x^2 - 1\right)$.

Solution: Solve by consolidating the two logarithms and then changing to exponential form.

- Collect and then consolidate the logarithm expressions on the left side of the equation:

$$\log_3 80x - \log_3\left(x^2 - 1\right) = 2, \text{ so } \log_3\left(\frac{80x}{x^2 - 1}\right) = 2.$$

- Change to exponential form: $\quad \dfrac{80x}{x^2 - 1} = 3^2 = 9.$

- Eliminate the fraction by multiplying both sides of the equation by $x^2 - 1$:

$$9\left(x^2 - 1\right) = 80x$$
$$9x^2 - 80x - 9 = 0$$
$$(9x + 1)(x - 9) = 0$$
$$x = -\frac{1}{9} \quad or \quad x = 9$$

- Check each root in the original equation. Reject $x = -\frac{1}{9}$ because the right side of the equation $\log_3 80x = \log_3\left(-\frac{80}{9}\right)$ is not defined. Hence, $x = -\frac{1}{9}$ is an extraneous root. You should verify that **x = 9** works and is, therefore, the only root of the equation.

Solving Exponential Equations with Different Bases

In solving an exponential equation, it may not be possible to write each side as a power of the same base, as in $2^{3x} = 6$. To eliminate the variable exponent, take the logarithm of each side of the equation:

$$\log\left(2^{3x}\right) = \log 6$$
$$3x \cdot \log 2 = \log 6$$
$$x = \frac{\log 6}{3 \log 2} \approx 0.8618$$

Exercise 6   **Solving an Exponential Equation Using Logarithms**

Solve $3^x = 21$, approximating x to the *nearest hundredth*. Confirm your answer graphically.

Solution: If $3^x = 21$, then $\log (3^x) = \log 21$ so

$$x \log 3 = \log 21 \quad and \quad x = \frac{\log 21}{\log 3} \approx \mathbf{2.77}.$$

Confirm graphically.

- Set $Y_1 = 3 \wedge x$ and $Y_2 = 21$.
- Display the graphs in an appropriate viewing rectangle, such as $[0,4.7] \times [0,30]$.
- Use the INTERSECT feature of your calculator to estimate the x-coordinate of the point of intersection of the two graphs, as shown in the accompanying figures.

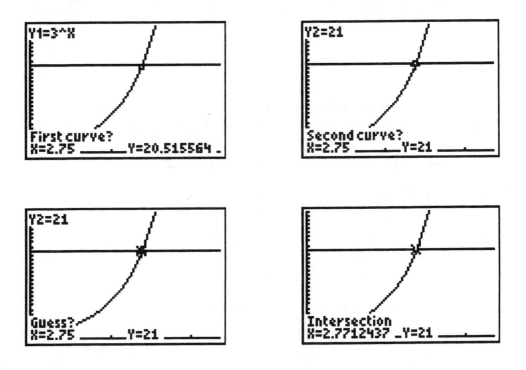

Exercise 7   **Modeling with an Exponential Function**

In a laboratory one amoeba divides into two amoebas every hour. Write an exponential function to model the reproduction of amoebas, letting N represent the total number present after t hours. Then find, to the *nearest tenth of an hour*, how long one amoeba will take to multiply to a total of 10,000 amoebas.

Solution: Make a table, and look for a pattern.

t	0	1	2	3	. . .	t
N	1	2	4	8	. . .	2^t

From the table, you now know that $N = 2^t$. Solve $N = 2^t$ for t when $N = 10,000$:

$$10,000 = 2^t$$
$$\log 10,000 = t \log 2$$
$$t = \frac{\log 10,000}{\log 2} = 4 \div \log 2 \approx \textbf{13.3}$$

Exponential and Logarithmic Models

KEY IDEAS

The growth patterns of many types of biological and physical quantities closely approximate exponential curves whose equations have the general form

$$y = \text{initial amount} \times \left(\text{constant growth factor}\right)^{\text{time variable}}.$$

When a quantity grows linearly over time, it increases by the same amount over equal intervals of time; when it grows exponentially, it increases by a fixed percent of its previous value.

Modeling Growth and Decay

If a quantity is increasing at a constant annual rate of $r\%$ of its previous value, then, after t years, an initial amount, A_0, of that quantity has grown to amount $A(t)$. For instance, suppose that \$320 is deposited in a bank account that earns 7% interest compounded annually.

- After 1 year, the balance in the account is

$$\underbrace{320}_{\text{Old amount}} + \underbrace{320 \times 0.07}_{\text{Interest}} = 320(1.07)^1.$$

- After the second year, the balance in the account is

$$\underbrace{320(1.07)}_{\text{Old amount}} + \underbrace{[320(1.07)] \times (0.07)}_{\text{Interest}} = 320(1.07)^2.$$

- After t years, the initial balance has increased by a factor of 1.07 for each of a total of t times, so

$$A(t) = 320\underbrace{(1.07)(1.07)\ldots(1.07)}_{t \text{ factors}} = 320\left(1.07\right)^t.$$

The function $A(t) = 320(1.07)^t$ that describes this growth process is called an **exponential growth model.** The initial amount is $A_0 = A(0) = 320$, and the annual growth factor is $1 + r = 1 + 0.07 = 1.07$. If a quantity is *decreasing* at a constant annual rate of $r\%$ of its previous value, then $r < 0$ and the exponential function that describes this process is referred to as an **exponential decay model.**

Exercise 1   Compounding Yearly Interest

Using the model $A(t) = 320(1.07)^t$, where $t = 0$ represents the year 2000, determine:

  a.   the balance to the nearest hundredth of a dollar in the account at the end of the year 2005.
  b.   the first year in which the account will total at least $700.

Solutions:

  a.   Since $t = 0$ represents the year 2000, $t = 5$ is 2005. Use your calculator to find that $320(1.07)^5 = 448.81655$, which, correct to the *nearest hundredth*, is **448.82**.

  b.   To find the first year in which the account will total at least $700, solve the exponential equation $320(1.07)^t = 700$:

$$\log\left[320(1.07)^t\right] = \log 700$$

$$\log 320 + t\log 1.07 = \log 700$$

$$t = \frac{\log 700 - \log 320}{\log 1.07} \approx 11.57$$

Since $t = 0$ represents the year 2000, any value of t between 11 and 12 is the year **2011**.

Exercise 2   Calculating Depreciation

Raymond buys a new car for $21,500. The car depreciates by about 11% per year. What is the value of the car after 5 years?

Solution: The problem situation can be modeled by the exponential decay function $A(t) = A(1+r)^t$, where $r = -11\% = -0.11$, $t = 5$, and $A = 21,500$:

$$A(5) = 21,500(1 - 0.11)^5$$

$$= 21,500(0.89)^5$$

$$\approx \mathbf{12,005.73}$$

TIPS

An exponential function of the form $A(t) = A_0 b^t$ can be used to model a growth or decay process in which $A_0 = A(0)$ is the initial amount of a substance, $A(t)$ is the amount present after t units of time, and b is the growth or decay factor.

- If $b > 1$, b is a *growth* factor. To find the rate of growth, given b, rewrite b as $1 + r$. For example, if $A(t) = 35(1.085)^t$ with t measured in years:

$$b = 1 + r = 1.085, \text{ so } 1.085 = 1 + r \text{ and } r = 0.085 = 8.5\%.$$

Hence, the annual rate of *growth* is 8.5%.

- If $0 < b < 1$, b is a *decay* factor. To find the rate of decay, given b, again rewrite b as $1 + r$. For example, if $A(t) = 47.3(.87)^t$ with t measured in years:

$$b = 0.87, \text{ so } 1 + r = 0.87 \text{ and } r = -0.13 = -13\%.$$

Because $r < 0$, the annual rate of *decay* is 13%.

Fitting an Exponential Curve to Data

The exponential function $y = ab^x$ may be a good fit to observed data that describe nonlinear growth. In Lesson 3-6, you learned how to fit a linear equation to data by using the regression feature of your graphing calculator. An exponential equation can be fitted to collected data by following a similar procedure.

Exercise 3   Fitting an Exponential Curve to Data

The data in the accompanying table show the growth of cellular phone subscriptions in the United States from 1995 to 2002. Fit an exponential curve, $y = ab^x$, to the data, where $x = 0$ represents the year 1990 and y is the number of cellular phone subscriptions. Approximate a and b to the *nearest thousandth*. Then use the model to predict the number of cellular phone subscriptions in the United States by 2006.

Year	Subscriptions (millions)
1995	33.8
1996	44.0
1997	55.3
1999	86.1
2001	99.6
2002	116.7

Solution: Enter the data by storing the six years in list L1 and the corresponding numbers of cellular phone subscriptions in list L2. Then press

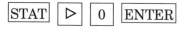

to find that the exponential regression equation, as seen from the accompanying figure, is approximately $y = 15.692(1.188)^x$. In 2006 there will be approximately $15.692(1.188)^{16} \approx$ **247.03 million** cellular phone subscriptions in the United States.

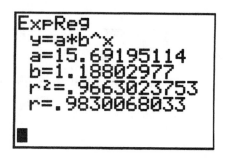

The Natural Logarithmic Function

The **natural logarithmic function**, denoted as $\ln x$, is the logarithmic function whose base is the special irrational number, e, called *Euler's constant*. The base of $\log x$ is understood to be 10 with $\log 10 = 1$. Similarly, the base of $\ln x$ is understood to be e with $\ln e = 1$. A graphing calculator has an $\boxed{\text{LN}}$ key with its inverse, e^x, above it. Use your calculator to verify that $e \approx 2.7182818. \ldots$ The same laws that work with common logarithmic functions apply also to natural logarithmic functions:

- $\ln\left(ab\right) = \ln a + \ln b$ **EXAMPLE:** $\ln\left(5x\right) = \ln 5 + \ln x$

- $\ln\left(\dfrac{a}{b}\right) = \ln a - \ln b$ **EXAMPLE:** $\ln\left(\dfrac{x}{5}\right) = \ln x - \ln 5$

- $\ln\left(a^n\right) = n \ln a$ **EXAMPLE:** $\ln\left(x^5\right) = 5 \ln x$

Exercise 4   Solving an Equation with e That Has a Quadratic Form

Solve for x: $e^{2x} - 2e^x - 3 = 0$.

Solution: Since the exponent of the first term is two times the exponent of the middle term, the equation has a quadratic form with $u = e^x$, so $u^2 = e^{2x}$:

$$u^2 - 2u - 3 = 0$$
$$\left(u - 3\right)\left(u + 1\right) = 0$$
$$u - 3 = 0 \quad or \quad u + 1 = 0$$
$$u = 3 \qquad\qquad e^x + 1 = 0$$
$$e^x = 3 \qquad\qquad e^x = -1 \leftarrow \text{Reject as impossible}.$$
$$x \ln e = \ln 3$$
$$x = \ln 3$$

The root $e^x = -1$ is rejected since e^x is always a positive number. The solution is $\boldsymbol{x = \ln 3}$.

Fitting a Natural Logarithmic Function to Data

When nonlinear data show a leveling off in growth over time, a logarithmic function of the form $y = a + b \ln x$ may be a good fit to the data.

| Exercise 5 | Fitting a Natural Logarithmic Function to Data |

The data in the accompanying table give the recommended numbers of hours of sleep for different ages. Fit a logarithmic regression equation to the data, estimating the regression coefficients to the *nearest tenth*. Then find, to the *nearest tenth of an hour*, the number of hours of sleep recommended for a 35-year-old.

Age, x (years)	Hours of Sleep Required, y
2	13
6	11
12	10
16	9
25	8
50	6

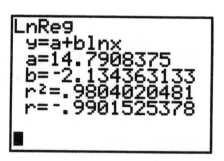

Solution: Enter the ages in list L1 and the corresponding hours of sleep in list L2. Choose option **9:LnReg** from the regression menu of your graphing calculator to obtain the regression display shown in the accompanying figure. If $x = 35$, then:

$$y = 14.8 - 2.1(\ln 35)$$
$$\approx 14.8 - (2.1)(3.56)$$
$$\approx \mathbf{7.3}$$

Modeling Continuous Growth and Decay

Things that grow or decay *continuously* over time at an exponential rate can be modeled by the function

$$A(t) = A_0 e^{kt},$$

where A_0 is the starting amount, $A(t)$ is the amount present after t units of time, e is the base of the natural logarithm function, and k is some constant specific to the particular growth or decay process that the function is modeling.

Exercise 6   Compounding Interest Continuously

If \$10,000 is invested for 5 years and interest is compounded continuously at an annual rate of 6%, what is the balance at the end of 5 years?

Solution: When an amount, A_0, is invested at an annual rate of $r\%$ with interest compounded continuously, the balance, $A(t)$, after t years is given by the function $A(t) = A_0 e^{rt}$. It is given that $A_0 = 10,000$, $r = 0.06$, and $t = 5$. Thus:

$$A(5) = 10,000e^{0.06 \times 5} = 10,000e^{0.30} \approx \mathbf{13,498.59}$$

Exercise 7   Finding the Growth Constant

Bacteria are growing continuously in a laboratory culture at an exponential rate. The initial number of bacteria was 200, which increased to 500 after 7 days. If the bacteria continue to grow at the same rate, how many will be in the culture after 10 days?

Solution: Use the exponential growth model $A(t) = A_0 e^{kt}$ to find k when $A(t) = 500$, $A_0 = 200$, and $t = 7$:

$$200e^{7k} = 500$$
$$e^{7k} = \frac{500}{200} = 2.5$$
$$\ln e^{7k} = \ln 2.5$$
$$7k = \ln 2.5$$
$$k = \frac{\ln 2.5}{7} \approx 0.1309$$

Now that you know k, use the same model to find $A(10)$:

$$A(10) = 200e^{0.1309 \times 10} = 200e^{1.309} \approx \mathbf{740}.$$

Exercise 8   Modeling Exponential Decay

The amount of a 20-milligram dose of a medicinal drug remaining in the bloodstream falls continuously at an exponential rate. After 2 hours, 17 milligrams remain in the bloodstream. Find, to the *nearest 10 minutes*, how much time is required for at least half of the original drug dose to leave the bloodstream.

Solution: Use the exponential growth model $A(t) = A_0 e^{kt}$ to find k when $A(t) = 17$, $A_0 = 20$, and $t = 2$:

$$20e^{2k} = 17$$

$$\ln e^{2k} = \frac{17}{20}$$

$$2k = \ln 0.85$$

$$k = \frac{\ln 0.85}{2} \approx -0.08126$$

Use the model $A(t) = A_0 e^{-0.08126t}$ to find t when $A(t) = 10$:

$$20e^{-0.08126t} = 10$$

$$\ln\left(e^{-0.08126t}\right) = \ln\left(\frac{10}{20}\right)$$

$$-0.08126t = \ln 0.5$$

$$t = \frac{\ln 0.5}{-0.08126} \approx 8.53$$

Because 0.53 hour $= 60 \times 0.53 \approx 32$ minutes, half of the original dose remains in the bloodstream after 8 hours and 32 minutes. After 8 hours and 30 minutes, *less than* half of the original dose has left the bloodstream. Hence, to the *nearest 10 minutes*, at least half of the original dose has left the bloodstream after **8 hours and 40 minutes**.

Modeling Radioactive Decay

The half-life of a radioactive element is the amount of time required for one-half of an initial amount of the element to decay. This process can be modeled by the function

$$A(t) = A_0 (0.5)^{\frac{\text{time, } t}{\text{half-life, } H}},$$

where A_0 is the initial amount of a radioactive substance with a half-life of H, and $A(t)$ is the amount of the original sample that remains after t units of time. For example, since the half-life of radioactive strontium-90 is 29 years, the decay of 250 grams of strontium-90 after t years can be modeled by the equation $A(t) = 250(0.5)^{\frac{t}{29}}$.

| Exercise 9 | Calculating Radioactive Decay |

In approximately how many years will 30% of a 250-gram mass of strontium-90 remain?

Solution: When 30% of the original amount of strontium-90 remains, $A(t) = 0.30 \times 250$. If x represents the number of years required for 30% of a 250-gram mass of strontium-90 to remain, then:

$$A(x) = 250(0.5)^{\frac{x}{29}}$$

$$0.30(250) = 250(0.5)^{\frac{x}{29}}$$

$$0.30 = (0.5)^{\frac{x}{29}}$$

$$\frac{x}{29} \log 0.5 = \log 0.30$$

$$x = \frac{29 \times \log 0.30}{\log 0.5} \approx \mathbf{50.4}$$

TIPS

- In Exercise 9, the equation $0.3 = (0.5)^{\frac{x}{29}}$ can also be solved by graphing $Y_1 = (0.5)^{\frac{x}{29}}$ and $Y_2 = 0.3$ in an appropriate viewing window, such as $[0,75.2] \times [0,0.5]$, and then using the INTERSECT feature to estimate the x-coordinate of the point of intersection of the two graphs.

- For the function $A(t) = A_0(0.5)^{\frac{time,\ t}{half\text{-}life,\ H}}$, the numerator and the denominator of the exponent must be expressed in the same units of time.

Chapter 8	CHECKUP EXERCISES

1–20. Multiple Choice

1. Solve for x: $64^{x-2} = 256^{2x}$.

   (1) $-\dfrac{6}{11}$ (2) $-\dfrac{6}{5}$ (3) $-\dfrac{1}{5}$ (4) 0

2. Which equation is the inverse of $y = -3x + 4$?

   (1) $y = \dfrac{1}{3}x - \dfrac{4}{3}$ (2) $y = -\dfrac{1}{4}x + \dfrac{3}{4}$ (3) $y = -\dfrac{1}{3}x + \dfrac{4}{3}$ (4) $y = -\dfrac{1}{4}x - \dfrac{3}{4}$

3. If $y = 7x + 2$ and $y = \dfrac{x}{7} + k$ are inverse functions, what is the value of k?

   (1) -2 (2) $-\dfrac{2}{7}$ (3) $\dfrac{2}{7}$ (4) $\dfrac{7}{2}$

4. If $g(x) = \dfrac{1}{x-2}$, where $x \neq 2$, what is $g^{-1}\left(\dfrac{1}{2}\right)$?

   (1) $-\dfrac{2}{3}$ (2) 2   (3) $\dfrac{3}{2}$ (4) 4

5. Solve for x: $4^{3x-4} = \left(\dfrac{1}{8}\right)^{x-1}$.

   (1) $\dfrac{5}{3}$ (2) $\dfrac{7}{9}$ (3) $\dfrac{11}{9}$ (4) $-\dfrac{1}{3}$

6. If $81^{x+2} = 27^{5x+4}$, then $x =$

   (1) $-\dfrac{2}{11}$ (2) $-\dfrac{3}{2}$ (3) $\dfrac{4}{11}$ (4) $-\dfrac{4}{11}$

7. If $f(x) = \dfrac{5x-1}{x+2}$, then $f^{-1}(x) =$

   (1) $\dfrac{x+2}{5x-1}$ (2) $\dfrac{2x+1}{x-5}$ (3) $\dfrac{2x+1}{5-x}$ (4) $\dfrac{2x-1}{x-5}$

8. If $\log a = 2$ and $\log b = 3$, what is the numerical value of $\log\left(\dfrac{\sqrt{a}}{b^3}\right)$?

   (1) 8   (2) –8   (3) 25   (4) –25

9. The expression $3\log x - \dfrac{1}{2}\log y$ is equivalent to

   (1) $\log\left(\dfrac{x^3}{y^2}\right)$ (2) $\log\left(\dfrac{x^3}{\sqrt{y}}\right)$ (3) $\log\sqrt{\dfrac{3x}{y}}$ (4) $\dfrac{\log 3x}{\dfrac{1}{2}\log y}$

10. If $27^x = 9^{y-1}$, what is y in terms of x?

    (1) $\dfrac{3}{2}x + 1$ (2) $\dfrac{3}{2}x + 2$ (3) $\dfrac{3}{2}x + \dfrac{1}{2}$ (4) $\dfrac{1}{2}x + \dfrac{2}{3}$

11. If $\log N = \dfrac{1}{2}(\log r - 2\log t) + \log s$, then $N =$

    (1) $\dfrac{\sqrt{rs}}{t}$ (2) $\dfrac{s\sqrt{r}}{t}$ (3) $\sqrt{\dfrac{r+s}{t^2}}$ (4) $\sqrt{\dfrac{r}{t^2}} + s$

12. If $f(x) = \log(x^2)$ and $g(x) = \dfrac{x}{10}$, then $f(g(x)) =$

    (1) $2\left(\log x - 1\right)$ (2) $\dfrac{\log x}{50}$ (3) $\dfrac{1}{2}\left(\log x + 1\right)$ (4) $2\log x + 1$

13. What is the domain of $f(x) = \log_b(x+2)$?

    (1) $\left\{x \mid x > -2\right\}$ (2) $\left\{x \mid x \geq -2\right\}$ (3) $\left\{x \mid 0 < x < 2\right\}$ (4) $\left\{x \mid -2 < x < 0\right\}$

14. If $\log 30 = b$, then $(b-1)^2$ is equivalent to

    (1) $\left(\log 29\right)^2$ (2) $2\left(\log 30 - 1\right)$ (3) $\left(\log 3\right)^2$ (4) $\log 899$

15. If $\log_3(x-2) = 2$, what is the value of x?

    (1) 7   (2) 8   (3) 11   (4) 9

16. If $\log_9 x = \dfrac{3}{2}$, what is the value of x?

    (1) $\dfrac{3}{2}$ (2) 8   (3) $\dfrac{27}{2}$ (4) 27

17. The expression $3\log_2 16 + \log_2 \dfrac{1}{16}$ is equivalent to

(1) 12   (2) 2   (3) 8   (4) 4

18. A culture of 5000 bacteria triples every 20 minutes. If $P(t)$ represents the size of the bacteria population after t minutes have elapsed, which equation can be used to model the exponential growth of these bacteria?

(1) $P(t) = 5000(20)^{3t}$

(2) $P(t) = 5000(3)^{20t}$

(3) $P(t) = 5000(3)^{\frac{t}{20}}$

(4) $P(t) = 5000\left(20^{\frac{t}{3}}\right)$

19. Which function is the inverse of $y = \log(x - 1)$?

(1) $y = x^{10} + 1$ (2) $y = 10^x - 1$ (3) $y = 10^x + 1$ (4) $y = \left(\dfrac{1}{10}\right)^x - 1$

20. When exposed to sunlight, the number of bacteria in a culture decreases exponentially at the rate of 10% per hour. What is the best approximation for the number of hours required for the initial number of bacteria to decrease by 50%?

(1) 6.6   (2) 6.0   (3) 5.4   (4) 4.6

21–28. Solve for x.

21. $9^{x^2 + x} = 3^4$

22. $\dfrac{1}{2}\ln(x + 2) = 2$

23. $\log_3(7x + 4) - \log_3 2 = 2\log_3 x$

24. $2^{1-3x} - \left(\dfrac{1}{4}\right)^2 = 0$

25. $\log_8 x + \log_8 (x + 6) = \dfrac{4}{3}$

26. $\log_4(x^2 + 3x) = 1 + \log_4(x + 5)$

27. $e^{2x} + 2 = 3e^x$

28. $\log_2(x^2 + 7) = 3 + \log_2(x + 5)$

29. Solve for x and y if $8^{\frac{x}{3}} = 16^{y+8}$ and $5^{x+y} = 125^{x-5}$.

30. If $\log_b 3 = p$ and $\log_b 5 = q$, express $\log_b\left(\dfrac{9}{\sqrt{5}}\right)$ and $\log_b \sqrt[3]{0.6}$ in terms of p and q.

31–33. Solve for x to the *nearest hundredth*:

31. $2^x = \dfrac{3}{16}$

32. $\log_7 75 = x$

33. $\log_3 x = 2 - \log_3(x+2)$

34. An archaeologist can determine the approximate ages of certain ancient specimens by measuring the amounts of radioactive carbon-14 they contain. The formula used to determine the age of a specimen is

$$A = A_0 \left(\frac{1}{2}\right)^{\frac{t}{5760}},$$

where A_0 is the original amount of carbon-14 that the specimen contained, A is the amount of carbon-14 that the specimen contains after t years, and 5760 is the half-life of carbon-14. A specimen that originally contained 120 milligrams of carbon-14 now contains 100 milligrams. What is the age of the specimen, to the *nearest hundred years*?

35. Bacteria are growing continuously in a laboratory culture at an exponential rate such that the number of bacteria triples after 2 hours. At this rate, how many hours will the culture require to increase to five times its original size?

36. The number of endangered species of animals, S, in the United States has grown since 1980 according to the exponential model $S(t) = 103e^{0.084t}$, where e is the base of the natural logarithmic function and t is the number of years since 1980.

   a. Use the model to estimate the number of endangered species at the end of the year 2000.

   b. Use a graph to estimate the year in which the number of endangered species will first reach 1000.

37. The Department of Health in a developing country has issued a warning that the number, N, of reported cases of a certain disease is increasing exponentially at an annual rate of 9%. The first year that statistics were collected, 5000 cases were reported.

   a. Estimate the number of cases reported after 7 years.

   b. Estimate the number of years, to the *nearest tenth*, required for the number of reported cases to rise to 11,000.

38. Radioactive iodine has a half-life of 60 days. A given sample of radioactive iodine decays according to the function $y(t) = 80(0.5)^{\frac{t}{60}}$, where y grams of the radioactive element remain after t days.

   a. To the *nearest tenth of a percent*, what percent of the original sample of the element will be present after 42 days?

   b. In approximately how many days, *correct to the nearest tenth*, will 15% of the original sample be present?

39 and 40. If P dollars are invested at an annual rate of $r\%$ compounded n times during the year, the amount, A, of the investment after t years is given by the equation

$$A = P\left(1 + \frac{r}{n}\right)^{nt}.$$

Mary invests $12,000 at an annual rate of 6.5%.

39. What is the *least* number of years and months needed for Mary's initial investment to double if interest is compounded quarterly?

40. What is the *least* number of years and months needed for Mary's initial investment to double if interest is compounded continuously?

41. Given $f(x) = 2x^2$, determine graphically the coordinates of the points of intersection of $y = f(x)$ and $y = f^{-1}(x)$ in the interval $0 \le x \le 2$.

42. In the equation $y = 0.5(1.21^x)$, y represents the number of snowboarders in millions and x represents the number of years since 1989.

   a. By what percent is the number of snowboarders increasing each year?

   b. In what year will the number of snowboarders reach 10 million for the first time?

43. Sean invests $12,000 at an annual rate of 6.75% compounded continuously. Find, to the *nearest year*, the *least* number of years required for his investment to exceed $28,000.

44. The accompanying table shows the average salaries of baseball players since 1994.

   a. Find the exponential regression equation for these data with the coefficient and base rounded to the *nearest hundredth*.

   b. Using your written regression equation, estimate the salary, to the *nearest thousand dollars*, of a baseball player in the year 2007.

Baseball Players' Salaries (thousands of dollars)

Numbers of Years Since 1994	Average Salary
0	290
1	320
2	400
3	495
4	600
5	700
6	820
7	1000
8	1250
9	1580

45. Prove that $\log_b x = \dfrac{\log x}{\log b}$ for $x, b > 0$.

46. Assume that the depreciation on a new car can be determined by the formula

$$V = C\left(1 - r\right)^t,$$

where V is the value of the car after t years, C is the original cost, and r is the rate of depreciation. A car purchased 6 years ago, when it was new, now has a value of $6000. If the car's original cost was $24,000, what is the rate of depreciation, correct to the *nearest tenth of a percent*?

47. An amount of a radioactive substance, A_0, decays exponentially according to the function

$$A\left(t\right) = A_0 e^{-0.025t},$$

where $A(t)$ is the amount that remains after t years. What is the half-life of this radioactive substance, correct to the *nearest year*?

48. Under laboratory conditions, the number of insects increases exponentially so that after 3 days the number of insects is 112, after 5 days the number of insects is 167, and after 8 days the number of insects is 302.

    a. Fit these data to an exponential curve, and find the regression coefficients correct to the *nearest tenth*. Then estimate the number of insects in the initial population.

    b. Find to the *nearest hour* when the number of insects is 210.

49. The accompanying table shows the number of milligrams, y, of a dose of a cold medication that remains in the body x hours after the dose has been ingested.

x (hr)	2	3	5	7	9
y (mg)	11	6.7	2.9	1.5	0.9

Determine an exponential function $y = ax^b$ that best fits the given data. Then estimate constants a and b to the *nearest tenth*. Find to the *nearest minute* when the amount of the drug that remains in the body first falls below 0.5 milligram.

STUDY UNIT III: TRIGONOMETRIC ANALYSIS

Chapter 9

TRIGONOMETRY

OVERVIEW

*T*rigonometry means "measurement of triangles." The study of trigonometry arose from the ancient need to understand the relationships between the sides and angles of triangles. With the development of calculus, trigonometry progressed from the study of ratios within right triangles to trigonometric *functions* that could be used to better represent the circular and repeating patterns of behavior that characterize a wide range of physical phenomena in the real world.

This chapter progresses from considering acute angles in right triangles to a more general view of angles as rotations about the origin in the coordinate plane. By fixing the vertex of such an angle at the origin and keeping one side of the angle aligned with the positive *x*-axis, we can give meaning to trigonometric functions of angles greater than 90° and less than 0°.

Lessons in Chapter 9

Degree and Radian Measures

KEY IDEAS

Angle measures can be expressed in units of *degrees* or in real-number units called *radians*. Degrees are based on fractional parts of a circular revolution. Radian measure compares the length of an arc that a central angle of a circle cuts off to the radius of the circle. The Greek letter θ (theta) is commonly used to represent an angle of unknown measure.

Measuring Angles in Degrees and Minutes

One **degree**, denoted as $1°$, is $\dfrac{1}{360}$ of one complete revolution about a fixed point. Each of the 60 equal parts of a degree is called a **minute**. The notation $28° 30'$ is read as "28 degrees, 30 minutes." Since 60 minutes is equivalent to 1 degree, dividing 30 minutes by 60 changes 30 minutes to a fractional part of a degree. Thus:

$$28° 30' = 28° + \left(\frac{30}{60}\right)° = 28.5°.$$

Changing Minutes to Degrees

$$x° y' = \left(x + \frac{y}{60}\right)°$$

Measuring Angles in Radians

In Figure 9.1, angle θ cuts off an arc of circle O that has the same length as the radius of the circle. Angle θ measures 1 radian. Unlike degrees, radians are real numbers.

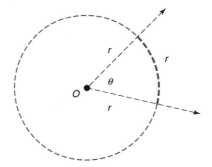

Figure 9.1 Defining a radian in a circle with radius r

Since the total number of radii that can be marked off along the circumference of the circle is 2π, the radian measure of a circle is 2π. The degree measure of a circle is $360°$. Hence, 2π radians $= 360°$, so π radians $= 180°$. This relationship provides the conversion factor for changing from one unit of angle measure to the other.

Radian and Degree Conversions

An angle of 1 radian is the angle at the center of a circle whose sides cut off an arc on the circle that has the same length as the radius of the circle.

- To convert from degrees to radians, multiply the number of degrees by $\dfrac{\pi}{180°}$.

- To convert from radians to degrees, multiply the number of radians by $\dfrac{180°}{\pi}$.

EXAMPLE: To convert $60°$ to radian measure, multiply it by $\dfrac{\pi}{180°}$:

$$60° = \overset{3}{\cancel{60°}} \times \frac{\pi}{\cancel{180°}} = \frac{\pi}{3} \text{ radians.}$$

EXAMPLE: To convert $\dfrac{7}{12}\pi$ radians to degrees, multiply it by $\dfrac{180°}{\pi}$:

$$\frac{7}{12}\pi \text{ radians} = \frac{7\cancel{\pi}}{\cancel{12}} \times \frac{\overset{15°}{\cancel{180}}°}{\cancel{\pi}} = 105°.$$

Lesson 9-2

Right-Triangle Trigonometry

KEY IDEAS

The **Pythagorean theorem** relates the measures of the *three* sides of a right triangle. A **trigonometric ratio** relates the measures of *two* sides and *one* of the acute angles of a right triangle.

The Three Basic Trigonometric Ratios

In Figure 9.2, $\triangle ABC$, $\triangle ADE$, and $\triangle AFG$ each contain a right angle and they have $\angle A$ in common.

When the ratio of the length of the leg opposite $\angle A$ to the length of the hypotenuse is computed, it is found to be the same for each right triangle:

$$\frac{\text{leg opposite } \angle A}{\text{hypotenuse}} = \frac{BC}{AB} = \frac{DE}{AD} = \frac{FG}{AF}.$$

This constant ratio is called the **sine** of $\angle A$. By means of a similar approach, the **cosine** and **tangent** ratios can also be defined in a right triangle.

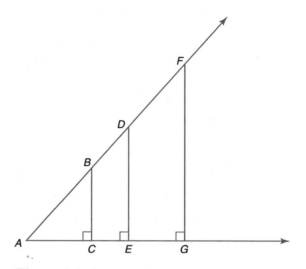

Figure 9.2 Comparing ratios in right triangles

Definitions of the Three Basic Trigonometric Functions		
$Sin\ A = \dfrac{\text{leg } Opposite \angle A}{Hypotenuse}$	$Cos\ A = \dfrac{\text{leg } Adjacent \text{ to} \angle A}{Hypotenuse}$	$Tan\ A = \dfrac{\text{leg } Opposite \angle A}{\text{leg } Adjacent \text{ to} \angle A}$
S O H	**C A H**	**T O A**

Writing consecutively the first letters of the three key words in the definitions of the three trigonometric ratios gives *SOH-CAH-TOA*. Remembering *SOH-CAH-TOA* can help you recall the definitions of sine, cosine, and tangent.

Viewing Trigonometric Ratios as Functions

For each possible acute angle of a right triangle, a trigonometric ratio of that angle corresponds to exactly one real number. Thus, each trigonometric ratio in a right triangle may be viewed as a function whose domain is the set of real numbers between $0\,(= 0°)$ and $\dfrac{\pi}{2}\,(= 90°)$.

Finding Trigonometric Function Values

To use your graphing calculator to find the value of $\sin\theta$, $\cos\theta$, or $\tan\theta$, where the value of θ is given in degrees:

1. Set the angular mode of your calculator to degrees, if necessary, by pressing $\boxed{\text{MODE}}$, highlighting **DEGREE**, and then pressing $\boxed{\text{ENTER}}$.

2. Return to the home screen by pressing $\boxed{\text{2nd}}$ $\boxed{\text{MODE}}$.

3. Press the key labeled with the name of the appropriate trigonometric function, enter the value of θ, and then press $\boxed{)}$ $\boxed{\text{ENTER}}$. Verify that $\sin 54° \approx 0.8090169944$ and $\cos 60° = 0.5$.

If evaluating a trigonometric function where the value of θ is given in radians, first set the angular mode of your calculator to **RADIAN** measure. Use your calculator to verify that $\tan 1.42 \approx 6.581119456$ and $\cos\left(\dfrac{\pi}{3}\right) = 0.5$.

Finding Angles Using Inverse Trigonometric Functions

Your graphing calculator has $\sin^{-1}$ printed above the $\boxed{\text{SIN}}$ key, $\cos^{-1}$ printed above the $\boxed{\text{COS}}$ key, and $\tan^{-1}$ printed above the $\boxed{\text{TAN}}$ key. These second or inverse calculator functions are used to find the measure of an angle when the value of a trigonometric function of that angle is known. For example, if $\tan x = 2.197$, then $x = \tan^{-1} 2.197$, which is read as "x is an angle whose tangent is 2.197."

To find the degree measure of $\angle x$, make sure that the angular mode of your calculator is set to **degree** measure. Then press:

$$\boxed{\text{2nd}} \quad \boxed{\text{TAN}} \quad \boxed{2} \quad \boxed{.} \quad \boxed{1} \quad \boxed{9} \quad \boxed{7} \quad \boxed{)} \quad \boxed{\text{ENTER}}.$$

You should verify that $x \approx 65.52657916°$.

Indirect Measurement and Trigonometry

Trigonometric functions are particularly useful when it is necessary to calculate the measure of a side or an angle of a right triangle that may be difficult, if not impossible, to measure directly.

Exercise 1   Finding the Measure of an Angle of a Right Triangle

A plane takes off from a runway and climbs while maintaining a constant angle with the ground. When the plane has traveled 1000 meters, its altitude is 290 meters, as shown in the accompanying figure. Find, correct to the *nearest degree*, the angle, x, at which the plane has risen with respect to the horizontal ground.

Solution: If $\angle A$ measures $x°$, then

$$\sin x = \frac{\text{length of leg opposite } \angle A}{\text{length of hypotenuse}}$$

$$= \frac{290}{1000}$$

$$x = \sin^{-1} 0.2900 \approx \mathbf{17}$$

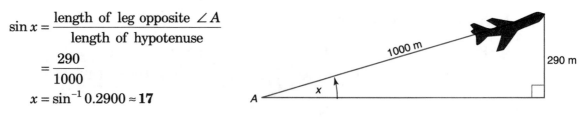

Exercise 2   Finding the Length of a Side of a Right Triangle

To determine the distance across a river, a surveyor marked two points, H and F, 65 meters apart on one riverbank. She also marked one point, K, on the opposite bank such that $\overline{KH} \perp \overline{HF}$, as shown in the accompanying figure. If $\angle HKF = 54°$, what is the width of the river, to the *nearest tenth of a meter*?

Solution: To find KH, the width of the river, use the tangent ratio:

$$\tan \angle HKF = \frac{\text{leg opposite } \angle HKF}{\text{leg adjacent to } \angle HKF}$$

$$\tan 54° = \frac{HF}{KH}$$

$$1.3764 = \frac{65}{KH}$$

$$KH = 65 \div 1.3764 \approx \mathbf{47.2}$$

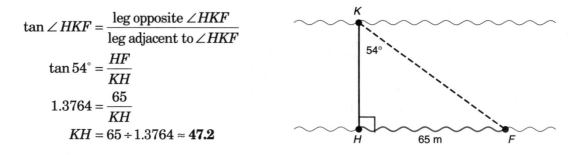

The Reciprocal Functions

Each of the three basic trigonometric functions has a reciprocal function. The reciprocal of sine is **cosec**ant (*csc*), the reciprocal of cosine is **sec**ant (*sec*), and the reciprocal of tangent is **cot**angent (*cot*):

$$\csc \theta = \frac{1}{\sin \theta}, \quad \sec \theta = \frac{1}{\cos \theta}, \quad \text{and } \cot \theta = \frac{1}{\tan \theta}.$$

The definitions of the six trigonometric functions are summarized in Figure 9.3.

Three Basic Trigonometric Functions	Reciprocal Trigonometric Functions	
$\sin \theta = \dfrac{a}{c}$	$\csc \theta = \dfrac{c}{a}$	
$\cos \theta = \dfrac{b}{c}$	$\sec \theta = \dfrac{c}{b}$	
$\tan \theta = \dfrac{a}{b}$	$\cot \theta = \dfrac{b}{a}$	

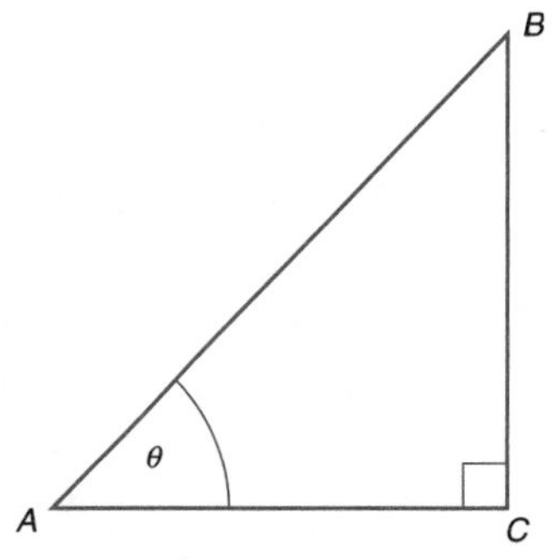

Figure 9.3 The six trigonometric functions of acute angle θ

To evaluate $\sec x$, $\csc x$, or $\cot x$, determine the value of its reciprocal function. Then find the reciprocal of that value by pressing the calculator's reciprocal key, $\boxed{x^{-1}}$.

EXAMPLE: To find $\sec 60°$, press

$$\boxed{\cos}\ \ 60\ \ \boxed{\ )\ }\ \ \boxed{x^{-1}}\ \ \boxed{\text{ENTER}}.$$

Thus, $\sec 60° = 2$.

EXAMPLE: To find $\csc 35° \, 20'$ correct to four decimal places, press

$$\boxed{\sin}\ \ 35.3333\ \ \boxed{\ )\ }\ \ \boxed{x^{-1}}\ \ \boxed{\text{ENTER}}.$$

Thus, $\csc 35° \, 20' \approx 1.7291$.

Cofunction Relationships

The prefix "*co*" in *co*sine, *co*secant, and *co*tangent represents *co*mplementary. Two angles are **complementary** when their measures add up to 90°. Pairs of cofunctions have equal values when their angles are complementary:

- $\sin \theta° = \cos(90 - \theta)°$ **EXAMPLE:** $\sin 50° = \cos 40°$

- $\sec \theta° = \csc(90 - \theta)°$ **EXAMPLE:** $\sec 24° = \csc 66°$

- $\tan \theta° = \cot(90 - \theta)°$ **EXAMPLE:** $\tan 20°50' = \cot 69°10'$

The General Angle

KEY IDEAS

By defining trigonometric functions of angles of rotation using coordinates, the domains of the six trigonometric functions can be expanded to include sets of real numbers. As a result we no longer are limited to positive acute angles. It will now be possible to consider trigonometric functions of angles greater than 90° or less than 0°.

Standard Position for a General Angle

An angle is in **standard position** when its vertex is at the origin and one of its sides, called the **initial side**, remains fixed on the positive x-axis. The side of the angle that rotates is called the **terminal side**.

- If the terminal side rotates in a counterclockwise direction, as shown in Figure 9.4, the angle of rotation is *positive*.
- If the terminal side rotates in a clockwise direction, as shown in Figure 9.5, the angle of rotation is *negative*.

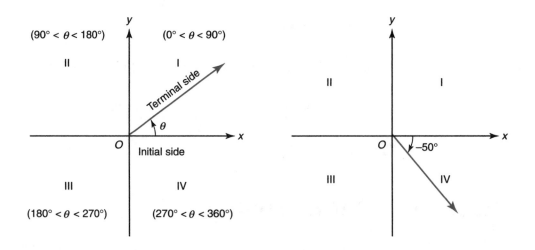

Figure 9.4 Positive angle   **Figure 9.5** Negative angle

Coterminal Angles

Angles of rotation of 50° and 410° are *coterminal angles*, as are angles 210° and –150°, as shown in Figures 9.6 and 9.7. **Coterminal angles** are angles in standard position whose terminal sides coincide. To find an angle that is coterminal with $\angle\theta$, add or subtract 360°, or a multiple of 360°, to or from the value of θ.

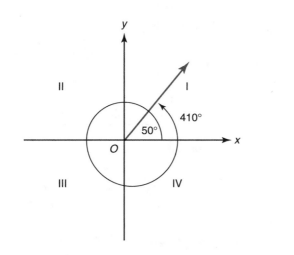

Figure 9.6 Coterminal angles

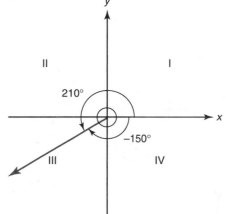

Figure 9.7 Coterminal angles

Reducing Angles Greater Than 360°

To reduce an angle greater than 360° to a coterminal angle between 0° and 360°, successively subtract 360° from the given angle until the difference is between 0° and 360°. Angles of rotation of 870° and 150° are coterminal because

$$870° - 360° = 510° \quad \text{and} \quad 510° - 360° = 150°.$$

Coterminal angles differ by a multiple of 360°. For example:

$$870° - 150° = 720° = 2 \times 360°.$$

Coordinate Trigonometric Definitions

Let $P(x,y)$ be any point on the terminal side of an angle in standard position. If r is the distance of P from the origin, the six trigonometric functions can be defined in terms of x, y, and r, as shown in Figure 9.8.

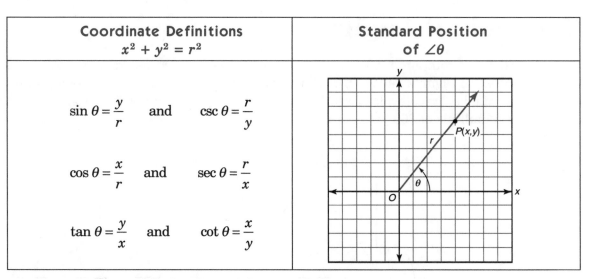

Coordinate Definitions $x^2 + y^2 = r^2$	Standard Position of $\angle\theta$
$\sin \theta = \dfrac{y}{r}$ and $\csc \theta = \dfrac{r}{y}$ $\cos \theta = \dfrac{x}{r}$ and $\sec \theta = \dfrac{r}{x}$ $\tan \theta = \dfrac{y}{x}$ and $\cot \theta = \dfrac{x}{y}$	

Figure 9.8 Coordinate definitions of the six trigonometric functions

When the terminal side of θ rotates $360°$, $P(x,y)$ is on a circle whose center is at $(0,0)$ and whose radius is r. According to the distance formula, $\sqrt{x^2 + y^2} = r$, so $x^2 + y^2 = r^2$. Since the coordinate definitions of the six trigonometric functions do not depend on a right triangle, they can be used to evaluate trigonometric functions of angles greater than $90°$ or less than $0°$.

Exercise 1   Applying the Coordinate Definitions

If the terminal side of $\angle\theta$ lies in Quadrant II and $\sin\theta = \dfrac{3}{5}$, find the exact values of the five remaining trigonometric functions.

Solution:

- It is given that $\sin\theta = \dfrac{3}{5} = \dfrac{y}{r}$, where $y = 3$ and $r = 5$. Because $3 - \underline{4} - 5$ is a Pythagorean triple and $\angle\theta$ lies in Quadrant II, where x is negative, $x = -4$. You can also find x by using the Pythagorean theorem:

$$x^2 + y^2 = r^2$$
$$x^2 + 3^2 = 5^2$$
$$x^2 = 16$$
$$x = \pm 4 \quad \leftarrow \text{Reject 4 because } x < 0 \text{ in Quadrant II.}$$
$$= -4$$

- The terminal side of $\angle\theta$ contains point $P(-4,3)$, as shown in the accompanying figure, where $r = 5$.
- Find the values of the five remaining trigonometric functions:

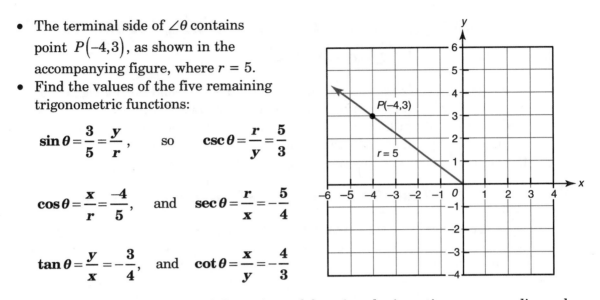

$$\sin\theta = \frac{3}{5} = \frac{y}{r}, \qquad \text{so} \qquad \csc\theta = \frac{r}{y} = \frac{5}{3}$$

$$\cos\theta = \frac{x}{r} = \frac{-4}{5}, \qquad \text{and} \qquad \sec\theta = \frac{r}{x} = -\frac{5}{4}$$

$$\tan\theta = \frac{y}{x} = -\frac{3}{4}, \qquad \text{and} \qquad \cot\theta = \frac{x}{y} = -\frac{4}{3}$$

You can also find the values of the reciprocal functions by inverting corresponding values of $\sin\theta$, $\cos\theta$, and $\tan\theta$.

| Exercise 2 | Using Coordinates to Find Sine and Cosine |

If $P(\sqrt{7},-3)$ is a point on the terminal side of $\angle\theta$, find the values of $\sin\theta$ and $\cos\theta$.

Solution: Determine the values of x, y, and r for the given point. Then use the coordinate definitions of sine and cosine.

- Let $P(\sqrt{7},-3) = P(x,y)$; then $x = +\sqrt{7}$ and $y = -3$. Because $x > 0$ and $y < 0$, $\angle\theta$ terminates in Quadrant IV, as shown in the accompanying figure.
- Find r using $x^2 + y^2 = r^2$, where $x = +\sqrt{7}$ and $y = -3$:

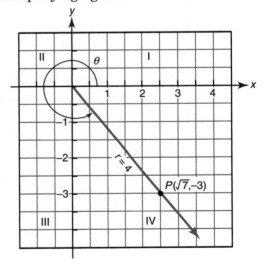

$$\left(\sqrt{7}\right)^2 + \left(-3\right)^2 = r^2$$
$$7 \quad + \quad 9 \quad = r^2$$
$$16 = r^2$$

Because r is always positive, $r = +\sqrt{16} = 4$.

- Thus, $\sin\theta = \dfrac{y}{r} = \dfrac{-3}{4}$ and $\cos\theta = \dfrac{x}{r} = \dfrac{+\sqrt{7}}{4}$.

Determining the Sign of a Trigonometric Function in a Quadrant

The signs of the six trigonometric functions of $\angle\theta$ depend on the signs of x and y in the quadrant in which the terminal side of θ lies, as shown in Table 9.1. Since x and y are both positive in Quadrant I, the first quadrant is the only quadrant in which the six trigonometric functions are all positive at the same time. In each of the other quadrants, only one of the basic trigonometric functions and its reciprocal function are positive.

Table 9.1 Signs of the Trigonometric Functions

Functions	Quadrant I	Quadrant II	Quadrant III	Quadrant IV
$\sin\theta$ and $\csc\theta$	+	+	—	—
$\cos\theta$ and $\sec\theta$	+	—	—	+
$\tan\theta$ and $\cot\theta$	+	—	+	—

Another way of remembering the quadrants in which the different trigonometric functions are positive is to memorize the sentence "**A**ll **S**tudents **T**ake **C**alculus," where the first letter of each word has special meaning:

> A = **A**ll are positive in Quadrant I.
> S = **S**ine (and cosecant) are positive in Quadrant II.
> T = **T**angent (and cotangent) are positive in Quadrant III.
> C = **C**osine (and secant) are positive in Quadrant III.

Exercise 3   Finding the Sine of an Angle, Given Its Cosine

If $\cos\theta = -\dfrac{4}{5}$ and $\tan\theta$ is positive, what is the exact value of $\sin\theta$?

Solution: Cosine is negative in Quadrants II and III. Tangent is positive in Quadrants I and III. Hence, $\angle\theta$ terminates in Quadrant III, where cosine is negative and, at the same time, tangent is positive.

- Since $\cos\theta = \dfrac{-4}{5} = \dfrac{x}{r}$, $x = -4$ and $r = 5$.

- Find y by recognizing $\underline{3}$-4-5 as a Pythagorean triple or by applying the Pythagorean theorem:

$$x^2 + y^2 = r^2$$
$$\left(-4\right)^2 + y^2 = 5^2$$
$$y = \pm\sqrt{9} = \pm 3.$$

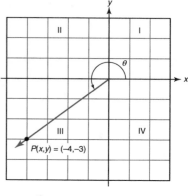

In Quadrant III y is negative, so $y = -3$. Hence, $\sin\theta = \dfrac{y}{r} = -\dfrac{3}{5}$.

Lesson 9-4

Working with Trigonometric Functions

KEY IDEAS

A trigonometric function of an angle whose terminal side lies in Quadrant II, III, or IV can be expressed as a trigonometric function of a positive *acute* angle. Familiar arithmetic operations can be performed with trigonometric functions by treating the trigonometric function of an angle as a single variable, as in $2 \tan x + 3 \tan x = 5 \tan x$.

Finding Reference Angles

For each $\angle \theta$ in standard position, there is a corresponding acute angle called the *reference angle*. The **reference angle,** θ_{Ref}, is the *acute* angle whose vertex is the origin and whose sides are the terminal side of $\angle \theta$ and the x-axis. If $\angle \theta$ is a Quadrant I angle, then θ and θ_{Ref} are the same angle. Figure 9.9 shows how to locate the reference angle when $\angle \theta$ terminates in Quadrants II, III, and IV. The right triangle that contains the reference angle is called the **reference triangle**.

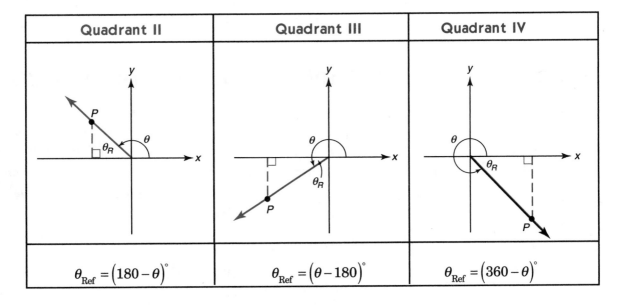

Figure 9.9 Locating reference angles

Exercise 1   Using Reference Triangles

If $\tan \theta = -\dfrac{5}{12}$ and $\sin \theta$ is positive, what is the value of $\cos \theta$?

Solution:

- First determine the quadrant in which $\angle\theta$ terminates: $\sin\theta > 0$ in Quadrants I and II, and $\tan\theta < 0$ in Quadrants II and IV. Hence, $\angle\theta$ terminates in the quadrant that satisfies both conditions, Quadrant II.

- Because $\tan\theta = -\dfrac{5}{12} = \dfrac{y}{x}$ and $y > 0$ in Quadrant II, attach the negative sign to 12 so that $P(-12,5)$ is a point on the terminal side of $\angle\theta$.

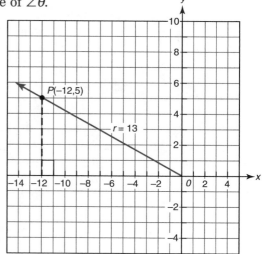

- Draw the reference triangle in Quadrant II, as shown in the accompanying figure. The reference triangle is a 5-12-_13_ right triangle, so $r = 13$.
  If you did not recognize 5-12-_13_ as a Pythagorean triple, use the Pythagorean theorem to find that the hypotenuse of the reference triangle is 13.

- Thus, $\cos\theta = \dfrac{x}{r} = \dfrac{-12}{13}$.

Reducing Trigonometric Functions of Angles

The trigonometric function of *any* angle θ can be expressed as either plus or minus the same trigonometric function of its reference angle, θ_{Ref}. Use the sign that corresponds to the sign of the function in the quadrant in which θ is located. To express $\cos 135°$, $\sin 135°$, and $\tan 135°$ as functions of a positive acute angle, first find the reference angle. As shown in Figure 9.10, $\theta_{\text{Ref}} = 180° - 135° = 45°$. Then determine the sign of the function in Quadrant II where the angle terminates:

- Cosine is negative in Quadrant II, so $\cos 135° = -\cos 45°$.
- Sine is positive in Quadrant II, so $\sin 135° = \sin 45°$.
- Tangent is negative in Quadrant II, so $\tan 135° = -\tan 45°$.

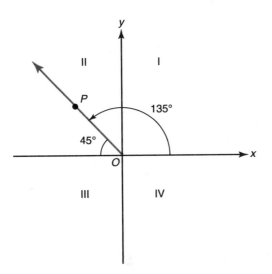

Figure 9.10 Evaluating a function of an angle greater than 90°

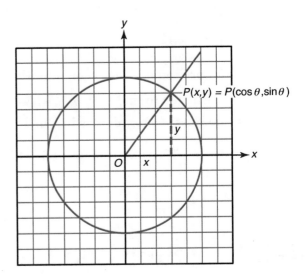

Figure 9.11 The unit circle

If $\angle\theta$ is greater than 360° or less than 0°, follow the same procedure after determining the quadrant in which the angle terminates.

EXAMPLE: $\cos 570° = \cos\left(570 - 360\right)° = \cos 210° = -\cos 30°$

EXAMPLE: $\sin\left(-140°\right) = \sin 220° = -\sin 40°$

EXAMPLE: $\tan 650° = \tan 290° = -\tan 70°$

The Unit Circle

The **unit circle** is the circle whose center is at the origin, O, and whose radius is 1, as shown in Figure 9.11. If the terminal side of $\angle\theta$ intersects the unit circle at $P(x,y)$, then

$$\cos\theta = \frac{x}{1} = x \quad \text{and} \quad \sin\theta = \frac{y}{1} = y, \text{ so } P(x,y) = P(\cos\theta, \sin\theta).$$

Exercise 2	Working in the Unit Circle

A point moves along the unit circle O in the counterclockwise direction from $A(1,0)$ to $P(-0.6,-0.8)$. Find, to the *nearest degree*, the angle of rotation.

Solution: Let θ represent the angle of rotation.

- $P(-0.6,-0.8)$ is on unit circle O, so $x = -0.6$, $y = -0.8$, and $r = 1$. Hence, $\angle\theta$ terminates in Quadrant III, where both x and y are negative.

- Let $\cos\theta = \dfrac{x}{r} = -\dfrac{0.6}{1} = -0.6$.

  Ignore the negative sign, and use your
  calculator to find that $\theta_{\text{Ref}} \approx 53°$.
- Because $\angle\theta$ terminates in Quadrant III:

  $$\theta = 180° + \theta_{\text{Ref}} \approx 180° + 53° = \mathbf{233°}.$$

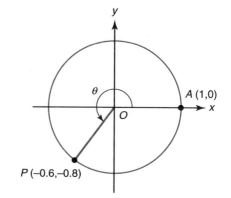

Relating Sine and Cosine to Tangent

In the unit circle, $\sin\theta = y$ and $\cos\theta = x$. Since

$$\frac{\sin\theta}{\cos\theta} = \frac{y}{x} \quad \text{and} \quad \tan\theta = \frac{y}{x},$$

it follows that $\tan\theta = \dfrac{\sin\theta}{\cos\theta}$ for all values of θ that do not make the denominator evaluate to 0.

Quotient Formulas

$$\tan\theta = \frac{\sin\theta}{\cos\theta} \quad \text{and} \quad \cot\theta = \frac{\cos\theta}{\sin\theta}$$

Operations with Trigonometric Functions

When a trigonometric function is raised to a power, the exponent is written next to the function and a half line above it, as in $\sin^2 x$. Thus, $\sin^2 x$ means $(\sin x)\cdot(\sin x)$, while $\sin x^2$ means $\sin(x^2)$, where the measure of the angle is x^2. Here are some examples that illustrate how to work with trigonometric functions:

- <u>Multiplying and Dividing Trigonometric Expressions</u>

EXAMPLE: $\left(2\cos^2 x\right)\cdot\left(4\cos x\right) = 8\cos^3 x$

EXAMPLE: $\dfrac{15\sin^3 x}{3\sin x} = 5\sin^2 x$

- Factoring Trigonometric Expressions

EXAMPLE: $\tan^2 - \tan x - 6 = \left(\tan x - 3\right)\left(\tan x + 2\right)$

EXAMPLE: $4\sin x - \sin^3 x = \sin x\left(4 - \sin^2 x\right) = \sin x\left(2 - \sin x\right)\left(2 + \sin x\right)$

- Combining Fractions Containing Trigonometric Expressions

EXAMPLE: $\dfrac{2\tan^2 x - 5}{\tan x + 1} + \dfrac{\tan^2 x + 2}{\tan x + 1} = \dfrac{3\tan^2 x - 3}{\tan x + 1}$

$$= \frac{3\left(\tan^2 x - 1\right)}{\tan x + 1}$$

$$= \frac{3\left(\tan x + 1\right)\left(\tan x - 1\right)}{\tan x + 1}$$

$$= 3\left(\tan x - 1\right)$$

- Solving a Trigonometric Equation

EXAMPLE: If $3\cos x + 7 = 5 - \cos x$, then $3\cos x + \cos x = 5 - 7$, so

$$4\cos x = -2 \quad \text{and} \quad \cos x = -\frac{1}{2}.$$

EXAMPLE: If $\cot^2 x - 5\cot x - 6 = 0$, then

$$\left(\cot x - 6\right)\left(\cot x + 1\right) = 0, \text{ so } \cot x = 6 \quad or \quad \cot x = -1.$$

| Exercise 3 | Applying the Quotient Formula |

Simplify: $\csc\theta \cdot \tan\theta \cdot \cos^2\theta$.

Solution: Change to sines and cosines, using the reciprocal and quotient function relationships. Then simplify.

$$\csc\theta \cdot \tan\theta \cdot \cos^2\theta = \frac{1}{\sin\theta} \cdot \frac{\sin\theta}{\cos\theta} \cdot \cos^2\theta$$

$$= \frac{1}{\sin\theta} \cdot \frac{\sin\theta}{\cos\theta} \cdot \cos^2\theta$$

$$= \boldsymbol{\cos\theta}$$

KEY IDEAS

You should be able to figure out the exact values of trigonometric functions of 30°, 45°, and 60°. These values are worth remembering. You should also remember the values of trigonometric functions of the *quadrantal angles*: 0°, 90°, 180°, 270°, and 360°.

Trigonometric Functions of 30°, 45°, and 60°

Trigonometric functions of 30°, 45°, and 60° can be determined geometrically by considering the special right triangles shown in Figures 9.12 and 9.13.

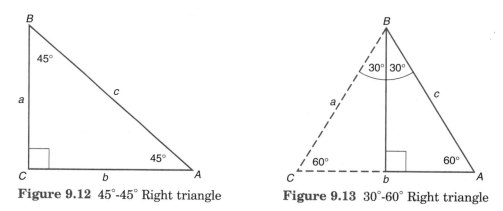

Figure 9.12 45°-45° Right triangle   **Figure 9.13** 30°-60° Right triangle

- In Figure 9.12, if $a = b = 1$, the Pythagorean theorem can be used to find that $c = \sqrt{2}$. Because the triangle is isosceles, each of the acute angles must have the same measure and, as a result, each measures 45°. The exact values of the sine, cosine, and tangent of 45° can be expressed in terms of the lengths of the sides of this right triangle, as shown in Table 9.2.

- In Figure 9.13, let $a = b = c = 1$. Because $\triangle ABC$ is equilateral, it is also equiangular, so each angle measures 60°. To form a right triangle, drop a perpendicular to the base that bisects the base and the angle from which it is drawn. The Pythagorean theorem can then be used to find that $CD = \sqrt{3}$. The exact values of the sine, cosine, and tangent of 30° and 60° can be expressed in terms of the lengths of the sides of the solid right triangle, as shown in Table 9.2.

Table 9.2 Trigonometric Function Values of 30°, 45°, and 60°

x	$\sin x$	$\cos x$	$\tan x$
30°	$\dfrac{1}{2}$	$\dfrac{\sqrt{3}}{2}$	$\dfrac{1}{\sqrt{3}}$ or $\dfrac{\sqrt{3}}{3}$
45°	$\dfrac{\sqrt{2}}{2}$	$\dfrac{\sqrt{2}}{2}$	1
60°	$\dfrac{\sqrt{3}}{2}$	$\dfrac{1}{2}$	$\sqrt{3}$

There is actually less to remember than may appear if you make use of the cofunction and quotient relationships:

- $\sin 30° = \cos 60° = \dfrac{1}{2}$ and $\cos 30° = \sin 60° = \dfrac{\sqrt{3}}{2}$

- $\sin 45° = \cos 45°$, so $\tan 45° = \dfrac{\sin 45°}{\cos 45°} = 1$

- $\tan 60° = \sin 60° \div \cos 60° = \dfrac{\sqrt{3}}{2} \div \dfrac{1}{2} = \sqrt{3}$, and $\tan 30° = \sin 30° \div \cos 30° = \dfrac{1}{2} \div \dfrac{\sqrt{3}}{2} = \dfrac{1}{\sqrt{3}}$

To find the values of cosecant, secant, and cotangent of 30°, 45°, and 60°, use the reciprocal relationships. Angles greater than 90° or less than 0° may have reference angles of 30°, 45°, or 60°. You should verify that:

- $\cos 120° = -\cos 60° = -\dfrac{1}{2}$

- $\tan 225° = \tan 45° = 1$

- $\csc 240° = -\dfrac{1}{\sin 60°} = -\dfrac{2}{\sqrt{3}}$

- $\sin 300° = -\sin 60° = -\dfrac{\sqrt{3}}{2}$

- $\cos(-30)° = \cos 30° = \dfrac{\sqrt{3}}{2}$

- $\sec 405° = \dfrac{1}{\cos 45°} = \sqrt{2}$

Trigonometric Functions of the Quadrantal Angles

An angle whose terminal side coincides with a coordinate axis is called a **quadrantal angle**. An angle of rotation of 0°, 90°, 180°, 270°, or 360° is a quadrantal angle. To figure out the values of sine, cosine, and tangent for each of the quadrantal angles, apply the coordinate definitions of these functions in the unit circle. For example, when $\theta = 90°$ in Figure 9.14, the terminal side of the angle is on the positive y-axis. Thus:

$$\sin 90° = y = 1$$

$$\cos 90° = x = 0$$

$$\tan 90° = \frac{y}{x} = \frac{1}{0} \leftarrow \text{Undefined}$$

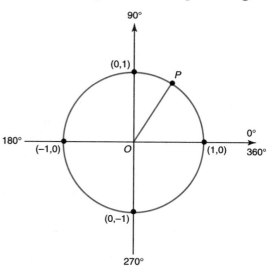

Figure 9.14 Quadrantal angles

The values of sine, cosine, and tangent of the other quadrantal angles can be determined in a similar way. See Table 9.3.

Table 9.3 Evaluating Trigonometric Functions of Quadrantal Angles

Trigonometric Function	$0°$ $360° = 2\pi$	$90° = \dfrac{\pi}{2}$	$180° = \pi$	$270° = \dfrac{3}{2}\pi$
$\sin x$	0	1	0	–1
$\cos x$	1	0	–1	0
$\tan x$	0	Undefined	0	Undefined

To find the values of cosecant, secant, and cotangent of quadrantal angles, use either the reciprocal or quotient function relationships.

EXAMPLE: $\sec 0° = \dfrac{1}{\cos 0°} = \dfrac{1}{1} = 1.$

EXAMPLE: $\csc \pi = \dfrac{1}{\sin \pi} = \dfrac{1}{0}$, so $\csc \pi$ is undefined.

EXAMPLE: $\cot 90° = \dfrac{\cos 90°}{\sin 90°} = \dfrac{0}{1} = 0$.

Exercise 1   Finding Where a Trigonometric Fraction Is Not Defined

For what value of x is $\dfrac{x-1}{1+\sin x}$, where $0 < x \le 2\pi$, not defined?

Solution: A fraction with a variable expression in the denominator is not defined for any value of the variable that makes the denominator evaluate to 0. Hence, the fraction is not defined when $1 + \sin x = 0$ or $\sin x = -1$. If $\sin x = -1$, then $\mathbf{x = \dfrac{3\pi}{2}}$.

Chapter 9	CHECKUP EXERCISES

1–20. Multiple Choice

1. In right triangle ABC with right angle C, $\cos A = \dfrac{5}{13}$. What is the value of $\sin A \cdot \tan A$?

   (1) $\dfrac{60}{65}$ (2) $\dfrac{144}{65}$ (3) $\dfrac{60}{156}$ (4) $\dfrac{144}{169}$

2. If $\sec(3x-10)^\circ = \csc(x+40)^\circ$, then a possible value of x is

   (1) 25 (2) 15 (3) 55 (4) 65

3. If $\dfrac{\sin(x-3)^\circ}{\cos(2x+6)^\circ} = 1$, then the value of x is

   (1) –9 (2) 26 (3) 29 (4) 64

4. At Slippery Ski Resort, the beginner's slope is inclined at an angle of 12.3°, while the advanced slope is inclined at an angle of 26.4°. If Rudy skis 1000 meters down the advanced slope, while Valerie skis the same distance on the beginner's slope, how much greater was the horizontal distance, in meters, that Valerie covered?

   (1) 81.3 (2) 231.6 (3) 895.7 (4) 977.0

5. The bottom of a pendulum traces an arc 3 feet in length when the pendulum swings through an angle of $\dfrac{1}{2}$ radian. Find the number of feet in the length of the pendulum.

   (1) 1.5 (2) 6 (3) $\dfrac{1.5}{\pi}$ (4) 6π

6. Through how many radians does the minute hand of a clock turn in 24 minutes?

   (1) 0.2π (2) 0.4π (3) 0.6π (4) 0.8π

7. If $\cos\theta = -\dfrac{3}{4}$ and $\tan\theta$ is negative, the value of $\sin\theta$ is

   (1) $-\dfrac{4}{5}$ (2) $-\dfrac{\sqrt{7}}{4}$ (3) $\dfrac{4\sqrt{7}}{7}$ (4) $\dfrac{\sqrt{7}}{4}$

8. If $\cos A = \dfrac{4}{5}$ and $\angle A$ is *not* in Quadrant I, what is the value of $\sin A$?

   (1) –0.2   (2) 0.75   (3) –0.6   (4) 0.6

9. In the accompanying diagram of the unit circle, $\overline{BA}$ is tangent to circle O at A, $\overline{CD}$ is perpendicular to the x-axis, and $\overline{OC}$ is a radius. Which distance represents $\sin\theta$?

   (1) OD (2) CD

   (3) BA (4) OB

10. In the accompanying diagram of a unit circle, $\overline{BA}$ is tangent to circle O at A, $\overline{CD}$ is perpendicular to the x-axis, and $\overline{OC}$ is a radius. Which distance represents $\tan\theta$?

   (1) OD (2) CD

   (3) BA (4) OB

Exercises 9 and 10

11. If θ is an angle in standard position and its terminal side passes through point $P(0.8,-0.6)$ on the unit circle, a possible value of $\angle\theta$, correct to the *nearest degree*, is

   (1) 127   (2) 143   (3) 307   (4) 323

12. If x is a positive acute angle and $\cos x = a$, an expression for $\tan x$ in terms of a is

   (1) $\dfrac{1-a}{a}$ (2) $\sqrt{1-a^2}$ (3) $\dfrac{\sqrt{1-a^2}}{a}$ (4) $\dfrac{1}{1-a}$

13. If $\tan x = -\dfrac{\sqrt{3}}{3}$ and $\cos x = -\dfrac{\sqrt{3}}{2}$, the measure of $\angle x$ is

   (1) $120°$ (2) $150°$ (3) $210°$ (4) $300°$

14. If $\dfrac{1}{2\sin x - 1}$ is undefined when $x = a°$, what is the value of $\cos\left(180+a\right)°$?

   (1) $-\dfrac{1}{2}$ (2) $\dfrac{1}{2}$ (3) $-\dfrac{\sqrt{3}}{2}$ (4) $-\dfrac{1}{3}$

15. The expression $\csc\left(-\dfrac{\pi}{2}\right)$ is equivalent to

   (1) $2\cos\dfrac{\pi}{3}$

   (2) $\tan\pi$

   (3) $2\sin\dfrac{5}{6}\pi$

   (4) $\cos\pi$

16. If θ is an angle in standard position and its terminal side passes through point $\left(\dfrac{\sqrt{3}}{2}, -\dfrac{1}{2}\right)$ on the unit circle, a possible value of $\angle\theta$ is

   (1) $210°$

   (2) $240°$

   (3) $300°$

   (4) $330°$

17. What is the image of $(1, 0)$ after a counterclockwise rotation of $135°$?

   (1) $\left(\dfrac{\sqrt{2}}{2}, -\dfrac{\sqrt{2}}{2}\right)$

   (2) $\left(-\dfrac{\sqrt{2}}{2}, \dfrac{\sqrt{2}}{2}\right)$

   (3) $\left(-\sqrt{2}, 1\right)$

   (4) $\left(-\dfrac{1}{2}, \dfrac{1}{2}\right)$

18. If $\sin A = b$, the value of $\sin A \cdot \cos A \cdot \tan A$ is

   (1) 1

   (2) $\dfrac{1}{b}$

   (3) b

   (4) b^2

19. The expression $\tan\theta\left(\cos\theta + \csc\theta\right)$ is equivalent to

   (1) $1 + \sin\theta$

   (2) $1 + \cos\theta$

   (3) $\dfrac{1}{\cos\theta\sin\theta}$

   (4) $\dfrac{1 + \cos\theta\sin\theta}{\cos\theta}$

20. The ordered pair (x,y) represents a point where the terminal side of $\angle\theta$ intersects the unit circle, as shown in the accompanying figure. If $\theta = -\dfrac{\pi}{3}$, what is the value of y?

   (1) $-\sqrt{3}$

   (2) $-\dfrac{\sqrt{2}}{2}$

   (3) $-\dfrac{\sqrt{3}}{2}$

   (4) $-\dfrac{1}{2}$

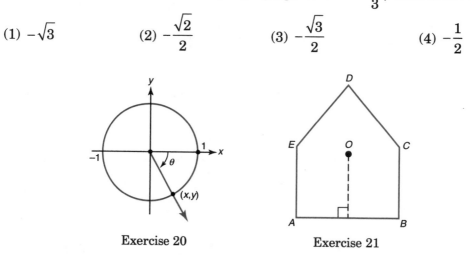

Exercise 20   Exercise 21

21. The perimeter of regular pentagon *ABCDE*, shown in the accompanying figure, is 100 inches.

   a. Find, to the *nearest tenth of an inch*, the length of the perpendicular from center *O* of the pentagon to side $\overline{AB}$.

   b. Find the area of the pentagon to the *nearest square inch*.

22–24. In the accompanying diagram, $\triangle ABC$ is an isosceles right triangle with $\angle C = 90°$.

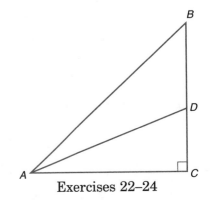

22. If $AB = 10$ and $\overline{AD}$ bisects $\angle BAC$, find BD to the *nearest tenth*.

23. If $AD = 41$ and $AC = 40$, find $\angle BAD$ to the *nearest tenth of a degree*.

24. If $BD = 87$ and $AC = 112$, find $\angle ADB$ to the *nearest tenth of a degree*.

Exercises 22–24

25. A triangular-shaped access ramp to an office building is constructed so that it rises $1\dfrac{1}{2}$ feet for every 8 feet of horizontal ground distance.

    a. Find, to the *nearest degree*, the angle of incline of the access ramp.

    b. Find, to the *nearest foot*, the length of the access ramp at the point at which the height of the ramp is 4 feet.

26. The equatorial diameter of Earth is approximately 8000 miles. A communications satellite makes a circular orbit around Earth at a distance of 1600 miles from Earth. If the satellite completes one orbit every 5 hours, how many miles does the satellite travel in 1 hour?

27. If $\tan \theta = -\dfrac{5}{12}$ and $\sin \theta$ is negative, determine the exact value of $\cos \theta$.

28. If $\sin \theta = -\dfrac{1}{\sqrt{5}}$ and $\cos \theta$ is negative, determine the exact value of $\tan \theta$.

29. If $\cot \theta = -\dfrac{15}{8}$ and $\csc \theta$ is positive, determine the exact value of $\sec \theta$.

30. If $\sin x = \dfrac{8}{17}$, where $0° < x < 90°$, determine the exact value of $\cos(x + 180°)$.

31–34. Write each of the following as an equivalent function of a positive acute angle:

31. $\tan 510°$

32. $\cos\left(\dfrac{4}{3}\pi\right)$

33. $\csc\left(-130°\right)$

34. $\cos\left(-20°\right)$

35. Write $\tan 110°$ as a function of a positive acute angle less than 45°.

36. Write $\cos 255°$ as a function of a positive acute angle less than 45°.

37–39. Factor completely.

37. $3\tan^3 x - 192\tan x$ 38. $2\sin^2 \theta - 18\cos^2 \theta$ 39. $6\sin^2 x + 3\sin x - 3$

40 and 41. Express in simplest form.

40. $\dfrac{3\sin y + 6}{4 - \sin^2 y} + \dfrac{4}{\sin y - 2}$ 41. $\dfrac{1 - \tan^2 x}{6\tan x + 6} \div \dfrac{\tan^4 x - 1}{6\tan^2 x + 6}$

42 and 43. Write each complex fraction in simplest form.

42. $\dfrac{\dfrac{3}{\cos^2 x} + \dfrac{1}{\cos x}}{1 - \dfrac{9}{\cos^2 x}}$ 43. $\dfrac{\dfrac{\sin x}{\cos x} - \dfrac{\cos x}{\sin x}}{\dfrac{1}{\cos x} - \dfrac{1}{\sin x}}$

44–46. Solve for the trigonometric function

44. $2\cos^2 x - 3\cos x = 2$ 45. $4\sin^2 x + 5\sin x + 1 = 0$ 46. $\dfrac{3}{\tan x + 3} + \dfrac{2}{\tan x - 4} = \dfrac{4}{3}$

Chapter 10

GRAPHING TRIGONOMETRIC FUNCTIONS

OVERVIEW

A unifying theme of this course is functions and their graphs. This chapter looks at trigonometric functions from a graphical point of view. Trigonometric functions are periodic because there is a repeating pattern of function values, which is easily observed from the graphs of the functions. Other important properties of trigonometric functions that may not be obvious from their equations are also apparent from their graphs.

Lessons in Chapter 10

Lesson 10-1
Periodic Functions and Their Graphs

KEY IDEAS

The values of a **periodic function** repeat over regular intervals. Each complete pattern of values is a **cycle**. When θ varies from $0°$ to $360°$, $\sin \theta$ and $\cos \theta$ take on values from -1 to $+1$. Since an angle formed by adding any multiple of $360°$ to θ is coterminal with θ, the same pattern of values for $\sin \theta$ and $\cos \theta$ repeats every $360°$. For example:

$$\cos 60° = \underbrace{\cos 420°}_{\cos(60+360)°} = \underbrace{\cos 780°}_{\cos(60+720)°} = \cdots = \frac{1}{2}.$$

Sine and cosine are periodic functions with periods of $360°$. The period of tangent is $180°$. When graphing trigonometric functions, we usually use radians rather than degrees as radians are real numbers.

Period and Amplitude

A function f is periodic if there exists some number p for which $f(x + p) = f(x)$ for all x.

- If p is the smallest such positive number, p is called the **period** of the function. The period of the function in Figure 10.1 is $8 - 3 = 5$, which is the length of the smallest interval of x-values that the function needs to complete one cycle.
- The **amplitude** of a periodic function is one-half of the difference between the maximum and minimum values of the function. For the periodic function in Figure 10.1, the amplitude is $\dfrac{3-(-2)}{2} = 2.5$.

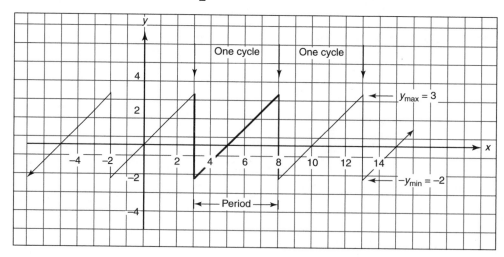

Figure 10.1 A periodic function

Sine and Cosine Curves

The graphs of $y = \sin x$ and $y = \cos x$ have the same basic shape except that the cosine curve is $\dfrac{\pi}{2}$ out of phase with the sine curve, as shown in Figures 10.2 and 10.3, where x varies from -2π to 2π. It's easy to see from the graphs that the period of each of the functions is 2π. The graph of the cosine curve has y-axis symmetry, while the graph of the sine curve has origin symmetry.

Frequency

The **frequency** of a trigonometric function is the number of cycles that its graph completes in an interval of 2π radians. Because the sine curve and the cosine curve each complete one cycle every 2π radians, the frequency of each curve is 1.

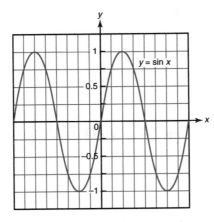

Figure 10.2 Graph of $y = \sin x$ Figure 10.3 Graph of $y = \cos x$

For both $y = \sin x$ and $y = \cos x$:

- period $= 2\pi$
- amplitude $= 1$

- domain $= \{$real numbers$\}$
- range $= \left\{ y \mid -1 \le y \le 1 \right\}$

Exercise 1   Comparing the Graphs of the Sine and Cosine Functions

In which quadrant is $y = \sin x$ decreasing and $y = \cos x$ increasing?

Solution: Sketch the graphs of $y = \cos x$ and $y = \sin x$ on the same set of axes, as shown in the accompanying figure. In Quadrant **III**, where $\pi < x < \dfrac{3\pi}{2}$, the sine curve is decreasing and the cosine curve is increasing.

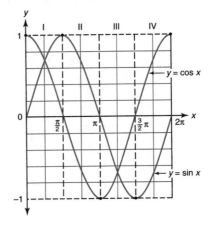

Unwrapping the Unit Circle

If point (x,y) is moving counterclockwise around the unit circle, the graph of $y = \sin x$ indicates how the height of the point changes as x varies from 0 to 2π, as shown in Figure 10.4. Similarly, the graph of $y = \cos x$ indicates how the horizontal position of the point changes.

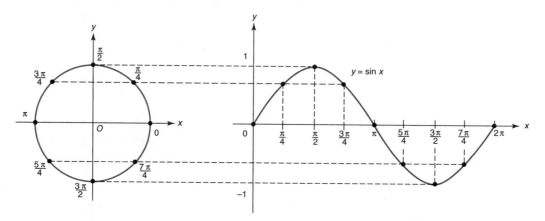

Figure 10.4 Mapping points of the unit circle onto the sine function

Amplitude and Period of $y = a\sin bx$ and $y = a\cos bx$

In the equations $y = a\sin bx$ and $y = a\cos bx$, the number a affects the amplitude and the number b determines the period. For each of these functions:

- The *amplitude* is $|a|$. For example, the maximum value of $y = 2\sin x$ is $+2$ and its minimum value is -2, so the amplitude of $y = 2\sin x$ is $\dfrac{+2 - (-2)}{2} = \dfrac{4}{2} = 2$. Figure 10.5 compares the graphs of $y = \sin x$, $y = 2\sin x$, and $y = \dfrac{1}{2}\sin x$ over the interval $0 \le x \le 2\pi$.

- The period is $\dfrac{2\pi}{|b|}$. If $y = \cos 2x$, then $b = 2$, so the period is $\dfrac{2\pi}{2} = \pi$. Therefore, the graph of $y = \cos 2x$ completes one full cycle in π radians. If $y = \cos \dfrac{1}{2}x$, then $b = \dfrac{1}{2}$, so the period is $\dfrac{2\pi}{1/2} = 4\pi$. Figure 10.6 compares the graphs of $y = \cos x$, $y = \cos 2x$, and $y = \cos \dfrac{1}{2}x$ over the interval $0 \le x \le 2\pi$. Because the period of $y = \cos \dfrac{1}{2}x$ is 4π, the graph completes one-half of a full cycle from 0 to 2π.

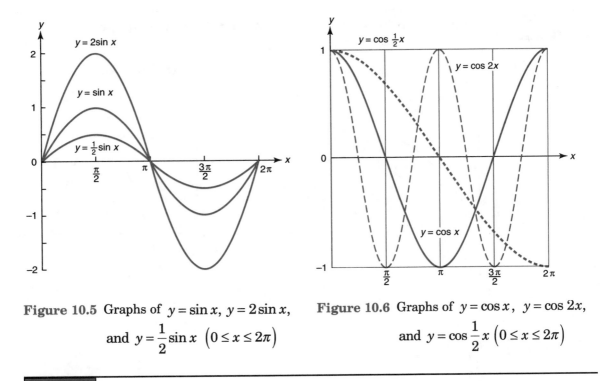

Figure 10.5 Graphs of $y = \sin x$, $y = 2\sin x$, and $y = \dfrac{1}{2}\sin x$ $\left(0 \le x \le 2\pi\right)$

Figure 10.6 Graphs of $y = \cos x$, $y = \cos 2x$, and $y = \cos \dfrac{1}{2}x$ $\left(0 \le x \le 2\pi\right)$

TIPS

For $y = a \sin bx$ and $y = a \cos bx$:

- $\left| a \right|$ is the highest point on the graph, and $-\left| a \right|$ is the lowest point on the graph. As can be seen in Figure 10.5, a, the number multiplying the trigonometric function, is a vertical scale factor for the graph. For example, $y = 2\sin x$ is a vertical stretching of the graph of $y = \sin x$ using a scale factor of 2.

- $\left| b \right|$ represents the frequency of the graph, which is the number of complete cycles that the curve completes in an interval of 2π radians. Figure 10.6 shows that b, the number multiplying x, is a horizontal scale factor. When $0 < b < 1$, the graph is stretched horizontally. For example, $y = \cos \dfrac{1}{2}x$ is a horizontal stretching of the graph of $y = \cos x$ using a scale factor of $\dfrac{1}{2}$, as the curve needs 4π radians, rather than 2π radians, to complete a full cycle.

| **Exercise 2** | Finding an Equation of a Sine or Cosine Function |

Determine an equation of the graph of each trigonometric function:

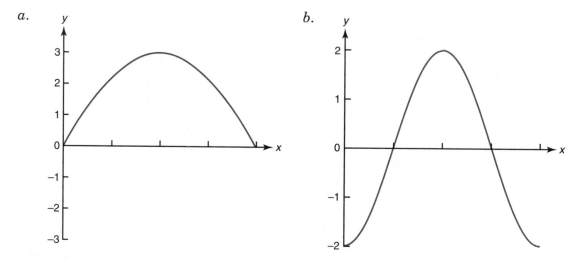

a.

b.

Solutions:

a. The curve has the shape of one arch of a sine curve, $y = a \sin bx$, whose amplitude is 3. The curve completes one-half of a complete cycle in 2π, so the frequency is $\frac{1}{2}$. Because $a = 3$ and $b = \frac{1}{2}$, an equation of the curve is $\boldsymbol{y = 3\sin \frac{1}{2}x}$.

b. The curve has the basic shape of a cosine curve, $y = a \cos x$, whose amplitude is 2 and whose period is 2π, except that it has been reflected in the x-axis. Thus, $a < 0$. The curve reaches a *minimum* of –2 at $x = 0$ and a *maximum* of +2 at $x = \pi$, so $a = -2$, and an equation of the curve is $\boldsymbol{y = -2\cos x}$.

| **Lesson 10-2** | **Graphing Trigonometric Functions** |

KEY IDEAS

The six trigonometric functions can be graphed by hand using graph paper or graphed using a calculator.

Graphing Sine and Cosine Curves by Hand

To sketch the graph of $y = \sin x$ or $y = \cos x$ from 0 to 2π:

1. Set up the coordinate axes on graph paper, using a convenient scale with quarter-period marks at $x = \dfrac{\pi}{4}$, π, $\dfrac{3}{4}\pi$, and 2π. The quarter-period marks correspond to the division points of the four quadrants.

2. "Frame" the graph by drawing broken horizontal lines through the maximum and minimum y-values of 1 and –1, and broken vertical lines through the endpoints of the interval on which the graph is being drawn.

3. Plot key points, such as the maximum point on the graph, the minimum point on the graph, and intercepts.

4. Use your knowledge of the basic shape of the curve to sketch the graph, as illustrated in Figure 10.7.

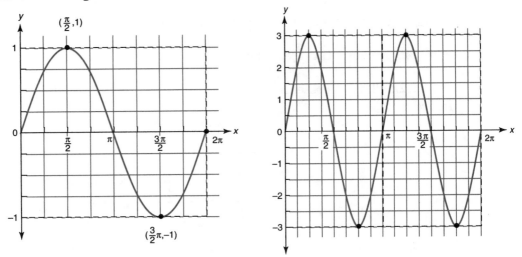

Figure 10.7 Graphing $y = \sin x$ **Figure 10.8** Graphing $y = 3\sin 2x$

A function of the form $y = a\sin bx$ or $y = a\cos bx$ can be graphed in a similar way by dividing the interval from $x = 0$ to $x = \dfrac{2\pi}{b}$ into four equal parts. For example, to sketch $y = 3\sin 2x$ from $x = 0$ to $x = 2\pi$:

1. Find the amplitude and the period. The amplitude of $y = 3\sin 2x$ is 3, and the frequency is 2. Hence, the graph completes two cycles from $x = 0$ to $x = 2\pi$.

2. Since the graph completes one cycle in the interval from $x = 0$ to $x = \pi$, set up the coordinate axes using a convenient scale with quarter-period marks from 0 to π. Also divide the interval from π to 2π into four equal parts.

3. Draw broken horizontal lines through the maximum and minimum y-values of +3 and –3, and broken vertical lines at $x = \pi$ and $x = 2\pi$, the right endpoints of the two cycles.

4. Use your knowledge of the basic shape of the sine curve to sketch one cycle of the graph from 0 to π, and another cycle of the graph from π to 2π, as shown in Figure 10.8.

Exercise 1   Hand Sketching a Sine and Cosine Function

On the same set of axes on graph paper, sketch and label the graphs of the equations $y = 2\sin\dfrac{1}{2}x$ and $y = \cos 2x$ in the interval $0 \le x \le 2\pi$.

What is the value of x in the interval $0 \le x \le 2\pi$ for which $2\sin\dfrac{1}{2}x - \cos 2x = 1$?

Solutions: The equation $y = 2\sin\dfrac{1}{2}x$ has the form $y = a\sin bx$, where $a = 2$ and $b = \dfrac{1}{2}$. Since the amplitude of this graph is 2, and its frequency is $\dfrac{1}{2}$. Since the amplitude is 2, the graph reaches a maximum height of 2 at $x = \pi$ and a minimum height of –2 at $x = 3\pi$. Because the frequency is $\dfrac{1}{2}$, the curve completes $\dfrac{1}{2}$ cycle in the interval $0 \le x \le 2\pi$. This represents one arch of the sine curve, as shown in the accompanying figure.

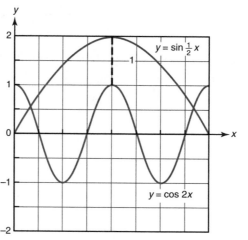

The equation $y = \cos 2x$ has the form $y = a\cos bx$, where $a = 1$ and $b = 2$. The amplitude of this graph is 1, and its frequency is 2. The curve reaches a maximum height of 1 and a minimum height of –1. Because the frequency is 2, the graph completes two cycles in the interval $0 \le x \le 2\pi$. On the x-axis mark off four quarter-period marks from 0 to π radians, and another four quarter-period marks from π to 2π radians. Using the quarter-period marks and the amplitude as guideposts, sketch one complete cycle of the cosine curve from 0 to π radians and another complete cycle from π to 2π radians, as shown in the accompanying figure.

According to the graph, at $x = \pi$ the height of the graph of $y = 2\sin \frac{1}{2}x$ is 2 and the height of the graph of $y = \cos 2x$ is 1. Therefore, $2\sin \frac{1}{2}x - \cos 2x = 1$ at $x = \boldsymbol{\pi}$.

Graphing a Trigonometric Function with a Calculator

To graph a trigonometric function using your graphing calculator, set the angular mode to radians, enter the function, and then press $\boxed{\text{ZOOM}}$ $\boxed{7}$. The graph will be displayed in the interval $-2\pi < x < 2\pi$ using the following preset values:

$$\text{Xmin} = -\left(\frac{47}{24}\right)\pi, \quad \text{Xmax} = \left(\frac{47}{24}\right)\pi, \quad \text{Xscl} = \frac{\pi}{2}, \text{ and Yscl} = 1.$$

Because Xscl is set to $\frac{\pi}{2}$, each tic mark on the x-axis is an integer multiple of $\frac{\pi}{2}$. If you need to view the graph in a different interval of x, press $\boxed{\text{WINDOW}}$ and selectively change the values of Xmin, Xmax, and Xscl.

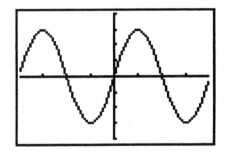

Figure 10.9
Graph of $y = 3\sin 2x \ (-\pi < x < \pi)$

For example, to graph $y = 3\sin 2x$ over the interval $-\pi < x < \pi$, press

$$\boxed{\text{ZOOM}} \quad \boxed{7} \quad \boxed{\text{WINDOW}}.$$

Divide Xmin, Xmax, and Xscl by 2 so that consecutive tic marks are $\frac{\pi}{4}$ radian apart in the interval $-\pi < x < \pi$. Because the period of $y = 3\sin 2x$ is π radians, one full cycle will be displayed on either side of the y-axis, as shown in Figure 10.9.

Graphing $y = \tan x$

The graph of the tangent function looks very different from the graphs of the sine and cosine functions. The function $y = \tan x$ is not defined at $x = \dfrac{\pi}{2}(=90°)$ or at any odd-integer multiple of $\dfrac{\pi}{2}$, so the graph has vertical asymptotes through these x-values, as shown in Figure 10.10.

Here are key facts about the tangent function that you can observe from the graph:

- period $= \pi$
- amplitude: none
- domain $=$ {real numbers except odd multiples of $\pm\dfrac{\pi}{2}$}
- range $=$ {real numbers y}
- x-intercepts: at integer multiples of π
- vertical asymptotes: at odd-integer multiples of $\pm\dfrac{\pi}{2}$. The graph completes one full cycle between consecutive vertical asymptotes.

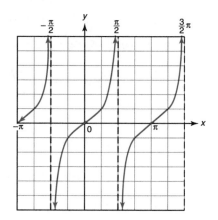

Figure 10.10 Graph of $y = \tan x$

Graphing $y = \csc x$

The graphs of $y = \csc x$ and its reciprocal, $y = \sin x$, are displayed on the same set of axes in Figure 10.11.

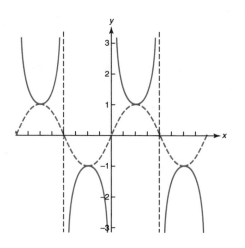

Figure 10.11 Graphs of $y = \csc x$ and $y = \sin x$ ($-2\pi \le x \le 2\pi$)

You can see from the graph that:

- The period of the cosecant function, like that of the sine function, is 2π.
- Whenever $\sin x = 1$, $\csc x = 1$; and whenever $\sin x = -1$, $\csc x = -1$.
- The graph of $y = \csc x$ has asymptotes at the values of x for which $\sin x = 0$.
- A local maximum of $y = \sin x$ coincides with a local minimum of $y = \csc x$, and a local minimum of $y = \sin x$ coincides with a local maximum of $y = \csc x$.

Graphing $y = \sec x$

The graphs of $y = \sec x$ and its reciprocal, $y = \cos x$, are displayed on the same set of axes in Figure 10.12.

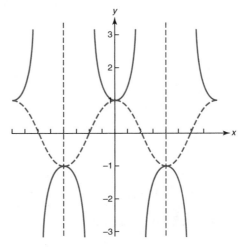

Figure 10.12 Graphs of $y = \sec x$ and $y = \cos x$ $(-2\pi \le x \le 2\pi)$

Here are four important features of the secant function that you can see from the graph:

- The period of the secant function, like that of the cosine function, is 2π.
- Whenever $\cos x = 1$, $\sec x = 1$.
- The graph of $y = \sec x$ has asymptotes at the values of x for which $\cos x = 0$.
- A local maximum of $y = \cos x$ coincides with a local minimum of $y = \sec x$, and a local minimum of $y = \cos x$ coincides with a local maximum of $y = \sec x$.

Graphing $y = \cot x$

The graphs of $y = \cot x$ and $y = \sin x$ are displayed on the same set of axes in Figure 10.13.

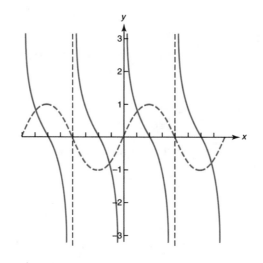

Figure 10.13 Graphs of $y = \cot x$ and $y = \sin x$ $(-2\pi \le x \le 2\pi)$

From the graph, you can see that:

- The period of the cotangent function, like that of the tangent function, is π.
- The graph of $y = \cot x$ has asymptotes at the values of x for which $\sin x = 0$. Because $\cot x = \dfrac{\cos x}{\sin x}$, cotangent is undefined when $\sin x = 0$.
- The cotangent function, like the tangent function, has no amplitude.

Lesson 10-3	**Transformations of Trigonometric Functions**

KEY IDEAS

The graphs of trigonometric functions can be *reflected* and *shifted* in the same way that these types of transformations are applied to the graphs of algebraic functions.

Reflecting Trigonometric Functions

The graphs of $y = -\sin x$ and $y = -\cos x$ are the reflections of the graphs of the basic sine and cosine curves in the x-axis, as shown in Figure 10.14, where the solid curves represent the graphs of $y = \sin x$ and $y = \cos x$.

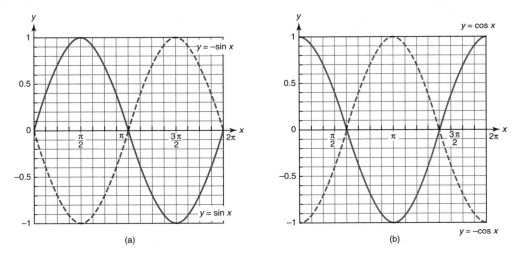

Figure 10.14 Reflecting sine (a) and cosine (b) curves in the x-axis

You can use your graphing calculator to verify that:

- The graph of $y = \sin(-x)$ is the reflection of the graph of $y = \sin x$ in the y-axis.

- The graph of $y = \cos x$ and its reflection in the y-axis, the graph of $y = \cos(-x)$, coincide.

Translating Trigonometric Functions

Trigonometric functions, like algebraic functions, can be shifted horizontally, vertically, or both horizontally and vertically. For example:

- Adding a constant, k, to the right side of the equation $y = a \sin bx$ or $y = a \cos bx$ shifts the graphs of these functions up when $k > 0$ and down when $k < 0$. For example, the graph of $y = \sin x + 3$ is the graph of $y = \sin x$ shifted up 3 units, as shown in Figure 10.15. Notice that the line $y = 3$ is a horizontal line of symmetry.

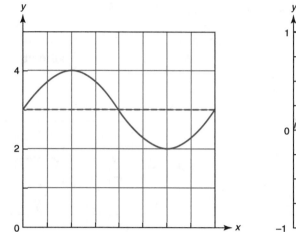

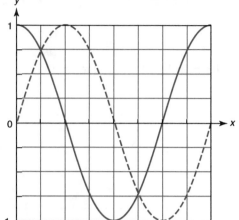

Figure 10.15 Graph of $y = \sin x + 3$

Figure 10.16 Horizontal shift of $y = \cos x$ by $\dfrac{\pi}{2}$

- The function $y = \cos\left(x - \dfrac{\pi}{2}\right)$ shifts the graph of $y = \cos x$ to the right by $\dfrac{\pi}{2}$ units, as shown in Figure 10.16, where the solid curve is the graph of $y = \cos x$. A horizontal translation of a periodic function is called a **phase shift**.

TIPS

- The line $y = k$ is a horizontal line of symmetry of the graphs of $y = a\sin(bx) + k$ and $y = a\cos(bx) + k$.
- The broken curve in Figure 10.16 represents the graph of $y = \cos\left(x - \dfrac{\pi}{2}\right)$, which coincides with the graph of $y = \sin x$. Thus, $\sin x = \cos\left(x - \dfrac{\pi}{2}\right)$.

Exercise 1   Finding an Equation of a Periodic Function Used as a Model

The number of visitors at a resort rises and falls during the year according to the accompanying graph. Determine an equation of this graph in terms of the month number, t.

Solution: Since the curve has the basic shape of the sine curve shifted vertically up relative to the horizontal axis, its equation has the form $y = a\sin(bt) + k$.

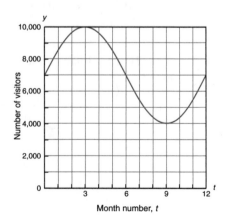

- Find a, the amplitude of the function:

$$a = \frac{\left(\text{maximum } y\text{-value}\right) - \left(\text{minimum } y\text{-value}\right)}{2}$$

$$= \frac{10,000 - 4000}{2}$$

$$= 3000$$

- Find k, the vertical shift of the function:

$$k = \frac{\left(\text{maximum } y\text{-value}\right) + \left(\text{minimum } y\text{-value}\right)}{2}$$

$$= \frac{10,000 + 4000}{2}$$

$$= 7000$$

- Find b. Since the graph completes one full cycle in 12 months, the period is 12 months. Hence, $\frac{2\pi}{b} = 12$, so $b = \frac{\pi}{6}$.

Because $a = 3000$, $b = \frac{\pi}{6}$, and $k = 7000$, an equation of the graph in terms of t is

$$\mathbf{y = 3000 \sin\left(\frac{\pi}{6}t\right) + 7000.}$$

Lesson 10-4 — Inverse Trigonometric Functions

KEY IDEAS

Because none of the graphs of the six trigonometric functions passes the horizontal-line test, none of these functions has an inverse function. Inverse trigonometric functions can be formed, however, by restricting the domains of the original trigonometric functions so that the graphs of the restricted functions pass the horizontal-line test.

Forming the Inverse Sine Function

Figure 10.17 shows that, in the interval $\left[-\frac{\pi}{2}, +\frac{\pi}{2}\right]$, the graph of $y = \sin x$ passes the horizontal-line test while $\sin x$ takes on its full range of values from $y = -1$ to $y = +1$.

Thus, over the restricted domain $\left[-\dfrac{\pi}{2}, +\dfrac{\pi}{2}\right]$, the sine function has an inverse function,

denoted as $y = \text{Sin}^{-1} x$ or $y = \text{Arc}\sin x$, read as "inverse sine of x." As you have seen with other pairs of inverse functions, the roles of x and y in the sine function and its inverse function are interchanged. Thus:

$$y = \text{Sin}^{-1} x \text{ and } x = \sin y$$

are equivalent for y-values in the restricted domain $\left[-\dfrac{\pi}{2}, +\dfrac{\pi}{2}\right]$ and for x-values in $[-1, +1]$, as shown in Figure 10.18.

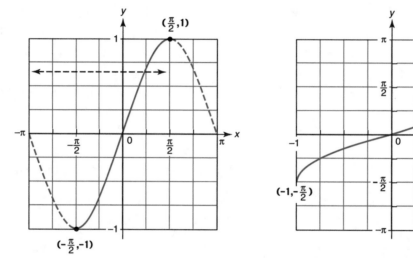

Figure 10.17 Graph of $y = \sin x$ **Figure 10.18** Graph of $y = \text{Arc}\sin x$

Evaluating Inverse Trigonometric Functions

Think of $y = \text{Sin}^{-1} x$ as the unique angle y whose sine is x, as in $\sin y = x$, for $-\dfrac{\pi}{2} \le y \le \dfrac{\pi}{2}$. If x is negative, then y must be between 0 and $-\dfrac{\pi}{2}$ (Quadrant IV).

EXAMPLE: If $y = \text{Arc}\sin\left(\dfrac{1}{2}\right)$, then $y = \dfrac{\pi}{6}$ because $\sin\left(\dfrac{\pi}{6}\right) = \sin 30° = \dfrac{1}{2}$.

EXAMPLE: If $y = \text{Sin}^{-1}\left(-\dfrac{\sqrt{3}}{2}\right)$, then $y = -\dfrac{\pi}{3}$ because $\sin\left(-\dfrac{\pi}{3}\right) = -\sin\left(\dfrac{\pi}{3}\right) = -\dfrac{\sqrt{3}}{2}$.

Inverse Cosine and Tangent Functions

With the same approach that was used in forming the inverse sine function, the inverse functions of cosine and tangent can be defined, as shown in the accompanying table.

Inverse Trigonometric Functions

INVERSE FUNCTION	RESTRICTED DOMAIN	RESTRICTED RANGE	EXAMPLE
$y = \text{Sin}^{-1} x$	$[-1, +1]$	$\left[-\dfrac{\pi}{2}, +\dfrac{\pi}{2} \right]$	$\text{Sin}^{-1}\left(\dfrac{\sqrt{2}}{2} \right) = \dfrac{\pi}{4}$
$y = \text{Cos}^{-1} x$	$[-1, +1]$	$[0, +\pi]$	$\text{Cos}^{-1}\left(-\dfrac{1}{2} \right) = \dfrac{2\pi}{3}$
$y = \text{Tan}^{-1} x$	$(-\infty, +\infty)$	$\left(-\dfrac{\pi}{2}, +\dfrac{\pi}{2} \right)$	$\text{Tan}^{-1}\left(\sqrt{3} \right) = \dfrac{\pi}{3}$

Think of $y = \text{Cos}^{-1} x$ as the unique angle y whose cosine is x, as in $\cos y = x$, for $0 \le y \le \pi$. If x is negative, then y must be between $\dfrac{\pi}{2}$ and π.

EXAMPLE: If $y = \text{Cos}^{-1}\left(\dfrac{\sqrt{3}}{2} \right)$, then $y = \dfrac{\pi}{6}$ since $\cos\dfrac{\pi}{6} = \dfrac{\sqrt{3}}{2}$.

EXAMPLE: If $y = \text{Arc}\cos\left(-1 \right)$, then $y = \pi$ since $\cos\pi = -1$.

Similarly, you can interpret $y = \text{Tan}^{-1} x$ as the unique angle y whose tangent is x for $-\dfrac{\pi}{2} < y < \dfrac{\pi}{2}$. If x is negative, then y must be between $0°$ and $-\dfrac{\pi}{2}$.

EXAMPLE: If $y = \text{Arc}\tan\left(-1 \right)$, then $y = -\dfrac{\pi}{4}$.

EXAMPLE: If $y = \sin\left(\tan^{-1}\left(\dfrac{\sqrt{3}}{2} \right) \right)$, then $\tan^{-1}\left(\dfrac{\sqrt{3}}{2} \right) = \dfrac{\pi}{6}$, so $y = \sin\left(\dfrac{\pi}{6} \right) = \dfrac{1}{2}$.

Exercise 1   **Applying an Inverse Trigonometric Function**

If $f(x) = \sin(\text{Arc}\tan x)$, what is the value of $f(1)$?

Solution: If $f(x) = \sin(\text{Arc}\tan x)$, then $f(1) = \sin(\text{Arc}\tan(1))$. Because $\text{Arc}\tan(1) = \dfrac{\pi}{4}$:

$$f(1) = \sin(\text{Arc}\tan(1)) = \sin\left(\frac{\pi}{4}\right) = \frac{\sqrt{2}}{2}.$$

Exercise 2   **Evaluating an Expression Involving Inverse Trigonometric Notation**

What is the value of $\tan\left(\text{Arc}\cos\left(-\dfrac{\sqrt{3}}{2}\right)\right)$?

Solution: Let $\theta = \text{Arc}\cos\left(-\dfrac{\sqrt{3}}{2}\right)$, where $0 \le \theta \le \pi$. Then:

- Because $\cos\theta = -\dfrac{\sqrt{3}}{2}$ and θ lies in Quadrant II, $\theta = \pi - \dfrac{\pi}{6} = \dfrac{5\pi}{6}$.

- Hence, $y = \tan\left(\text{Arc}\cos\left(-\dfrac{\sqrt{3}}{2}\right)\right) = \tan\left(\dfrac{5\pi}{6}\right) = -\dfrac{\sqrt{3}}{3}$.

Exercise 3   **Evaluating an Expression Involving Inverse Trigonometric Notation**

What is the value of $\cot\left[\text{Sin}^{-1}\left(-\dfrac{12}{13}\right)\right]$?

Solution: Let $\theta = \text{Sin}^{-1}\left(-\dfrac{12}{13}\right)$, where $-\pi \le \theta \le \pi$.

- Because $\sin\theta = \dfrac{-12}{13} = \dfrac{y}{r}$, locate θ in Quadrant IV so that $y = -12$ and $r = 13$, as shown in the accompanying figure.

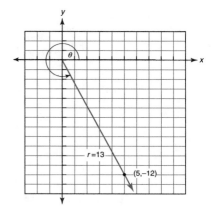

- Use the relationship $x^2 + y^2 = r^2$ to find x:

$$x^2 + (-12)^2 = 13^2$$
$$x^2 = 169 - 144$$
$$= 25$$
$$x = \sqrt{25} = 5$$

- In Quadrant IV, $\cot \theta = \dfrac{x}{y} = \dfrac{5}{-12}$.

- Hence:

$$\cot\left[\operatorname{Sin}^{-1}\left(\frac{-12}{13}\right)\right] = \cot \theta = -\frac{5}{12}.$$

Chapter 10 — CHECKUP EXERCISES

1–25. Multiple Choice

1. In which quadrant is the terminal side of $\angle x$ located if the graphs of $y = \sin x$ and $y = \cos x$ are both decreasing when $\angle x$ is increasing?

   (1) I   (2) II   (3) III   (4) IV

2. What is the range of the function $y = 2 \sin 3x$?

   (1) all real numbers   (3) $-2 \le y \le 2$

   (2) $-2\pi \le x \le 2\pi$ (4) $-3 \le y \le 3$

3. The graph of which equation has an amplitude of $\dfrac{1}{2}$ and a period of π?

   (1) $y = 2 \sin \dfrac{1}{2} x$ (2) $y = \dfrac{1}{2} \sin \dfrac{1}{2} x$ (3) $y = 2 \sin 2x$ (4) $y = \dfrac{1}{2} \sin 2x$

4. The graph of which equation has a period of π radians and passes through the origin?

   (1) $y = \cos 2x$ (2) $y = \sin 2x$ (3) $y = \cos \dfrac{1}{2} x$ (4) $y = \sin \dfrac{1}{2} x$

5. The graph of the equation $y = 2 \cos 2x$, $0 \le x \le 2\pi$, has a line of symmetry at

   (1) $x = \pi$ (2) $x = \dfrac{\pi}{4}$ (3) $y = 2$ (4) the x-axis

6. In the interval $0 < \theta \le 2\pi$, the number of solutions of the equation $\sin \theta - \cos \theta = 0$ is

   (1) 1   (2) 2   (3) 3   (4) 4

7. What phase shift of the cosine function will translate it onto the sine function?

   (1) 1   (2) $\dfrac{\pi}{2}$ (3) $-\dfrac{\pi}{2}$ (4) -1

8. If $y = 2 \cos \dfrac{1}{2} x - 3$, what is the minimum value of y?

   (1) -3 (2) -2 (3) -1 (4) -5

9. Which is an equation of the image of the graph of $y = 3 \sin \dfrac{1}{2} x$ under a size transformation with a scale factor of 2?

   (1) $y = 6 \sin x$ (2) $y = 6 \sin \dfrac{1}{4} x$ (3) $y = 1.5 \sin \dfrac{1}{4} x$ (4) $y = 1.5 \sin x$

10. Which equation could represent the graph in the accompanying figure?

(1) $y = 3 \sin 2x$ (2) $y = 2 \sin 3x$ (3) $y = 3 \sin x$ (4) $y = 2 \sin 4x$

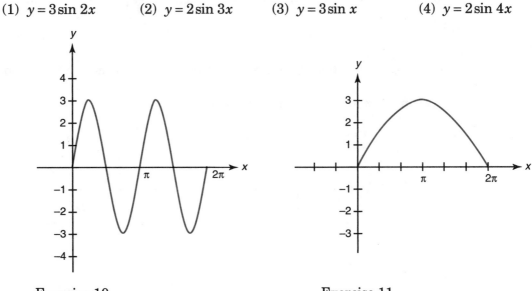

Exercise 10   Exercise 11

11. Which equation could represent the graph in the accompanying figure?

(1) $y = 3 \sin 2x$ (2) $y = 3 \sin \dfrac{1}{2} x$ (3) $y = 2 \sin 3x$ (4) $y = \dfrac{1}{2} \sin 3x$

12. Which equation could represent the graph in the accompanying figure?

(1) $y = 2 \sin \dfrac{1}{2} x$ (3) $y = 2 \cos \dfrac{1}{2} x$

(2) $y = \dfrac{1}{2} \sin 2x$ (4) $y = \dfrac{1}{2} \cos 2x$

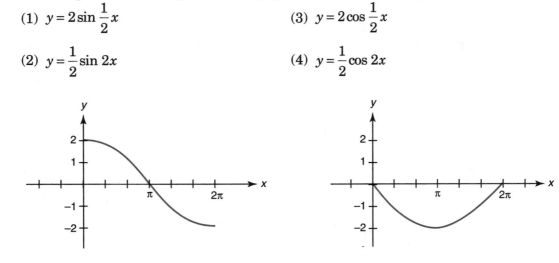

Exercise 12   Exercise 13

13. Which equation could represent the graph in the accompanying figure?

(1) $y = -2 \sin \dfrac{1}{2} x$ (3) $y = \dfrac{1}{2} \sin 2x$

(2) $y = -\dfrac{1}{2} \sin 2x$ (4) $y = 2 \sin \dfrac{1}{2} x$

14. At which value of x does the graph of $y = \tan\left(x + \dfrac{\pi}{4}\right)$ have a vertical asymptote?

  (1) $-\dfrac{\pi}{4}$ (2) 0   (3) $\dfrac{\pi}{4}$ (4) $\dfrac{\pi}{2}$

15. If $f(x) = \cos x$, which graph represents the composite function formed by a reflection

   of $f(x)$ in the x-axis followed by a reflection in the y-axis?

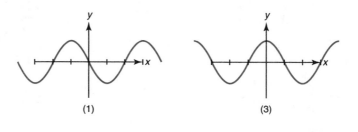

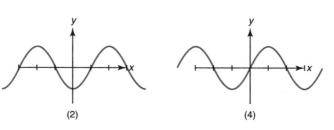

16. If $\theta = \operatorname{Arc\,cos}\left(-\dfrac{\sqrt{2}}{2}\right)$, what is the value of $\tan\theta$?

  (1) 1   (2) -1 (3) $-\dfrac{1}{\sqrt{3}}$ (4) $\sqrt{3}$

17. If $f(x) = \sin(\operatorname{Arc\,cos} x)$, what is the value of $f\left(\dfrac{1}{2}\right)$?

  (1) 1   (2) $\dfrac{1}{2}$ (3) $\dfrac{\sqrt{3}}{2}$ (4) $\dfrac{\sqrt{2}}{2}$

18. If $y = \sec\left(\operatorname{Arc\,sin}\dfrac{1}{2}\right)$, what is the value of y?

  (1) $30°$ (2) 2   (3) $\dfrac{2}{\sqrt{3}}$ (4) $\dfrac{1}{2}$

19. What is the value of $\csc\left[\operatorname{Arc\,tan}\left(\dfrac{4}{3}\right)\right]$?

  (1) $\dfrac{5}{3}$ (2) $\dfrac{5}{4}$ (3) $\dfrac{3}{5}$ (4) $\dfrac{4}{5}$

20. If $y = \text{Arc} \sin\left(\dfrac{\sqrt{2}}{2}\right) + \text{Arc} \cos\left(\dfrac{\sqrt{2}}{2}\right)$, then $y =$

    (1) $\dfrac{\pi}{4}$ (2) $\dfrac{\pi}{2}$ (3) $\dfrac{2\pi}{3}$ (4) π

21. What is the value of $\cos\left[\cos^{-1}\left(-\dfrac{1}{2}\right) + \sin^{-1}\left(\dfrac{1}{2}\right)\right]$?

    (1) $-\dfrac{\sqrt{3}}{2}$ (2) $-\dfrac{1}{2}$ (3) $\dfrac{1}{2}$ (4) $\dfrac{\sqrt{3}}{2}$

22. If $y = \cos\left(\text{Arc} \tan \dfrac{24}{7}\right)$, then $y =$

    (1) $\dfrac{7}{25}$ (2) $\dfrac{24}{25}$ (3) $\dfrac{25}{24}$ (4) $\dfrac{25}{7}$

23. What is the value of $\sin\left[\text{Arc} \sec\left(-\dfrac{41}{9}\right)\right]$?

    (1) $-\dfrac{41}{40}$ (2) $\dfrac{40}{9}$ (3) $-\dfrac{9}{40}$ (4) $\dfrac{40}{41}$

24. What is the value of $\cot\left[\text{Arc} \sin\left(-\dfrac{3}{\sqrt{13}}\right)\right]$?

    (1) $\dfrac{2}{3}$ (2) $-\dfrac{3}{2}$ (3) $\dfrac{3}{2}$ (4) $-\dfrac{2}{3}$

25. What is the value of $\sin\left(\text{Arc} \cos \dfrac{1}{x}\right)$ where $x \neq 0$?

    (1) $\dfrac{\sqrt{1-x^2}}{x}$ (2) $\dfrac{\sqrt{1+x^2}}{x}$ (3) $\dfrac{\sqrt{x^2-1}}{x}$ (4) $\dfrac{x}{\sqrt{1+x^2}}$

26. A normal breathing cycle consists of inhaling followed by exhaling. Assume that one complete breathing cycle occurs every 5 seconds, with a maximum airflow rate of 0.6 liter per second. If the function $f(t) = A \sin Bt$ is used to model this process, where $f(t)$ gives the airflow at time t, what are the values of A and B?

27. If the graph in the accompanying figure represents a vertical translation of a basic sine curve, write an equation of the translated graph shown.

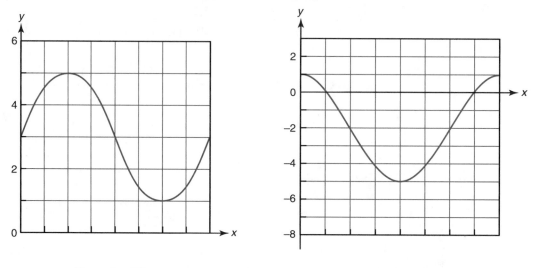

Exercise 27   Exercise 28

28. If the graph in the accompanying figure represents a vertical translation of a basic cosine curve, write an equation of the translated graph shown.

29 The peak demand for water in a certain town occurs during the summer months and can be modeled by the function $W(t) = 1600\sin\left(\dfrac{\pi}{90}t\right) + 5000$, where $W(t)$ represents the number of cubic liters of water needed for day number t $(0 \le t \le 90)$. June has 30 days, and the months of July and August each have 31 days. Assume that $t = 1$ corresponds to June 1. On what date is the demand for water the greatest, and how many cubic liters of water are demanded on that day?

30. A student attaches one end of a rope to a wall at a fixed point 3 feet above the ground, as shown in the accompanying diagram, and moves the other end of the rope up and down, producing a wave described by the equation $y = a\sin bx + c$. The height of the rope above the ground varies from 1 to 5 feet. The period of the wave is 4π. Write an equation that represents the wave.

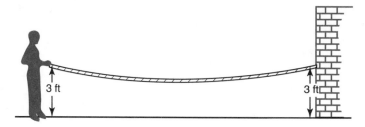

Exercise 30

31. *a.* On the same set of axes on graph paper, sketch the graphs of $y = 2\sin x$ and

$y = \cos\dfrac{1}{2}x$ for all values of x in the interval $0 \le x \le 2\pi$.

b. Each graph drawn in part *a* is symmetric about which of the following?

(1) line $x = \pi$ (2) x-axis   (3) $(\pi, 0)$ (4) $(0, 0)$

32. *a.* On the same set of axes on graph paper, sketch and label the graphs of the equations $y = -2\sin x$ and $y = 3\cos\dfrac{1}{2}x$ in the interval $0 \le x \le 2\pi$.

b. In the interval $0 \le x \le 2\pi$, which value of x satisfies the equation

$3\cos\dfrac{1}{2}x + 2\sin x = 0$?

33. *a.* On the same set of axes on graph paper, sketch and label the graphs of the equations $y = \sin 2x$ and $y = 3\cos x$ in the interval $-\pi \le x \le \pi$.

b. From the graphs drawn in part *a*, find all values of x in the interval $-\pi \le x \le \pi$ that satisfy the equation $\sin 2x = 3\cos x$.

34. A person who has just finished exercising has a respiratory cycle modeled by the

equation $v = 1.8\sin\left(\dfrac{2}{3}\pi t\right)$, where v is the velocity of airflow after t seconds. If one

respiratory cycle consists of an inhalation ($v > 0$) and an exhalation ($v < 0$):

a. Determine the number of seconds to the *nearest tenth* for one respiratory cycle.

b. Determine the number of full respiratory cycles completed in 1 minute.

c. Sketch the graph of $v = 1.8\sin\left(\dfrac{2}{3}\pi t\right)$ for one complete respiratory cycle.

35. A ball attached to the end of a spring is bobbing up and down according to the equa-

tion $d = 20\sin\left(\dfrac{1}{2}\pi t\right)$, where d is the distance in centimeters from the position of the

ball at rest at time t, which is measured in seconds. Assume that the motion of the ball is not affected by friction or air resistance, and continues to move up and down in a uniform manner. Oscillating motion of this type is called **simple harmonic motion**.

a. What is the distance between the lowest and highest positions of the bobbing ball?

b. What is the least number of seconds that the ball takes to travel from its lowest to its highest position?

c. Sketch the graph of $d = 20\sin\left(\dfrac{1}{2}\pi t\right)$ from $t = 0$ second to $t = 6$ seconds.

36. *a.* Write an equation of the form $d = a \sin bt$ that describes the simple harmonic motion of a ball attached to a spring, if the motion of the ball starts at its highest position of 8 inches above its rest point, bounces down to its lowest position of 8 inches below its rest point, and then bounces back to its highest position in a total of 6 seconds.

    b. Sketch the graph of the equation obtained in part *a* from $t = 0$ second to $t = 12$ seconds.

37. The tide at a boat dock can be modeled by the equation $y = -2\cos\left(\dfrac{\pi}{6}t\right) + 8$, where t is the height of the tide, in feet. For how many hours between $t = 0$ and $t = 12$ is the tide at least 7 feet high?

38. A building's temperature, T, varies with time of day, t, during the course of 1 day, according to the function $T(t) = 8\cos t + 78$. The air-conditioning operates when $T \geq 80°\,F$. Graph this function for $6 \leq t < 17$. Determine, to the *nearest tenth of an hour*, the amount of time in 1 day that the air-conditioning operates in the building.

39. The population of a certain fish in a river, over a 12-month interval, as shown in the accompanying graph, can be modeled by the equation $y = A\cos(Bx) + D$. Determine the values of A, B, and D, and explain how you arrived at these values.

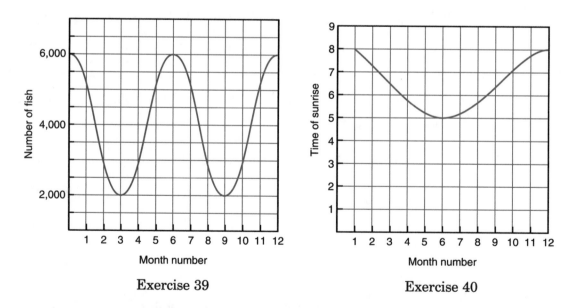

Exercise 39   Exercise 40

40. The times of average monthly sunrise, as shown in the accompanying diagram, over the course of a 12-month interval can be modeled by the equation $y = A\cos(Bx) + D$. Determine the values of A, B, and D, and explain how you arrived at these values.

41. *a.* On the same set of axes, sketch the graphs of $y = \cos 2x$ and $y = \tan x$ as x varies

from $-\dfrac{\pi}{2}$ to $\dfrac{\pi}{2}$ radians.

b. From the graph in part *a*, determine the number of points between $-\dfrac{\pi}{2}$ and $\dfrac{\pi}{2}$
radians for which $\tan x - \cos 2x = 0$.

Chapter 11

TRIGONOMETRIC IDENTITIES AND EQUATIONS

OVERVIEW

Important trigonometric formulas, called *identities*, appear so frequently when working with trigonometric expressions and equations that they need to be studied. You have already encountered one such identity, $\tan\theta = \dfrac{\sin\theta}{\cos\theta}$. Recognizing when a trigonometry identity appears in an expression can often simplify your work. Equations that contain different trigonometric functions can usually be reduced to one equation with a single trigonometric function by applying an appropriate identity.

Lesson 11-1   # Pythagorean Trigonometric Identities

KEY IDEAS

The equation $\tan\theta = \dfrac{\sin\theta}{\cos\theta}$ is an example of a *trigonometric identity* since it is true for all possible replacements of θ for which the expressions are defined. *Pythagorean trigonometric identities* can be derived using the unit circle.

Conditional Equation Versus an Identity

A **conditional equation** is true only for particular values of the variable, while an **identity** is an equation that is true for each value in the domain of the variable.

- The equation $2x = x + 1$ is a conditional equation. The graphs of $y = 2x$ and $y = x + 1$ intersect in exactly one point. Therefore, the left and right members of the equation $2x = x + 1$ agree for only one value of x, namely, $x = 1$.

- The equation $x^2 + 1 = (x+1)(x-1) + 2$ is an identity. The graphs of $y = x^2 + 1$ and $y = (x+1)(x-1) + 2$ coincide. Therefore, the left and right members of the equation $x^2 + 1 = (x+1)(x-1) + 2$ agree for all possible replacement values for x.

Basic Trigonometric Identities

Let $P(\cos\theta, \sin\theta)$ represent any point on the unit circle, $x^2 + y^2 = 1$. Substituting $\cos\theta$ for x and $\sin\theta$ for y produces the fundamental Pythagorean identity.

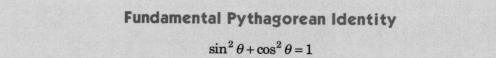

Fundamental Pythagorean Identity

$$\sin^2\theta + \cos^2\theta = 1$$

Dividing each term of $\sin^2\theta + \cos^2\theta = 1$ by either $\sin^2\theta$ or $\cos^2\theta$ leads to two additional Pythagorean identities which are listed below. Each of the three Pythagorean trigonometric identities may be written in more than one way.

Pythagorean Trigonometric Identities

THREE PYTHAGOREAN TRIGONOMETRIC IDENTITIES	SOME EQUIVALENT FORMS
<ul><li>$\sin^2\theta + \cos^2\theta = 1$</li><li>$\tan^2\theta + 1 = \sec^2\theta$</li><li>$\cot^2\theta + 1 = \csc^2\theta$</li></ul>	<ul><li>$\sin^2\theta = 1 - \cos^2\theta$ *or* $\cos^2\theta = 1 - \sin^2\theta$</li><li>$\tan^2\theta = \sec^2\theta - 1$ *or* $\sec^2\theta - \tan^2\theta = 1$</li><li>$\cot^2\theta = \csc^2\theta - 1$ *or* $\csc^2\theta - \cot^2\theta = 1$</li></ul>

Exercise 1 **Using a Pythagorean Identity to Simplify an Expression**

If $y = \cos A(\sec A - \cos A)$, then which choice is equivalent to y?

(1) $\cos^2 A$ (2) $\cos A - \sin A$ (3) $\sin^2 A$ (4) $\cot A - 1$

Solution: Replace $\sec A$ with $\dfrac{1}{\cos A}$:

$$y = \cos A\left(\sec A - \cos A\right)$$
$$= \cos A\left(\frac{1}{\cos A} - \cos A\right)$$
$$= 1 - \cos^2 A$$
$$= \mathbf{\sin^2 A}$$

The correct choice is **(3)**.

Exercise 2 **Using a Pythagorean Identity to Simplify an Expression**

If $\csc x \neq -1$, which expression is equivalent to $\dfrac{\cot^2 x}{1 + \csc x}$?

(1) $1 - \csc x$ (2) $\csc x - 1$ (3) $-\csc x$ (4) $\csc x - \cot x$

Solution: Rewrite the numerator using the Pythagorean identity, $\cot^2 x = \csc^2 x - 1$:

$$\frac{\cot^2 x}{1 + \csc x} = \frac{\csc^2 x - 1}{1 + \csc x}$$

Factor:
$$= \frac{\left(\csc x + 1\right)\left(\csc x - 1\right)}{1 + \csc x}$$

Divide:
$$= \frac{\left(\cancel{\csc x + 1}\right)\left(\csc x - 1\right)}{\cancel{1 + \csc x}}$$
$$= \mathbf{\csc x - 1}$$

The correct choice is **(2)**.

Proving Trigonometric Identities

Proving that a trigonometric equation is an identity involves showing that the two sides of the equation can be made to look exactly alike. Start with the more complicated side of the equation, and express it using only sines and cosines. Continue to work on the same side of the equation by doing one or more of the following:

- Factoring.
- Combining and simplifying fractional terms.
- Making a substitution using a known trigonometric identity such as a quotient, reciprocal, or Pythagorean identity.

If the two sides of the equation still do not look exactly the same, it may help to change the other side of the equation to sines and cosines, as illustrated in Exercise 3.

Exercise 3   Proving a Trigonometric Identity

Prove that the following equation is an identity for all values of θ for which the expressions are defined:

$$\sec \theta - \cos \theta = \sin \theta \tan \theta.$$

Solution: Draw a vertical boundary line separating the left and right sides of the equation. Working independently on each side of the equation, change secant and tangent to sine and cosine. Then simplify.

$$\sec \theta - \cos \theta \ = \ \sin \theta \, \tan \theta$$

Use the reciprocal identity:
$$\frac{1}{\cos \theta} - \cos \theta$$

Rewrite the second term, $\cos \theta$, as a fraction having $\cos \theta$ as it denominator:
$$\frac{1}{\cos \theta} - \frac{\cos^2 \theta}{\cos \theta}$$

Combine the fractions:
$$\frac{1 - \cos^2 \theta}{\cos \theta}$$

Use the Pythagorean identity in the numerator:
$$\frac{\sin^2 \theta}{\cos \theta}$$

Change the right side to sines and cosines:
$$\frac{\sin^2 \theta}{\cos \theta} \qquad \sin \theta \left(\frac{\sin \theta}{\cos \theta} \right)$$

Multiply on the right side:
$$\frac{\sin^2 \theta}{\cos \theta} \overset{\checkmark}{=} \frac{\sin^2 \theta}{\cos \theta}$$

Lesson 11-2

Solving Trigonometric Equations

KEY IDEAS

Trigonometric equations that are not identities are solved in much the same way as algebraic equations, but with two important exceptions:

- A trigonometric equation is generally solved in two stages. First the function is solved for, and then the angle. There may be more than one solution since a trigonometric function is positive in two quadrants and negative in two quadrants.
- A substitution using a Pythagorean, quotient, or reciprocal identity may be needed.

Solving Trigonometric Equations

To solve a trigonometric equation, first solve for the trigonometric function. Then find the angle, keeping in mind that function is positive or negative in more than one quadrant.

Exercise 1 **Solving a Linear Trigonometric Equation**

Solve $2\sin x + 1 = 0$ for x, where $0 \le x < 2\pi$.

Solution:

- Solve for $\sin x$: If $2\sin x + 1 = 0$, then $\sin x = -\dfrac{1}{2}$.

- Solve for $\angle x$: Because $x = \sin^{-1}\left(-\dfrac{1}{2}\right)$, the reference angle is $30°$. Sine is negative in Quadrants III and IV, so there are two possible solutions:

$$Q_{\text{III}}: x_1 = 180° + 30° = 210° \quad or \quad Q_{\text{IV}}: x_2 = 360° - 30° = 330°.$$

The two solutions are **210°** and **330°**.

Exercise 2 **Solving a Quadratic Trigonometric Equation**

Solve $\tan^2 x = \tan x + 2$, where $0° \le x < 360°$, for x to the *nearest tenth of a degree*.

Solution: Rewrite $\tan^2 x = \tan x + 2$ in standard form. Then solve by factoring.

- Solve for $\tan x$:

$$\tan^2 x - \tan x - 2 = 0$$
$$(\tan x + 1)(\tan x - 2) = 0$$
$$\tan x + 1 = 0 \qquad or \qquad \tan x - 2 = 0$$

- Solve for $\angle x$:

If $\tan x = -1$, the reference angle is $45°$. Tangent is negative in Quadrants II and IV. Thus:

$$Q_{II} : x_1 = 180° - 45° = 135° \quad or \quad Q_{IV} : x_2 = 360° - 45° = 315°.$$

If $\tan x = 2$, $x = \tan^{-1} 2$. Use a calculator to find that the reference angle, to the *nearest tenth of a degree*, is $63.4°$. Because tangent is positive in Quadrants I and III:

$$Q_I : x_3 \approx 63.4° \quad or \quad Q_{III} : x_4 \approx 180° + 63.4° = 243.4°.$$

Hence, there are four solutions: **63.4°**, **135°**, **243.4°**, and **315°**.

Solving Trigonometric Equations Requiring Substitutions

If a trigonometric equation contains two different trigonometric functions, use a trigonometric identity to transform the equation into an equivalent equation that contains the same trigonometric function.

Exercise 3   Solving a Trigonometric Equation Using a Substitution

Solve for x to the *nearest tenth of a degree*: $3\cos^2 x + 5\sin x = 4$ $(0° \le x \le 360°)$.

Solution: Because the equation contains two different trigonometric functions, one should be eliminated. Transform the original equation into an equivalent equation that contains only the sine function by replacing $\cos^2 x$ with $1 - \sin^2 x$:

$$3\cos^2 x \quad + 5\sin x = 4$$
$$3\left(1 - \sin^2 x\right) \quad + 5\sin x = 4$$
$$3 - 3\sin^2 x + 5\sin x = 4$$
$$3\sin^2 x - 5\sin x + 1 = 0$$

- Solve for $\sin x$ using the quadratic formula:

$$\sin x = \frac{-b \pm \sqrt{b^2 - 4ac}}{2a}$$

Let $a = 3$, $b = -5$, $c = 1$:

$$= \frac{-(-5) \pm \sqrt{(-5)^2 - 4(3)(1)}}{2(3)}$$

$$= \frac{5 \pm \sqrt{13}}{6}$$

$$\approx \frac{5 \pm 3.60555}{6}$$

- Solve for $\angle x$:

$$\sin x \approx \frac{5 - 3.60555}{6}$$
$$\approx 0.2324$$
$$x \approx \sin^{-1} 0.2324$$
$$x_{\text{Ref}} \approx 13.5°$$
$$Q_{\text{I}}: \quad x_1 = 13.5°$$
$$Q_{\text{II}}: \quad x_2 = 180° - 13.5° = 166.5°$$

or

$$\sin x \approx \frac{5 + 3.60555}{6}$$
$$\approx 1.39343$$

Reject since the maximum value of $\sin x$ is 1.

The equation has two solutions: **13.5°** and **166.5°**.

Sum and Difference Identities

KEY IDEAS

You can easily verify that $\sin(30 + 60)° \neq \sin 30° + \sin 60°$. Although $\sin(A + B) \neq \sin A + \sin B$, there are identities or formulas that allow trigonometric functions of the sum or difference of two angles to be expressed in terms of combinations of trigonometric functions of the individual angles.

Functions of the Sum of Two Angles

The sine, cosine, and tangent of the sum or difference of two angles can be written in terms of trigonometric functions of the individual angles.

Sum and Difference Formulas

- $\sin(A+B) = \sin A \cos B + \cos A \sin B$
- $\sin(A-B) = \sin A \cos B - \cos A \sin B$

- $\cos(A+B) = \cos A \cos B - \sin A \sin B$
- $\cos(A-B) = \cos A \cos B + \sin A \sin B$

- $\tan(A+B) = \dfrac{\tan A + \tan B}{1 - \tan A \cdot \tan B}$
- $\tan(A-B) = \dfrac{\tan A - \tan B}{1 + \tan A \cdot \tan B}$

Exercise 1   Recognizing a Sum-of-Two-Angles Identity

What is the exact value of $\sin 17° \cos 13° + \cos 17° \sin 13°$?

Solution: The given expression has the same form as the right side of the formula for the sum of the sines of $\angle A$ and $\angle B$, where $A = 17°$ and $B = 13°$. Hence:

$$\sin 17° \cos 13° + \cos 17° \sin 13° = \sin(17+13)° = \sin 30° = \frac{1}{2}.$$

Exercise 2   Using a Sum-of-Two-Angles Formula

If $\sin A = \dfrac{4}{5}$, $\cos B = \dfrac{12}{13}$, $\angle A$ is obtuse, and $\angle B$ is acute, what is the value of $\cos(A+B)$?

Solution: Before the formula for the cosine of the sum of two angles can be used, the values of $\cos A$ and $\sin B$ need to be determined by first locating the reference triangles.

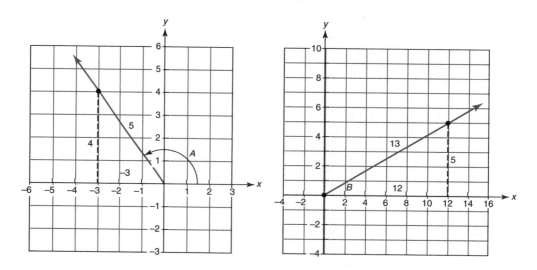

- $\sin A = \dfrac{4}{5} = \dfrac{y}{r}$. The reference triangle is a $\underline{3} - 4 - 5$ right triangle, as shown in the left figure on page 302. Since $\angle A$ lies in Quadrant II, where $x < 0$, $x = -3$. Then $\cos A = \dfrac{x}{r} = \dfrac{-3}{5}$.

- $\sin B = \dfrac{y}{r} = \dfrac{5}{13}$. Since the reference triangle is a $5 - \underline{12} - 13$ right triangle in Quadrant I, as shown in the right figure on page 302, $x = 12$. Then $\cos B = \dfrac{x}{r} = \dfrac{12}{13}$.

- Evaluate the formula for $\cos(A + B)$:

$$\cos(A + B) = \cos A \cos B - \sin A \sin B$$

$$= \left(\frac{-3}{5} \right)\left(\frac{12}{13} \right) - \left(\frac{4}{5} \right)\left(\frac{5}{13} \right)$$

$$= \left(-\frac{36}{65} \right) \quad - \left(\frac{20}{65} \right)$$

$$= -\frac{56}{65}$$

Exercise 3 Using a Difference-of-Two-Angles Formula

If $A = 30°$ and $B = \text{Arc}\cos \dfrac{3}{5}$, what is the exact value of $\sec(A - B)$?

Solution: Since $\sec(A - B) = \dfrac{1}{\cos(A - B)}$, first calculate $\cos(A - B)$.

- Since $A = 30°$, $\sin A = \dfrac{1}{2}$ and $\cos A = \dfrac{\sqrt{3}}{2}$.

- If $B = \text{Arc}\cos \dfrac{3}{5}$, then $\angle B$ is in Quadrant I, where $\cos B = \dfrac{3}{5} = \dfrac{x}{r}$ and $x = 3$, $y = 4$, and $r = 5$. Then $\sin B = \dfrac{y}{r} = \dfrac{4}{5}$.

- Use the formula for $\cos(A-B)$:

$$\cos(A-B) = \cos A \cos B + \sin A \sin B$$

$$= \left(\frac{\sqrt{3}}{2}\right)\left(\frac{3}{5}\right) + \left(\frac{1}{2}\right)\left(\frac{4}{5}\right)$$

$$= \frac{3\sqrt{3}+4}{10}$$

Since $\cos(A-B) = \dfrac{3\sqrt{3}+4}{10}$, $\sec(A-B) = \dfrac{10}{3\sqrt{3}+4}$.

Negative Angle Formulas

Exercise 4 asks you to prove a formula for rewriting the tangent of a negative angle as a function of a positive angle.

Exercise 4   Proving a Trigonometric Relationship

If $x > 0$, show that $\tan(-x) = -\tan x$.

Solution: Use the formula for $\tan(A-B)$, where $A = 0$ and $B = x$:

$$\tan(0-x) = \frac{\tan 0 - \tan x}{1 + \tan 0 \cdot \tan x}$$

$$= \frac{0 - \tan x}{1 + 0 \cdot \tan x}$$

$$\mathbf{\tan(-x) = -\tan x}$$

With an approach similar to the one demonstrated in Exercise 4, formulas for the sine and cosine of a negative angle can also be derived.

Negative Angle Formulas

The following formulas show how to rewrite a function of a negative angle as the same function of a positive angle where $A > 0$:

- $\sin(-A) = -\sin A$

- $\cos(-A) = \cos A$

- $\tan(-A) = -\tan A$

Double-Angle Identities

KEY IDEAS

Because the graphs of $y = \sin 2x$ and $y = 2\sin x$ do not coincide, you know that $\sin 2x = 2\sin x$ is not an identity. For example, $\sin 2(30°) \neq 2\sin 30°$. There are formulas, however, that allow a trigonometric function of a double angle to be expressed in terms of a trigonometric function of a single angle.

Functions of the Double Angle

The double-angle formulas $\sin 2A$, $\cos 2A$, and $\tan 2A$ can be derived from the corresponding formulas for $\sin(A + B)$, $\cos(A + B)$, and $\tan(A + B)$ by letting $B = A$.

Double-Angle Formulas		
$\sin 2A = 2\sin A \cos A$	$\cos 2A = \cos^2 A - \sin^2 A$ *or* $\cos 2A = 2\cos^2 A - 1$ *or* $\cos 2A = 1 - 2\sin^2 A$	$\tan 2A = \dfrac{2\tan A}{1 - \tan^2 A}$

Exercise 1   Using the Double-Angle Formula for Sine

Find the value of $\sin 2A$ if $\sin A = \dfrac{3}{5}$ and $\angle A$ is obtuse.

Solution: First find $\cos A$. Because $\sin A = \frac{3}{5} = \frac{y}{r}$ and $\angle A$ is obtuse, as shown in the accompanying figure, $\cos A = \frac{x}{r} = -\frac{4}{5}$. Use the double-angle identity for sine:

$$\sin 2A = 2\sin A \cos A$$
$$= 2\left(\frac{3}{5}\right)\left(\frac{-4}{5}\right)$$
$$= -\frac{24}{25}$$

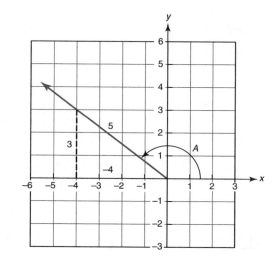

Exercise 2 Using the Double-Angle Formula for Cosine

If $\sin x = \frac{3}{4}$, what is the value of $\cos 2x$?

Solution: Since the value of $\sin x$ is given, choose the form of the identity for $\cos 2x$ that involves only sine: $\cos 2x = 1 - 2\sin^2 x$. Then:

$$\cos 2x = 1 - 2\sin^2 x = 1 - 2\left(\frac{3}{4}\right)^2 = -\frac{1}{8}.$$

Exercise 3 Solving an Equation Using a Double-Angle Formula

Find all values of x in the interval $0 \le x < 2\pi$ that satisfy the equation $\cos 2x - \cos x = 0$. Express answers in terms of π.

Solution: Transform the given equation into an equation that does not contain a double angle. Since the equation also contains $\cos x$, choose the form of the identity for $\cos 2x$ that is expressed only in terms of $\cos x$:

$$\cos 2x - \cos x = 0$$

Replace $\cos 2x$ with $2\cos^2 x - 1$:

$$\left(2\cos^2 x - 1\right) - \cos x = 0$$

Factor:

$$\left(2\cos x + 1\right)\left(\cos x - 1\right) = 0$$

- If $2\cos x + 1 = 0$, then $\cos x = -\dfrac{1}{2}$. The reference angle is $\dfrac{\pi}{3}$. Since cosine is negative in Quadrants II and III:

$$Q_{\text{II}} : x_1 = \frac{2\pi}{3} \quad or \quad Q_{\text{III}} : x_2 = \frac{4\pi}{3}.$$

- If $\cos x - 1 = 0$, then $\cos x = 1$, so $x_3 = 0$.

The solutions are **0, $\dfrac{2\pi}{3}$, and $\dfrac{4\pi}{3}$.**

Exercise 4   Proving an Identity Using a Double-Angle Formula

Prove that the following equation is an identity for all values of A for which the expressions are defined:
$$\frac{1 - \cos 2A}{\sin 2A} = \tan A.$$

Solution: Substitute $1 - 2\sin^2 A$ for $\cos 2A$ in the numerator, and replace $\sin 2A$ with $2\sin A \cos A$ in the denominator:

$$\frac{1 - \cos 2A}{\sin 2A} \overset{?}{=} \tan A$$

$$\frac{1 - \left(1 - 2\sin^2 A\right)}{2\sin A \cos A}$$

$$\frac{2\sin^2 A}{2\sin A \cos A}$$

$$\frac{\sin A}{\cos A}$$

$$\tan A = \tan A \;\checkmark$$

Exercise 5   Proving an Identity Using a Double-Angle Formula

Prove that the following equation is an identity for all values of x for which the expressions are defined:
$$\frac{\left(\sin x + \cos x\right)^2 - 1}{\cos x} = \left(\sin 2x\right)\left(\tan x\right)\left(\csc x\right).$$

Solution: Square the binomial on the left side of the equation, and simplify. Working independently on the right side of the equation, change to sines and cosines and simplify.

$$\frac{(\sin x + \cos x)^2 - 1}{\cos x} \overset{?}{=} (\sin 2x)(\tan x)(\csc x)$$

Square the binomial:

$$\frac{\sin^2 x + (2\sin x \cos x) + \cos^2 x - 1}{\cos x}$$

Substitute 1 for $\sin^2 x + \cos^2 x$:

$$\frac{2\sin x \cos x + 1 - 1}{\cos x}$$

Simplify:

$$\frac{2\sin x \cancel{\cos x}}{\underset{1}{\cancel{\cos x}}}$$

$$2\sin x$$

Change the right hand member to sines and cosines:

$$(2\sin x \cos x)\left(\frac{\sin x}{\cos x}\right)\left(\frac{1}{\sin x}\right)$$

$$\left(2\sin x \overset{1}{\cancel{\cos x}}\right)\left(\frac{\sin x}{\cancel{\cos x}}\right)\left(\frac{1}{\cancel{\sin x}}\right)$$

Simplify:

$$1$$

$$2\sin x \quad \overset{\checkmark}{=} \quad 2\sin x$$

Power-Reducing Formulas

It is sometimes useful to rewrite previously learned identities in equivalent ways. You already know that $\cos 2A = 1 - 2\sin^2 A$. Solving for $\sin^2 A$ gives

$$\sin^2 A = \frac{1 - \cos 2A}{2}.$$

This formula expresses the sine function raised to the second power in terms of the cosine function raised to the first power. Power-reducing formulas for $\cos^2 A$ and $\tan^2 A$ can be obtained in a similar manner.

Power-Reducing Formulas		
$\sin^2 A = \dfrac{1 - \cos 2A}{2}$	$\cos^2 A = \dfrac{1 + \cos 2A}{2}$	$\tan^2 A = \dfrac{1 - \cos 2A}{1 + \cos 2A}$

Lesson 11-5   Half-Angle Identities

KEY IDEAS

The graphs of $y = \sin\left(\dfrac{1}{2}x\right)$ and $y = \dfrac{1}{2}\sin x$ do not coincide indicating that

$\sin\left(\dfrac{1}{2}x\right) \neq \dfrac{1}{2}\sin x$. For example, $\sin\left(\dfrac{1}{2} \cdot 60°\right) \neq \dfrac{1}{2}\sin 60°$ since $\sin\left(\dfrac{1}{2} \cdot 60°\right) = \sin 30° = \dfrac{1}{2}$ but

$\dfrac{1}{2}\sin 60° = \dfrac{1}{2} \cdot \dfrac{\sqrt{3}}{2} = \dfrac{\sqrt{3}}{4}$. There are formulas, however, that allow a trigonometric function of a half angle to be expressed in terms of a trigonometric function of a single angle.

Functions of a Half Angle

The half-angle formulas for sine, cosine, and tangent can be developed from the corresponding power-reducing formulas for these functions. For example, you learned in Lesson 11-4 that $\sin^2 A = \dfrac{1 - \cos 2A}{2}$. Let $2A = x$; then $A = \dfrac{1}{2}x$, which makes $\sin^2 \dfrac{1}{2}x = \dfrac{1 - \cos x}{2}$. Taking the square root of each side of the equation gives

$$\sin \frac{1}{2}x = \pm\sqrt{\frac{1 - \cos x}{2}}.$$

Half-Angle Formulas		
$\sin \dfrac{1}{2}x = \pm\sqrt{\dfrac{1 - \cos x}{2}}$	$\cos \dfrac{1}{2}x = \pm\sqrt{\dfrac{1 + \cos x}{2}}$	$\tan \dfrac{1}{2}x = \pm\sqrt{\dfrac{1 - \cos x}{1 + \cos x}}$

The choice of a positive or negative sign in front of each radical depends on the sign of the trigonometric function in the quadrant in which $\dfrac{1}{2}x$ lies.

Exercise 1   Working with Half-Angle Formulas

If $\cos x = \dfrac{7}{8}$ and $\dfrac{3\pi}{2} \leq x < 2\pi$, find the values of $\sin \dfrac{1}{2}x$ and $\cos \dfrac{1}{2}x$.

Solutions: First determine the quadrant in which $\frac{1}{2}x$ lies.

- It is given that $\frac{3\pi}{2} \leq x < 2\pi$ or, equivalently, $270° \leq x < 360°$. To determine the quadrant in which $\frac{1}{2}x$ lies, divide each member of the inequality by 2, which gives $135° \leq \frac{1}{2}x < 180°$. Hence, $\frac{1}{2}x$ lies in Quadrant II.

- Because sine is positive in Quadrant II, use the positive value of the radical in the formula for $\sin\frac{1}{2}x$, where $\cos x = \frac{7}{8}$:

$$\sin\frac{1}{2}x = +\sqrt{\frac{1-\cos x}{2}} = \sqrt{\frac{1-\left(\frac{7}{8}\right)}{2}} = \sqrt{\frac{1}{16}} = \frac{1}{4}.$$

- Because cosine is negative in Quadrant II, use the negative value of the radical in the formula for $\cos\frac{1}{2}x$, where $\cos x = \frac{7}{8}$:

$$\cos\frac{1}{2}x = -\sqrt{\frac{1+\cos x}{2}} = -\sqrt{\frac{1+\frac{7}{8}}{2}} = -\sqrt{\frac{15}{16}} = -\frac{\sqrt{15}}{4}.$$

Exercise 2   **Working with a Half-Angle Formula**

If $\angle A$ is obtuse and $\sin A = \frac{\sqrt{5}}{3}$, what is the exact value of $\tan\frac{1}{2}A$?

Solution: If $\sin A = \frac{\sqrt{5}}{3}$, then

$$\cos^2 A = 1 - \sin^2 A = 1 - \left(\frac{\sqrt{5}}{3}\right)^2 = 1 - \frac{5}{9} = \frac{4}{9}.$$

Because $\angle A$ is obtuse, $\cos A$ is negative. Thus, $\cos A = -\sqrt{\frac{4}{9}} = -\frac{2}{3}$. Find $\tan\frac{1}{2}A$ using the half-angle formula for tangent where $\frac{1}{2}A$ is a Quadrant I angle so $\tan\frac{1}{2}A$ is positive:

$$\tan\frac{1}{2}A = \sqrt{\frac{1-\cos A}{1+\cos A}} = \sqrt{\frac{1-\left(-\frac{2}{3}\right)}{1+\left(-\frac{2}{3}\right)}} = \sqrt{\frac{5}{3} \div \frac{1}{3}} = \sqrt{5}.$$

TIP

The half-angle identities work whenever the two angles in the identity are in the

ratio of 1 to 2. For example, the identity $\sin \dfrac{1}{2} x = \pm \sqrt{\dfrac{1 - \cos x}{2}}$ can be expressed as

$\sin x = \pm \sqrt{\dfrac{1 - \cos 2x}{2}}$, which, of course, could have been obtained also from the power-

reducing identity for $\sin^2 x$.

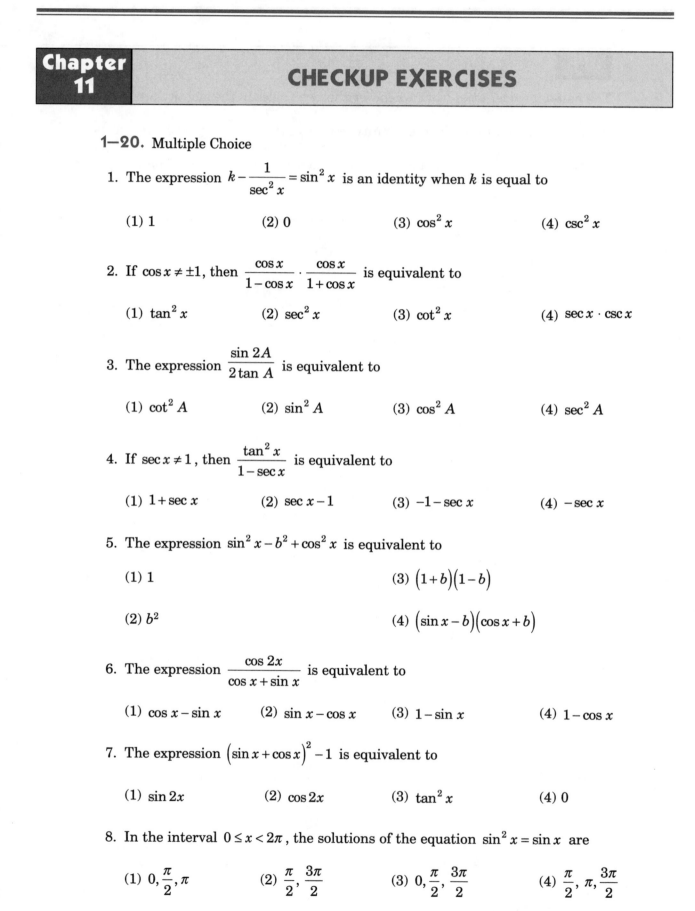

| Chapter 11 | CHECKUP EXERCISES |

1–20. Multiple Choice

1. The expression $k - \dfrac{1}{\sec^2 x} = \sin^2 x$ is an identity when k is equal to

   (1) 1   (2) 0   (3) $\cos^2 x$ (4) $\csc^2 x$

2. If $\cos x \neq \pm 1$, then $\dfrac{\cos x}{1 - \cos x} \cdot \dfrac{\cos x}{1 + \cos x}$ is equivalent to

   (1) $\tan^2 x$ (2) $\sec^2 x$ (3) $\cot^2 x$ (4) $\sec x \cdot \csc x$

3. The expression $\dfrac{\sin 2A}{2 \tan A}$ is equivalent to

   (1) $\cot^2 A$ (2) $\sin^2 A$ (3) $\cos^2 A$ (4) $\sec^2 A$

4. If $\sec x \neq 1$, then $\dfrac{\tan^2 x}{1 - \sec x}$ is equivalent to

   (1) $1 + \sec x$ (2) $\sec x - 1$ (3) $-1 - \sec x$ (4) $-\sec x$

5. The expression $\sin^2 x - b^2 + \cos^2 x$ is equivalent to

   (1) 1   (3) $(1 + b)(1 - b)$

   (2) b^2 (4) $(\sin x - b)(\cos x + b)$

6. The expression $\dfrac{\cos 2x}{\cos x + \sin x}$ is equivalent to

   (1) $\cos x - \sin x$ (2) $\sin x - \cos x$ (3) $1 - \sin x$ (4) $1 - \cos x$

7. The expression $(\sin x + \cos x)^2 - 1$ is equivalent to

   (1) $\sin 2x$ (2) $\cos 2x$ (3) $\tan^2 x$ (4) 0

8. In the interval $0 \leq x < 2\pi$, the solutions of the equation $\sin^2 x = \sin x$ are

   (1) $0, \dfrac{\pi}{2}, \pi$ (2) $\dfrac{\pi}{2}, \dfrac{3\pi}{2}$ (3) $0, \dfrac{\pi}{2}, \dfrac{3\pi}{2}$ (4) $\dfrac{\pi}{2}, \pi, \dfrac{3\pi}{2}$

9. If θ is an angle in Quadrant I and $\cot^2 \theta - 4 = 0$, what is the value of θ to the *nearest degree*?

   (1) 45   (2) 2   (3) 27   (4) 63

10. In $\triangle ABC$, $\sin(A+B) = \dfrac{3}{5}$. What is the value of $\sin C$?

    (1) $\dfrac{2}{5}$ (2) $\dfrac{2}{3}$ (3) $\dfrac{3}{5}$ (4) $\dfrac{11}{18}$

11. If $270° < x < 360°$ and $\cos x = 0.28$, what is the value of $\sin \dfrac{1}{2} x$?

    (1) 0.56   (2) 0.6   (3) –0.56   (4) – 0.6

12. If $\cos \dfrac{1}{2} x = \dfrac{\sqrt{5}}{4}$, what is the value of $\cos^2 x$?

    (1) $\dfrac{1}{64}$ (2) $\dfrac{1}{4}$ (3) $\dfrac{9}{64}$ (4) $\dfrac{16}{25}$

13. If $1 - \cos y = 0.2$ and $\angle y$ is acute, what is the value of $\tan \dfrac{1}{2} y$?

    (1) $\dfrac{1}{3}$ (2) $\dfrac{\sqrt{3}}{3}$ (3) $\dfrac{2}{3}$ (4) $\dfrac{1}{2}$

14. If $1 + \cos A = 0.38$ and $\angle A$ is obtuse, what is the value of $\sin \dfrac{1}{2} A$?

    (1) 0.44   (2) 0.1   (3) $\dfrac{\sqrt{19}}{10}$ (4) 0.9

15. If $A = \operatorname{Arc} \tan \dfrac{2}{3}$ and $B = \operatorname{Arc} \tan \dfrac{1}{2}$, what is the value of $\tan(A+B)$?

    (1) $\dfrac{1}{8}$ (2) $\dfrac{7}{8}$ (3) $\dfrac{1}{4}$ (4) $\dfrac{7}{4}$

16. The expression $\dfrac{\sin(90° + x)}{\sin(-x)}$ is equivalent to

    (1) –1   (2) 1   (3) –cot x (4) cot x

17. If $\tan \theta = a$, the expression $\tan\left(45° - \theta\right)$ is equivalent to

    (1) $\dfrac{1-a}{1+a}$ (2) $\dfrac{a-1}{a+1}$ (3) $\dfrac{1}{a}$ (4) a

18. If $\cos x = \dfrac{12}{13}$, $\sin y = \dfrac{4}{5}$, and x and y are acute angles, what is the value of $\sin\left(x - y\right)$?

    (1) $\dfrac{72}{65}$ (2) $\dfrac{56}{65}$ (3) $-\dfrac{16}{65}$ (4) $-\dfrac{33}{65}$

19. If $\sin A = \dfrac{3}{5}$ and $\sin B = \dfrac{2}{3}$, and A and B are acute angles, what is the value of $\cos\left(A - B\right)$?

    (1) $-\dfrac{2}{3}$ (2) $\dfrac{4\sqrt{5}-6}{15}$ (3) $\dfrac{4\sqrt{5}+2}{5}$ (4) $\dfrac{4\sqrt{5}+6}{15}$

20. If $\tan A = \dfrac{1}{2}$, what is the value of $\tan 2A$?

    (1) 1 (2) $\dfrac{1}{4}$ (3) $\dfrac{3}{4}$ (4) $\dfrac{4}{3}$

21. If $\sin \theta = \dfrac{\sqrt{5}}{3}$, what is the value of $\cos 2\theta$?

22. Prove the identity: $\cot 2\theta = \dfrac{\cot^2 \theta - 1}{2\cot \theta}$.

23–28. Find all values of x in the interval $0 \le x < 2\pi$ that satisfy the equation.

23. $\sqrt{4\cos^2 x - 1} - 1 = 0$

24. $\left|2\sin x - 1\right| = 1$

25. $\cos 2x = -3\sin x - 1$

26. $3\tan^2 x + \sqrt{3}\tan x = 0$

27. $\csc^2 x + \cot x = 1$

28. $2\cos x - 3 = 2\sec x$

29. Navigators aboard ships and airplanes use nautical miles to measure distance. The length of a nautical mile varies with latitude. The length of a nautical mile, L, in feet, on the latitude line θ is given by the formula $L = 6077 - 31\cos 2\theta$. Find, to the *nearest degree*, the angle, θ ($0° \leq \theta \leq 90°$), at which the length of a nautical mile is approximately 6075 feet.

30. Find all values of θ in the interval $0° \leq \theta < 360°$ that satisfy the equation $\dfrac{\sin^2 \theta}{1 - \cos \theta} = \dfrac{1}{2}$.

31–36. Find, correct to the *nearest tenth of a degree*, all values of x in the interval $0° \leq x < 360°$ that satisfy the equation.

31. $4\sin^2 x = 5\sin x + 6$

34. $5\sin^2 x - 9\cos x = 3$

32. $4\cos^2 x = 5(\sin x + 1)$

35. $2\tan x(\tan x - 1) = 3$

33. $\sec^2 x + 2\tan x = 4$

36. $\dfrac{3\sin x}{6\sin x + 1} = \dfrac{\csc x}{3}$

37. If $\angle x$ is acute, $\angle y$ is obtuse, $\cos x = \dfrac{15}{17}$, and $\sin y = \dfrac{3}{5}$, what is the exact value of $\sin(x + y)$?

38. When $\sin x = -\dfrac{8}{17}$ and x terminates in Quadrant III, and $\cos y = -\dfrac{4}{5}$ and y terminates in Quadrant II, what is the exact value of $\cos(x - y)$?

39. If $\operatorname{Arc}\sin x = -0.6$ and $\operatorname{Arc}\cos y = 0.5$, what is the exact value of $\cot(x - y)$?

40. If $A = 45°$ and $B = \operatorname{Arc}\sin \dfrac{3}{5}$, what is the exact value of $\csc(A - B)$?

41 and 42. Given $\tan A = 8$ and $\tan(A + B) = -\dfrac{17}{6}$.

41. What is the exact value of $\tan B$?

42. What is the exact value of $\sec^2 A + \sec^2 B$?

43–46. Find, to the *nearest tenth of a degree*, all nonnegative values of θ less than $360°$ that satisfy the equation.

43. $2\sin 2\theta + \cos\theta = 0$

44. $2\cos 2\theta + \cos\theta = 1$

45. $\sin\theta + 3\cos 2\theta = 1$

46. $4\cos 2\theta + 3 = 2\cos\theta$

47. If $\tan A = \dfrac{24}{7}$ and $180° < A < 270°$, determine the exact value of $\cos\dfrac{1}{2}A$.

48. If $\sin A = \dfrac{\sqrt{15}}{8}$ and $90° < A < 180°$, determine the exact value of $\cot\dfrac{1}{2}A$.

49. If $\sin A = -\dfrac{9\sqrt{19}}{50}$ and $180° < A < 270°$, determine the exact value of $\csc\dfrac{1}{2}A$.

50. If $\tan\dfrac{1}{2}\theta = \dfrac{2}{5}$ and $\angle\theta$ is acute, determine the exact value of $\sin\theta$.

51–60. Prove that each equation is an identity for all values for which the equation is defined.

51. $(\cot\theta + \csc\theta)(1 - \cos\theta) = \sin\theta$

52. $\dfrac{\tan\theta}{\cot\theta} + 1 = \sec^2\theta$

53. $\dfrac{\sin\theta}{\sin^2\theta + \cos 2\theta} = \dfrac{\sec\theta}{\cot\theta}$

54. $\dfrac{\csc x(\cos x + \sin x)^2}{1 + \sin 2x} = \dfrac{\cot x}{\cos x}$

55. $\dfrac{\sin 2A}{1 + \cos 2A} = \tan A$

56. $\dfrac{1 + \csc A}{\sec A} = \cos A + \cot A$

57. $\cot A - \dfrac{\cos 2A}{\sin A\cos A} = \tan A$

58. $\dfrac{\sec A\sec B}{\tan A - \tan B} = \csc(A - B)$

59. $\dfrac{\tan\theta - \cot\theta}{\tan\theta + \cot\theta} = 2\sin^2\theta - 1$

60. $\dfrac{\sin A}{1 - \cos A} + \dfrac{\sin A}{1 + \cos A} = 2\csc A$

Chapter 12

SOLVING TRIANGLES

OVERVIEW

Solving a triangle means finding the unknown measures of its sides and angles. To solve a triangle that does not contain a right angle, you need to know the length of at least one side and the measures of any two other parts of the triangle. The *Law of Sines* relates the lengths of two sides of a triangle to the sines of the angles opposite those sides. The *Law of Cosines* relates the lengths of the three sides of a triangle to the cosine of the angle opposite one of the sides.

Lessons in Chapter 12
Lesson 12-1: The Area of a Triangle
Lesson 12-2: The Law of Sines
Lesson 12-3: The Law of Cosines

Lesson 12-1

The Area of a Triangle

KEY IDEAS

The area of a triangle can be expressed in terms of the lengths of two sides of the triangle and the sine of the included angle.

Side-Angle-Side (SAS) Formula for the Area of a Triangle

When a line segment is drawn from one of the vertices of a triangle perpendicular to the opposite side, the perpendicular segment is the **height** of the triangle and the side to which it is drawn is the **base**. In Figure 12.1, $\overline{AC}$ is the base and $\overline{BH}$ is the height. The area of the triangle is $\frac{1}{2} \times AC \times BH = \frac{1}{2}bh$.

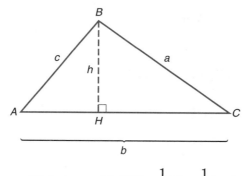

Figure 12.1 Area of $\triangle ABC = \frac{1}{2}bh = \frac{1}{2}bc \sin A$

To eliminate the height from the formula for the area of a triangle, consider right triangle AHB, where $\sin A = \frac{h}{c}$. Because $h = c \times \sin A$:

$$\text{Area of } \triangle ABC = \frac{1}{2}bh = \frac{1}{2}bc \sin A.$$

Choosing either of the other two sides as the base leads to a similar formula.

Formula for Area of a Triangle

In $\triangle ABC$, a, b and c are the lengths of the sides opposite vertices A, B, and C, respectively.

$$\text{Area of } \triangle ABC = \frac{1}{2}bc \sin A = \frac{1}{2}ac \sin B = \frac{1}{2}ab \sin C$$

Exercise 1 Finding the Area of a Triangle, Given SAS

If $\angle A = 150°$, $b = 8$ centimeters, and $c = 10$ centimeters, as shown in the accompanying diagram, find the area of $\triangle ABC$.

Solution:

Area of $\triangle ABC = \dfrac{1}{2} b c \sin A$

$\qquad = \dfrac{1}{2}(8)(10)\sin 150°$

$\qquad = 40(0.5)$

$\qquad = \mathbf{20\ cm^2}$

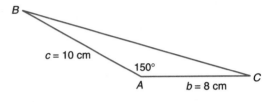

Exercise 2 Finding the Area of a Parallelogram

In parallelogram $ABCD$, $AB = 20$, $AD = 10$, and $\angle A = 45°$, as shown in the accompanying diagram. What is the exact area of parallelogram $ABCD$?

Solution:

- Draw diagonal $\overline{DB}$, as shown in the diagram. Then find the area of $\triangle DAB$:

  Area of $\triangle DAB = \dfrac{1}{2}(AD)(AB)\sin 45°$

  $\qquad = \dfrac{1}{2}(10)(20)\dfrac{\sqrt{2}}{2}$

  $\qquad = 50\sqrt{2}$

- Because a diagonal of a parallelogram divides the parallelogram into two equal triangles, area of $\triangle DCB = 50\sqrt{2}$.

- Hence, area of parallelogram $ABCD = 50\sqrt{2} + 50\sqrt{2} = \mathbf{100\sqrt{2}}$.

Lesson 12-2 — The Law of Sines

KEY IDEAS

The *Law of Sines* is a proportion that relates the lengths of any two sides of a triangle to the sines of the angles opposite those sides.

Deriving the Law of Sines

The Law of Sines can be obtained using the formula for the area of a triangle. Since

$$\text{area of } \triangle ABC = \frac{1}{2}ac\sin B \quad \text{and} \quad \text{area of } \triangle ABC = \frac{1}{2}bc\sin A,$$

then:

$$\frac{1}{2}\, a\, c\, \sin B = \frac{1}{2}\, b\, c\, \sin A$$

$$a\sin B = b\sin A$$

$$\frac{a}{\sin A} = \frac{b}{\sin B}$$

Following a similar approach, you can easily show that $\dfrac{a}{\sin A} = \dfrac{c}{\sin C}$. These proportions are referred to collectively as the **Law of Sines**.

The Law of Sines

$$\frac{\text{side of } \triangle ABC}{\text{sine of angle opposite side}} = \frac{a}{\sin A} = \frac{b}{\sin B} = \frac{c}{\sin C}$$

Using the Law of Sines, Given Angle-Angle-Side (AAS)

If AAS measurements of a triangle are given, the Law of Sines can be used to find the length of the side that is not given.

Exercise 1   **Using the Law of Sines, Given AAS**

In $\triangle ABC$, $a = 12$, $\sin A = 0.6$ and $\sin B = 0.4$. What is the length of side b?

Solution: Since AAS measurements are given, solve for $b\,(= AC)$ by using the Law of Sines.

$$\frac{a}{\sin A} = \frac{b}{\sin B}$$

$$\frac{12}{0.6} = \frac{b}{0.4}$$

$$0.6b = 0.4(12)$$

$$b = \frac{4.8}{0.6} = \mathbf{8}$$

Before using the Law of Sines, you may need to find the measure of the third angle of the triangle, so that you have angle-side-angle (ASA) measurements.

Exercise 2   Using the Law of Sines, Given ASA

In $\triangle ABC$, $\angle A = 59°$, $\angle B = 74°$, and $b = 100$ meters, as shown in the accompanying diagram. Find c to the *nearest tenth of a meter*.

Solution: Since ASA measurements are given, first find the measure of the third angle of the triangle:

$$\angle C = 180° - \left(59° + 74°\right) = 47°.$$

Then find c, using the Law of Sines:

$$\frac{b}{\sin B} = \frac{c}{\sin C}$$

$$\frac{100}{\sin 74°} = \frac{c}{\sin 47°}$$

$$c \sin 74° = 100 \sin 47°$$

$$c = \frac{100 \sin 47°}{\sin 74°} \approx \mathbf{76.1}$$

TIPS

When performing a calculation involving several calculator operations, such as calculating c using the proportion in Exercise 2:

- First solve for the variable without calculating approximations for the trigonometric functions of the given angles or performing any intermediate calculations.
- Perform all of the required calculator operations using the full accuracy of the calculator by storing intermediate results in the calculator's memory until the final answer is displayed.
- Round the final answer to the required number of digits.

Using the Law of Sines, Given Side-Side-Angle (SSA)

If SSA measurements of a triangle are given, the Law of Sines can be used to find the measure of the angle that is not given.

Exercise 3   Using the Law of Sines, Given SSA

In $\triangle JKL$, $j = 88.2$, $k = 100$, $\angle J = 26°10'$, and $\angle K$ is acute, as shown in the accompanying diagram. Find $\angle L$ correct to the *nearest hundredth of a degree*.

Solution: Since SSA measurements are given, first find $\angle K$ using the Law of Sines. Then use the measures of $\angle J$ and $\angle K$ to find $\angle L$.

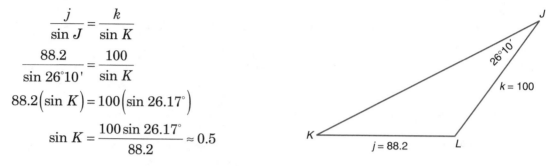

$$\frac{j}{\sin J} = \frac{k}{\sin K}$$

$$\frac{88.2}{\sin 26°10'} = \frac{100}{\sin K}$$

$$88.2 \left(\sin K \right) = 100 \left(\sin 26.17° \right)$$

$$\sin K = \frac{100 \sin 26.17°}{88.2} \approx 0.5$$

If $\sin K = 0.5$, then $\angle K$ measures 30° or 150°. However, since it is given that $\angle K$ is acute, $\angle K = 30°$. The sum of the degree measures of the three angles of a triangle is 180:

$$\angle J + \angle K + \angle L = 180°$$

$$\angle L = 180° \quad - \left(26°10' + 30° \right)$$

$$= 180° \quad - 56°10'$$

$$= 179°60' - 56°10'$$

$$= 123°50' = \left(123 + \frac{50}{60} \right)° = \mathbf{123.83°}$$

Solving "Double-Triangle" Problems

Two triangles may overlap so that they share an angle or a side. To use a trigonometric relationship in one triangle, it may be necessary to first find the measure of the shared angle or side by working in the other triangle.

Exercise 4   Solving a "Double Triangle"

The angle of elevation from a ship at point A to the top of a lighthouse at point B is 43°. When the ship reaches point C, 300 meters closer to the lighthouse, the angle of elevation is 56°. Find, to the *nearest meter*, the height of the lighthouse.

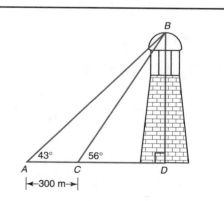

Solution: In the accompanying diagram of right triangle BDC, if the length of hypotenuse $\overline{BC}$ is known, BD, the height of the lighthouse, can be determined by using the sine ratio. To solve for BC, first find $\angle ABC$ and then use the Law of Sines in $\triangle ABC$.

- Find the measure of $\angle ABC$. Since an exterior angle of a triangle is equal to the sum of the two nonadjacent interior angles, $\angle ABC + 43° = 56°$, so $\angle ABC = 13°$.
- In $\triangle ABC$, use the Law of Sines:

$$\frac{300}{\sin 13°} = \frac{BC}{\sin 43°}$$

$$BC\left(\sin 13°\right) = 300\left(\sin 43°\right)$$

$$BC = \frac{300 \sin 43°}{\sin 13°} \approx 909.33$$

- In $\triangle BDC$, calculate BD using the sine ratio.

$$\sin 56° = \frac{BD}{909.33}$$

$$BD = 909.33 \sin 56° \approx \mathbf{754}$$

Ambiguous Case: Given Angle Is Acute

When Side-Side-Angle (SSA) measurements of a triangle are given, a unique triangle is not necessarily determined. In fact, it may be possible to construct no triangle, one triangle, or two triangles with the given dimensions. For this reason, the situation is ambiguous.

If the given angle is acute, four situations are possible. In Figure 12.2, a and b are the given sides of the triangle, h is the height (altitude), and A is the given angle.

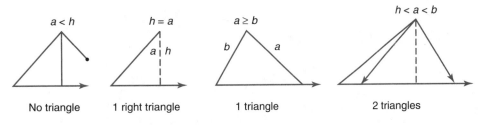

Figure 12.2 The ambiguous case where $\angle A$ is acute and $h = b \sin A$

In the two-triangle figure, when $h < a < b$, one triangle will be obtuse and the other acute.

Ambiguous Case: Given Angle Is Obtuse

Figure 12.3 summarizes the two possible situations when the given angle is obtuse.

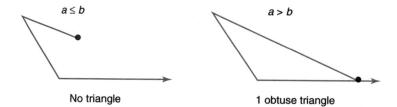

Figure 12.3 The ambiguous case, given sides a and b and obtuse $\angle A$

Counting Possible Triangles, Given SSA

When Side-Side-Angle (SSA) measurements are given, you can figure out the possible number of triangles, if any, that have these measurements. Suppose $a = 7$, $b = 10$, and $\angle A = 37°$, as shown in the accompanying diagram.

- Find the height, h, of a possible triangle:

$$h = b \sin A$$
$$= 10 \sin 37°$$
$$\approx 6.02$$

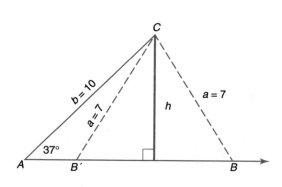

- Compare a, h and b. Since $a = 7$, $h = 6.02$, and $b = 10$, $h < a < b$.

- Hence, *two* triangles are possible: $\triangle ABC$ and $\triangle AB'C$.

You can arrive at the same conclusion if you start by assuming a triangle is possible and then apply the Law of Sines:

$$\frac{7}{\sin 37°} = \frac{10}{\sin B}$$

$$7 \sin B = 10 \sin 37°$$

$$\sin B = \frac{10 \sin 37°}{7} \approx 0.8597$$

Therefore, $\angle B \approx 59°$ *or* $121°$.

At least *one* triangle is possible. Can $\angle B$ be obtuse? Since $\angle A + \angle B = 37° + 121° = 158°$, and 158° is less than 180°, an obtuse triangle in which $\angle C = 180° - 158° = 22°$ *is* possible. Hence, *two* triangles are possible.

Exercise 5   Determining the Number of Possible Triangles, Given SSA

Find the number of triangles that can be constructed if $a = 2$, $b = 5$, and $\angle A = 30°$, as shown in the accompanying diagram.

Solution: The *shortest* distance from C to the base of a possible triangle is $h = 5\sin 30° = 2.5$.

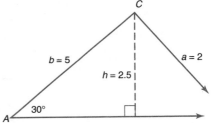

Thus, a must be *greater than* 2.5. But this contradicts the given fact that $a = 2$. Hence, *no* triangle can be constructed with the given measurements.

Trying to apply the Law of Sines leads to the same conclusion. If it is assumed that $\triangle ABC$ is possible, then:

$$\frac{a}{\sin A} = \frac{b}{\sin B}$$

$$\frac{2}{\sin 30°} = \frac{5}{\sin B}$$

$$2\sin B = 5(0.5)$$

$$\sin B = \frac{2.5}{2} = 1.25 \qquad \leftarrow \text{Impossible!}$$

Since $\sin A$ cannot be greater than 1, **no triangle is possible**.

Exercise 6   Determining the Number of Possible Triangles, Given SSA

Given $r = 7$, $s = 10$, and $\angle S = 40°$. How many distinct triangles can be constructed? If a triangle can be constructed, is it acute, obtuse, or right?

Solutions: Here the given angle is $\angle S$. As shown in the accompanying figure, s is the length of the side opposite the given angle and is greater than r. Hence, exactly **one** triangle is possible. When two sides of a triangle are unequal in length, the angle opposite the longer side is greater than the angle opposite the shorter side. Since $s > r$, $\angle S > \angle R$, so $\angle R$ must be less than 45°,  making $\angle S + \angle R$ less than 90°. Because the sum of the measures of the three angles of a triangle is 180°, the measure of the remaining angle of the triangle is greater than 90° and the triangle is **obtuse**.

Lesson 12-3 — The Law of Cosines

KEY IDEAS

The *Law of Cosines* relates the cosine of any angle of a triangle to the lengths of the three sides of the triangle. The Law of Cosines is used when **Side-Angle-Side (SAS)** or **Side-Side-Side (SSS)** triangle measurements are given. The Law of Sines is needed when **Angle-Angle-Side (AAS)**, **Side-Side-Angle (SSA)**, or **Angle-Side-Angle (ASA)** triangle measurements are given.

Using the Law of Cosines, Given SAS

The accompanying table summarizes the formulas for finding the length of a side of a triangle when the measures of the other two sides and their included angle are given. In the formula $c^2 = a^2 + b^2 - 2ab \cos C$, when $\angle C$ is a right angle, $\cos C = 0$, so the Law of Cosines reduces to $c^2 = a^2 + b^2$. Hence, the Law of Cosines can be considered to be a generalization of the Pythagorean theorem since it works in any triangle.

Law of Cosines, Given SAS

$\triangle ABC$	Given SAS	To Find	Law of Cosines
	$b, \angle A, c$	a	$a^2 = b^2 + c^2 - 2bc \cos A$
	$a, \angle B, c$	b	$b^2 = a^2 + c^2 - 2ac \cos B$
	$a, \angle C, b$	c	$c^2 = a^2 + b^2 - 2ab \cos C$

Exercise 1   Using the Law of Cosines, Given SAS

In $\triangle ABC$, $a = 6$, $b = 10$ and $\angle C = 120°$, as shown in the accompanying diagram. What is the length of side c?

Solution: Since SAS measurements are given, use the Law of Cosines to find the length of side c.

$$c^2 = a^2 + b^2 - 2ab \cos C$$
$$= 6^2 + 10^2 - 2(6)(10)\cos 120°$$
$$= 36 + 100 - 120(-0.5)$$
$$= 136 + 60$$
$$= 196$$
$$c = \sqrt{196} = 14$$

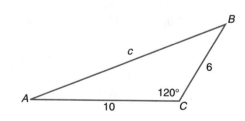

The length of side c is **14**.

Using the Law of Cosines, Given SSS

If the lengths of the three sides of $\triangle ABC$ are known (SSS), then the measure of any angle of the triangle can be determined by using the form of the Law of Cosines that involves that angle.

Exercise 2   Using the Law of Cosines, Given SSS

Peter (P) and Jamie (J) have computer factories that are 132 miles apart. They both ship computer parts to Diane (D), who is 72 miles from Peter and 84 miles from Jamie, as shown in the accompanying diagram. If points P, D, and J are located on a map drawn to scale, what is the measure of the largest angle of $\triangle PDJ$ to the *nearest tenth of a degree*?

Solution: The largest angle of a triangle lies opposite the longest side of the triangle. The longest side of $\triangle PDJ$ is $\overline{PJ}$. Since SSS measurements are given, use the Law of Cosines to find $\angle D$.

$$(PJ)^2 = (PD)^2 + (JD)^2 - 2(PD)(JD)\cos D$$
$$(132)^2 = (72)^2 + (84)^2 - 2(72)(84)\cos D$$
$$17,424 = 12,240 \qquad - 12,096\cos D$$
$$12,096\cos D = 12,240 \qquad - 17,424$$
$$\cos D = -\frac{5184}{12,096} \approx -0.42857$$
$$\angle D = \cos^{-1}(-0.42857) \approx \mathbf{115.4°}$$

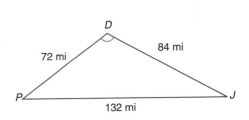

Finding the Resultant Force

If two forces act simultaneously on a body, then the single force that has the same effect as these two forces, called the **resultant**, is the diagonal of a parallelogram whose adjacent sides represent the two forces. For example, suppose that a 20-pound force and a 30-pound force act on a body at an angle of 60° with each other. The magnitude of the resultant is the length of the diagonal of the parallelogram whose adjacent sides are the two forces, as shown in Figure 12.4. To calculate the magnitude of the resultant to the *nearest tenth of a pound*, find AC using the Law of Cosines.

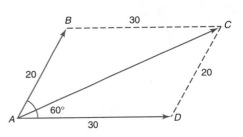

Figure 12.4 Resultant force

- $\angle B = 180° - 60° = 120°$ since consecutive angles of a parallelogram are supplementary.
- Since opposite sides of a parallelogram have the same length, $DC = AB = 20$.
- In $\triangle ABC$, use the Law of Cosines:

$$
\begin{aligned}
(AC)^2 &= (BC)^2 + (AB)^2 - 2(BC)(AB)\cos B \\
&= (30)^2 + (20)^2 - 2(30)(20)\cos 120° \\
&= 900 + 400 - 1200(-0.5) \\
&= 1900 \\
AC &= \sqrt{1900} \approx 43.6 \text{ pounds}
\end{aligned}
$$

Exercise 3   Finding the Angle Between Two Forces

Two forces of 437 pounds and 876 pounds act on a body at an acute angle with each other. The angle between the resultant and the 437-pound force is 41°10'. Find, to the *nearest ten minutes* and also to the *nearest tenth of a degree*, the measure of the angle between the original two forces.

Solution:

- Draw parallelogram $ABCD$ with $AB = 437$, $AD = 876$, and $\angle CAB = 41°10'$, as shown in the accompanying diagram. Find $\angle DAB$ by first finding $\angle DAC$. Let $x = \angle DAC = \angle BCA$.
- In $\triangle ABC$, use the Law of Sines:

$$\frac{876}{\sin 41°10'} = \frac{437}{\sin x}$$

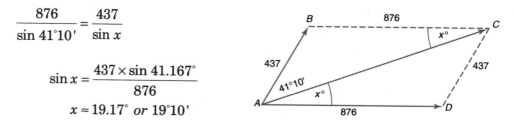

$$\sin x = \frac{437 \times \sin 41.167°}{876}$$

$$x \approx 19.17° \text{ or } 19°10'$$

- $\angle DAB \approx 41°10' + 19°10' = \mathbf{60°20'}$ *or*, equivalently, **60.3°**.

Exercise 4   Finding the Magnitude of the Resultant of Two Forces

In Exercise 4, find the magnitude of the resultant force to the *nearest pound*.

Solution: Use either the Law of Sines or the Law of Cosines.

<u>Method 1</u>: Use the Law of Sines.

- In $\triangle ABC$, find AC using the Law of Sines.
- $\angle B = 180° - \angle DAB = 179°60' - 60°20' = 119°40'$.

- $\dfrac{AC}{\sin 119°40'} = \dfrac{876}{\sin 41°10'}$, so $AC = \dfrac{\sin 119.667° \times 876}{\sin 41.167°} \approx \mathbf{1156}$.

<u>Method 2:</u> Use the Law of Cosines.

$$(AC)^2 = (437)^2 \quad +(876)^2 \quad -2\,(437)(876)\cos 119°40'$$
$$= \; 190{,}969 + 767{,}376 + 378{,}984$$
$$= 1{,}337{,}329$$
$$AC = \sqrt{1{,}337{,}329} \; \approx \mathbf{1156}$$

| Chapter 12 | CHECKUP EXERCISES |

1–15. Multiple Choice

1. In $\triangle ABC$, $a = 8$, $b = 9$, and $\angle C = 135°$. What is the area of $\triangle ABC$?

   (1) 18   (2) 36   (3) $18\sqrt{2}$ (4) $36\sqrt{2}$

2. An angle of a parallelogram has a measure of 150°. If the sides of the parallelogram measure 10 and 12 centimeters, what is the area of the parallelogram?

   (1) $30\,\text{cm}^2$ (2) $60\,\text{cm}^2$ (3) $60\sqrt{2}\,\text{cm}^2$ (4) $60\sqrt{3}\,\text{cm}^2$

3. In $\triangle ABC$, $a = 8$, $b = 2$, and $c = 7$. What is the value of $\cos C$?

   (1) $-\dfrac{19}{32}$ (2) $-\dfrac{11}{28}$ (3) $\dfrac{109}{112}$ (4) $\dfrac{19}{32}$

4. If $\angle A = 30°$, $BC = 10$, and $AC = 12$, then $\angle C$ in $\triangle ABC$ can be

   (1) an acute angle, only   (3) a right angle

   (2) an obtuse angle, only   (4) either an acute angle or an obtuse angle

5. In $\triangle DEF$, if $d = \sqrt{3}$, $e = 4$, and $\angle F = 30°$, the length of side f is

   (1) 7   (2) $\sqrt{17}$ (3) $\sqrt{7}$ (4) $\sqrt{3}$

6. Sasha is designing a triangular piece for a metal sculpture. He tells Vanessa that the length of two of the sides of the piece are 40 inches and 15 inches, and the angle opposite the 40-inch side measures 120°. Vanessa decides to sketch the piece that Sasha described. How many different triangles can she sketch that match Sasha's description?

   (1) 1   (2) 2   (3) 3   (4) 4

7. If $a = 4$, $b = 6$, and $\sin A = \dfrac{3}{5}$ in $\triangle ABC$, then $\sin B$ equals

   (1) $\dfrac{3}{20}$ (2) $\dfrac{6}{10}$ (3) $\dfrac{8}{10}$ (4) $\dfrac{9}{10}$

8. If $a = 5$, $c = 12$, and $\angle A = 30°$, what is the total number of distinct triangles that can be constructed?

   (1) 1   (2) 2   (3) 3   (4) 0

9. For which set of measurements can more than one triangle be constructed?

    (1) $r = 6$, $s = 5$, and $\angle R = 100°$

    (2) $r = 6$, $s = 5$, and $\angle R = 30°$

    (3) $r = 3$, $s = 5$, and $\angle R = 30°$

    (4) $r = 3$, $s = 6$, and $\angle R = 30°$

10. If $\sin A = 0.75$ and $b = 8$, for which value of a is it possible to construct two distinct triangles?

    (1) 5   (2) 6   (3) 7   (4) 9

11. If $\angle A = 48°$, $BC = 7$, and $AC = 9$, then $\angle C$ in $\triangle ABC$ can be

    (1) an acute angle, only

    (2) an obtuse angle, only

    (3) a right angle

    (4) either an acute angle or an obtuse angle

12. In $\triangle ABC$, $a = 4$, $b = 3$, and $c = \sqrt{37}$. What is the degree measure of the largest angle of the triangle?

    (1) 60   (2) 120   (3) 135   (4) 150

13. The sides of a triangle measure 6, 7, and 9. What is the value of the cosine of the largest angle?

    (1) $-\dfrac{4}{84}$ (2) $\dfrac{2}{21}$ (3) $\dfrac{4}{84}$ (4) $-\dfrac{1}{81}$

14. In isosceles triangle ABC, $BC = 1$ and $\angle C = 120°$. The length of $\overline{AB}$ is

    (1) 1   (2) $\sqrt{2}$ (3) $\sqrt{2.5}$ (4) $\sqrt{3}$

15. In $\triangle ABC$, $\angle B = 35°$ and side $c = 10$. Which value of side b will make it possible for two different triangles to be formed?

    (1) $b = 5$ (2) $b = 8$ (3) $b = 10$ (4) $b = 13$

16. In isosceles triangle ABC, $\angle C = 30°$, $AB = 8$, and $BC = 10$. What is the greatest possible area of $\triangle ABC$?

17. A ski lift begins at ground level 0.75 mile from the base of a mountain whose face has a 50° angle of elevation, as shown in the accompanying diagram. The ski lift ascends in a straight line at an angle of 20°. Find, to the *nearest hundredth of a mile*, the length of the ski lift from its beginning to the top of the mountain.

Exercise 17

18. Two equal forces act on a body at an angle of 80°. If the resultant force is 100 newtons, find, to the *nearest hundredth of a newton,* the magnitude of one of the two equal forces.

19. Gregory wants to build a garden in the shape of an isosceles triangle with the length of one of the congruent sides equal to 12 yards. If the area of the garden will be 55 square yards, find, to the *nearest tenth of a degree,* the measures of the three acute angles of the triangle.

20. A 54-foot entrance ramp makes an angle of 4.3° with the level ground. In order to comply with the most recent wheelchair-accessibility guidelines, the ramp must be extended, as indicated in the accompanying diagram, so that it makes an angle of, at most, 3° with the level ground. What is the minimum distance from the point on the ground at which the incline of the old ramp begins to the point on the ground where the incline of the new ramp must begin? Approximate your answer correct to the *nearest inch.*

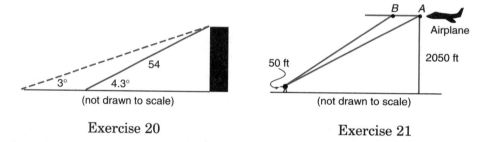

Exercise 20   Exercise 21

21. An airplane traveling at a level altitude of 2050 feet sights the top of a 50-foot tower at an angle of depression of 28° from point *A,* as shown in the accompanying diagram. After continuing in level flight to point *B,* the angle of depression to the same tower is 34°. Find, to the *nearest foot,* the distance that the plane traveled from point *A* to point *B.*

22. Engineers are designing a straight tunnel through a hill from point *A* to point *B,* which are on opposite sides of the hill and at the same level. Point *C* is chosen on the top of the hill in such a way that points *A, B,* and *C* lie in the same vertical plane. From points *A* and *B,* the angles of elevation to *C* are 26°40' and 38°10', respectively. If the distance from *A* to *C* is 400 meters, find the length of tunnel $\overline{AB}$ to the *nearest tenth of a meter.*

23. A forest preserve has the shape of quadrilateral *ABCD,* shown in the accompanying figure, where *AB* = 3.6 kilometers, *AD* = 4.8 kilometers, ∠*DAB* = 90°, ∠*DBC* = 50°, and *BC* = 2.8 km. Find, to the *nearest tenth,* the number of square kilometers in the area of the forest preserve.

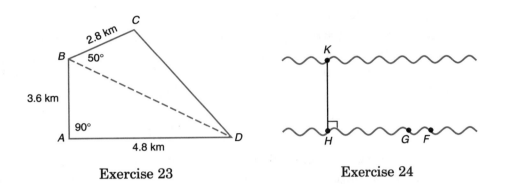

Exercise 23   Exercise 24

24. To determine the distance across a river, a surveyor marked three points along a riverbank: H, G, and F, as shown in the accompanying diagram. She also marked one point, K, on the opposite bank in such a way that $\overline{KH} \perp \overline{HGF}$, $\angle KGH = 41°$, and $\angle KFH = 37°$. The distance between G and F is 45 meters. Find KH, the width of the river, to the *nearest tenth of a meter*.

25. A hiking trail is planned in the shape of a triangle with sides measuring 2.3 miles, 8.1 miles, and 6.2 miles. Find, to the *nearest tenth of a square mile*, the area of the triangular region bounded by the hiking trail.

26. Kieran is traveling from city A to city B. As the accompanying diagram indicates, Kieran could drive directly from city A to city B along County Route 21 at an average speed of 55 miles per hour, or travel on the interstates, 45 miles along I-85 and then 20 miles along I-64. The two interstates intersect at an angle of 150° at C and have a speed limit of 65 miles per hour. How much time will Kieran save by traveling along the interstates at an average speed of 65 miles per hour?

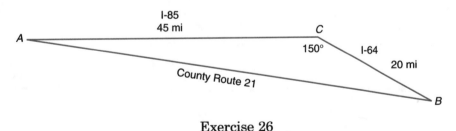

Exercise 26

27. A farmer has determined that a crop of strawberries yields a yearly profit of $1.50 per square yard. Strawberries are planted on a triangular plot of land whose sides measure 50 yards, 75 yards, and 100 yards. How much profit, to the *nearest hundred dollars*, can the farmer expect to make from this plot of land during the next harvest?

28. Two forces of 50 and 68 pounds act on a body to produce a resultant force of 70 pounds. Find, to the *nearest tenth of a degree*, the measure of the angle formed between the resultant force and the smaller force.

29. A metal frame is constructed in the form of an isosceles trapezoid, with diagonals acting as braces to strengthen the frame. Each base angle measures 73°30', the length of the shorter base is 8.0 feet, and the length of each of the nonparallel sides is 5.0 feet. Find, to the *nearest tenth of a foot*, the length of a diagonal brace of this frame.

30. Patricia and Quentin are separated by a wall. To calculate the distance between them, Akim positions himself 50 meters from Patricia and 75 meters from Quentin. If Akim's horizontal line of sight must change 120°40' when he switches his view from Patricia to Quentin, find, to the *nearest tenth of a meter*, the distance between Patricia and Quentin.

31. Home plate and the three bases on a Little League baseball field are at the vertices of a square in which each side has a regulation length of 60 feet. If the pitcher's mound is 46 feet from home plate, what is the number of feet, to the *nearest tenth*, from the pitcher's mound to first base?

32. Two forces acting on a body make an angle of 104° with each other. The magnitude of the first force is 390 pounds. If the resultant makes an angle of 47° with the first force, what is the magnitude of the resultant, to the *nearest tenth of a pound*?

33. A ship at sea heads directly toward a cliff on the shoreline. The accompanying diagram shows the top of the cliff, D, sighted from two locations, A and B, separated by distance S. If $\angle DAC = 30°$, $\angle DBC = 45°$, and $S = 30$ feet, what is the height of the cliff, to the *nearest foot*?

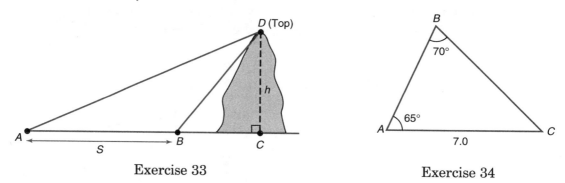

Exercise 33   Exercise 34

34. In the accompanying diagram of $\triangle ABC$, $\angle A = 65°$, $\angle B = 70°$, and the length of the side opposite vertex B is 7.0. Find the area of $\triangle ABC$ correct to the *nearest tenth*.

35. In $\triangle ABC$, the lengths of sides a, b, and c are in the ratio 4:6:8. Find the ratio of the cosine of $\angle C$ to the cosine of $\angle A$.

36. If the vertices of $\triangle ZAP$ are $Z(6,5)$, $A(-6,0)$, and $P(10,5)$, find the measure of the greatest angle of the triangle to the *nearest tenth of a degree*.

37. Carmen and Jamal are standing 5280 feet apart on a straight, horizontal road. They observe a hot-air balloon between them directly above the road. The angle of elevation from Carmen to the sighting of the balloon is 60°, and from Jamal it is 75°. Find the height of the balloon, correct to the *nearest foot*.

38. A parcel of land is shaped like an isosceles triangle in which the angle included between the congruent sides measures 53°10'. The area of the parcel of land is 1 acre. Find the perimeter of the parcel of land, correct to the *nearest foot*, if 1 acre is equal to 43,560 square feet.

39. Two forces of 130 and 150 pounds yield a resultant force of 170 pounds. Find, to the *nearest tenth of a degree*, the measure of the angle between the original two forces.

40. Two forces of 30 pounds and 40 pounds act on a body, forming an acute angle with each other. The angle between the resultant force and the 30-pound force is 35.2°. Find, to the *nearest tenth of a degree*, the angle between the two original forces.

41. A force of at least 3000 pounds is needed to pull a car out of a deep ditch. Two tow trucks try to pull a car out of the ditch. One truck applies a force of 1500 pounds, and the other truck applies a force of 2000 pounds. What is the greatest angle, rounded to the *nearest degree*, at which the forces can be applied in order for the tow trucks to be able to pull the car out of the ditch?

42. Two forces of 25 newtons and 38 newtons act on a body at an angle of 74.5°. Find, to the *nearest tenth* of a degree, the angle formed by the resultant and the larger of the two forces.

43. Circles P, Q, and R with radii 2, 4, and 9, respectively, are tangent to each other, as shown in the accompanying figure. Find, to the nearest tenth of a square unit, the area of the triangle formed by connecting their centers.

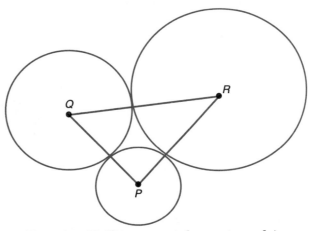

Exercise 43 (figures not drawn to scale)

STUDY UNIT IV:
POLAR
COORDINATES AND
CONIC SECTIONS

Chapter 13

POLAR COORDINATES AND PARAMETRIC EQUATIONS

OVERVIEW

Up to now, a curve has been defined by a single equation in two variables and then graphed using rectangular coordinates. In certain situations, however, a curve may not provide a complete picture of the real-world situation it is modeling. For example, when the vertical position of a moving object is plotted against the horizontal position of the object, we can tell the position of the object from the graph, but not the time at which the object was at that position. In such situations, it is helpful to define a curve using *two* equations by writing x as a function of a third variable, say t, and also writing y as a function of t. The equations $x = g(t)$ and $y = g(t)$ define a curve **parametrically** in the coordinate plane.

The **polar coordinate system** locates a point P in the plane using the ordered pair (r, θ), where the initial side of $\angle\theta$ is on the positive x-axis and its vertex, O, is fixed at the origin. The terminal side of $\angle\theta$ is the line segment OP, where $r = OP$, as shown in the accompanying figure.

Connections between polar coordinates and graphs of complex numbers lead to relationships that allow us to find the nth roots of a complex number.

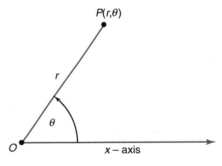

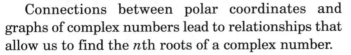

Lessons in Chapter 13

Parametric Equations

KEY IDEAS

For some situations, introducing a third variable when graphing a plane curve can provide additional insight into the process or function that the curve describes.

Defining a Curve Parametrically

Suppose that a particle is moving within the coordinate plane in such a way as to trace out the graph of $y = x^2 - 2x$. From this function we know that at, say, $x = 5$, $y = 15$. However, we do not know from this function *when* the particle was at $(5,15)$. Although the function allows us to determine the points *where* the particle has been, it does not tell us *when* the particle was at those locations.

It is helpful in this situation to introduce a third variable that represents time. By expressing both x and y as a function of time, t, as in $x = 2t + 1$ and $y = 4t^2 - 1$, we can obtain the position of the particle when the time is given. At time $t = 2$, for instance, the particle is located at

$$(2t + 1, 4t^2 - 1) = \left(2(2) + 1,\ 4(2)^2 - 1\right) = (5,15).$$

Over any specified interval of t-values, the pair of functions $x = 2t + 1$ and $y = 4t^2 - 1$ defines a plane curve consisting of the set of ordered pairs $(2t + 1, 4t^2 - 1)$.

Parametric Equations

Let f and g represent functions of the same variable, t, called the **parameter**, defined on some specified interval of t-values. The equations $x = f(t)$ and $y = g(t)$ are the **parametric equations** for the plane curve that consists of the set of ordered pairs $(f(t), g(t))$.

Graphing Parametric Equations

Most graphing calculators have a parametric mode that allows you to easily graph a pair of parametric equations. To graph $x = 2t + 1$ and $y = 4t^2 - 1$ for $-3 \le t \le 4$:

1. Change the mode to PARametric, and then set $X1_T = 2T + 1$ and $Y1_T = 4T \wedge 2 - 1$.
2. Select the viewing window, and set the minimum and maximum values of x, y, and t. Because $-3 \le t \le 4$, set Tmin $= -3$ and Tmax $= 4$. Also, choose a value for TSTEP. For example, letting TSTEP $= 0.1$ makes the calculator plot the points when $t = -3, -2.9, -2.8$, and so forth.

3. Display the graph. In Figure 13.1, the dimensions of the window are $[-10,10]\times[-15, 70]$. Using the TRACE feature of the calculator, you can determine graphically that, when $t = 2$, the particle is located at $(5,15)$, as shown in Figure 13.2.

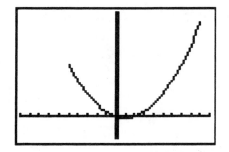   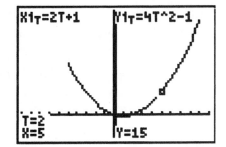

      Figure 13.1 Graph in parametric mode   **Figure 13.2** Graph with TRACE on

Eliminating the Parameter

To change from parametric to rectangular form, eliminate the parameter by solving for t in one of the parametric equations. Then use that equation to replace t in the other equation. The pair of equations $x = 2t + 1$ and $y = 4t^2 - 1$, where $-3 \le t \le 4$, can be changed to rectangular form by proceeding as follows:

1. Solve for t in the first equation: $t = \dfrac{x-1}{2}$.

2. Substitute $\dfrac{x-1}{2}$ for t in the second equation:

$$y = 4\left(\frac{x-1}{2}\right)^2 - 1$$
$$= \frac{4}{4}\left(x^2 - 2x + 1\right) - 1$$
$$= x^2 - 2x$$

3. Determine the domain and the range of $y = x^2 - 2x$. Notice that, by tracing along the parametric curve from $t = -3$ to $t = +4$, as shown in Figure 13.3, $a - c$, x increases from -5 (a) to 9 (c), while y varies from a low of -1 (b) to a high of 63 (c). Hence, the domain of $y = x^2 - 2x$ is restricted to $-5 \le x \le 9$, while the range is limited to $-1 \le y \le 63$.

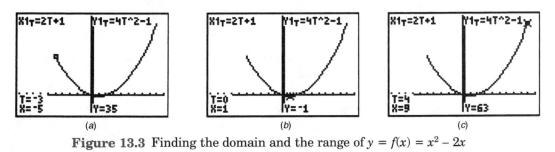

       (a)   (b)   (c)

      Figure 13.3 Finding the domain and the range of $y = f(x) = x^2 - 2x$

Changing from parametric to rectangular form can help you identify the type of curve that a set of parametric equations represents, as illustrated in Exercise 1.

Exercise 1   Finding the Rectangular Equation of a Curve Defined Parametrically

Find the rectangular equation of the plane curve defined parametrically by the equations $x = 4\cos t$ and $y = 3\sin t$, where $0 \le t \le 2\pi$.

Solution: Solve for $\cos t$ and $\sin t$. Then use the Pythagorean trigonometric identity.

$$\cos^2 t + \sin^2 t = 1$$

$$\left(\frac{x}{4}\right)^2 + \left(\frac{y}{3}\right)^2 = 1$$

$$\frac{x^2}{16} + \frac{y^2}{9} = 1 \leftarrow \text{Equation of an ellipse centered at the origin}$$

Exercise 2   Finding the Rectangular Equation of a Curve Defined Parametrically

a. Find the rectangular equation of the plane curve defined parametrically by the equations $x = \cos^2 t$ and $y = \sin^2 t$, where $0 \le t \le 2\pi$.

b. Describe the graph of the equation.

Solutions:

a. Because $x + y = \cos^2 t + \sin^2 t = 1$, the rectangular equation of the plane curve is **$x + y = 1$**.

b. Since $x = \cos^2 t \ge 0$ and $y = \sin^2 t \ge 0$, the graph consists of only that part of the line $x + y = 1$ in Quadrant I with endpoints $(0,1)$ and $(1,0)$.

Exercise 3   Finding the Parametric Equations of a Curve

Find the pair of parametric equations that represent the graph of $y = \frac{1}{2}x^2$ using the parameter $t = \sqrt{x} - 2$, where $x \ge 0$ and $0 \le y \le 18$.

Solution: When $t = \sqrt{x} - 2$, $x = (t+2)^2$, so

$$y = \frac{1}{2}(t+2)^2 = \frac{1}{2}(t^2 + 2t + 4) = \frac{1}{2}t^2 + t + 2.$$

- Because $t = \sqrt{x} - 2$ and $x \ge 0$, $t \ge -2$.

- It is also given that $0 \le y \le 18$, so $y = \frac{1}{2}(t+2)^2 \le 18$ and $(t+2)^2 \le 36$, implying that the maximum value of t is 4.

- Hence, the parametric equations that represent the graph are:

$$x(t) = (t+2)^2 \quad \text{and} \quad y(t) = \frac{1}{2}t^2 + t + 2,$$

where $-2 \le t \le 4$.

The Polar Coordinate System

KEY IDEAS

The **polar coordinate system** locates a point, P, in a plane relative to a fixed point, O, called the **pole**; a horizontal ray whose endpoint is the pole, called the **polar axis**; and the directed angle, θ, whose initial side is on the polar axis and whose vertex is on the pole. The directed distance, r, is marked off on the terminal side of $\angle\theta$ in such a way that $r = OP$, as shown in the accompanying figure. The polar coordinates of P are (r,θ).

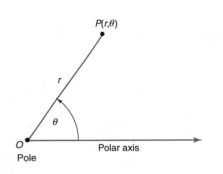

Locating Points in the Polar Coordinate System

When locating $P(r,\theta)$:

- If $\angle\theta$ is positive, the terminal side of the angle is rotated counterclockwise, and if $\angle\theta$ is negative, the terminal side of the angle is rotated clockwise, as illustrated in Figures 13.4 and 13.5.

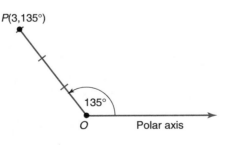

Figure 13.4 Graph of $P(3,135°)$

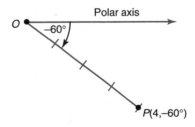
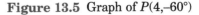

Figure 13.5 Graph of $P(4,-60°)$

- If r is positive, P is r units along the terminal side of $\angle\theta$, as illustrated in Figure 13.6. If r is negative, first locate $\left(\left|r\right|,\theta\right)$. Then extend the terminal side of $\angle\theta$ its own length in the opposite direction, as illustrated in Figure 13.7.

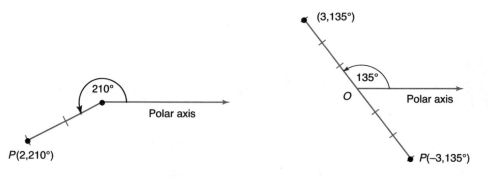

Figure 13.6 Graph of $P(2,210°)$ **Figure 13.7** Graph of $P(-3,135°)$

Multiple Representations in Polar Form

Unlike the case in the rectangular coordinate system, the representation of a point in the polar coordinate system is not unique. You can easily verify that the polar coordinates $(7,200°)$, $(7,-160°)$, $(7,560°)$, and $(-7,20°)$ all refer to the same point.

Coordinate System Conversions

The polar coordinate system can be superimposed on the rectangular coordinate system by aligning the pole with the origin and the polar axis with the positive x-axis, as shown in Figure 13.8. When the same point is represented in rectangular and polar coordinates, formulas that relate the two coordinate systems can be easily derived. For example, using the sine and cosine ratios in right triangle OAP gives:

$$\cos\theta = \frac{x}{r}, \text{ so } x = r\cos\theta,$$

and

$$\sin\theta = \frac{y}{r}, \text{ so } y = r\sin\theta.$$

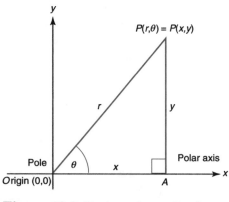

Figure 13.8 Rectangular and polar coordinate systems

> ## Polar-Rectangular Conversion Formulas
>
> Let point P have rectangular coordinates (x, y) and polar coordinates (r, θ).
>
> - To obtain (x, y) from (r, θ), use these relationships:
>
> $$x = r \cos \theta \quad \text{and} \quad y = r \sin \theta.$$
>
> - To obtain (r, θ) from (x, y), use these relationships:
>
> $$x^2 + y^2 = r^2 \quad \text{and} \quad \tan \theta = \frac{y}{x}.$$

Exercise 1   Converting from Polar to Rectangular Coordinates

Convert each point from polar to rectangular coordinates.

a. $(4, 150°)$ b. $(-2, 270°)$

Solutions:

a. To convert $(4, 150°)$ into rectangular coordinates (x, y), use the conversion formulas with $r = 4$ and $\theta = 150°$.

$$x = r \cos \theta = 4 \cos 150° = 4\left(-\cos 30°\right) = 4\left(-\frac{\sqrt{3}}{2}\right) = -2\sqrt{3}.$$

$$y = r \sin \theta = 4 \sin 150° = 4\left(\frac{1}{2}\right) = 2.$$

The equivalent rectangular coordinates are $\left(x, y\right) = \left(\mathbf{-2\sqrt{3}, 2}\right)$.

b. To convert $(-2, 270°)$ into rectangular coordinates (x, y), use the conversion formulas with $r = -2$ and $\theta = 270°$:

$$x = r \cos \theta = -2 \cos 270° = -2(0) = 0.$$

$$y = r \sin \theta = -2 \sin 270° = -2(-1) = 2.$$

The equivalent rectangular coordinates are $(x, y) = \mathbf{(0, 2)}$.

Exercise 2   Converting from Rectangular to Polar Coordinates

Convert each point from rectangular to polar coordinates.

a. $\left(3, \sqrt{3}\right)$ b. $\left(2, -2\right)$ c. $\left(-1, 0\right)$ d. $\left(-5, -12\right)$

Solutions:

a. To convert $\left(3, \sqrt{3}\right)$ into polar coordinates, find θ and r when $x = 3$ and $y = \sqrt{3}$.

- Find the reference angle:

$$\theta_{\text{Ref}} = \tan^{-1}\left(\left|\frac{\sqrt{3}}{3}\right|\right) = 30.$$

- Determine θ. Because x and y are both positive in Quadrant I, θ is a Quadrant I angle, so $\theta = \theta_{\text{Ref}} = 30°$.
- Find r:

$$r = \sqrt{x^2 + y^2} = \sqrt{3^2 + \left(\sqrt{3}\right)^2} = \sqrt{12} = 2\sqrt{3}.$$

The equivalent polar coordinates are $\left(r, \theta\right) = \left(\mathbf{2\sqrt{3}, 30°}\right)$.

b. To convert $(2, -2)$ into polar coordinates, find θ and r when $x = 2$ and $y = -2$.

- Find the reference angle:

$$\theta_{\text{Ref}} = \tan^{-1}\left(\left|\frac{-2}{2}\right|\right) = \tan^{-1}\left(1\right) = 45.$$

- Determine θ. Because x is negative and y is positive in Quadrant IV, θ is a Quadrant IV angle. Hence:

$$\theta = 360° - \theta_{\text{Ref}} = 360° - 45° = 315°.$$

- Find r:

$$r = \sqrt{x^2 + y^2} = \sqrt{2^2 + \left(-2\right)^2} = \sqrt{8} = 2\sqrt{2}.$$

The equivalent polar coordinates are $\left(r, \theta\right) = \left(\mathbf{2\sqrt{2}, 315°}\right)$.

c. To convert $(-1, 0)$ into polar coordinates, find θ and r when $x = -1$ and $y = 0$.

- Find the reference angle:

$$\theta_{\text{Ref}} = \tan^{-1}\left(\left|\frac{0}{-1}\right|\right) = \tan^{-1}\left(0\right) = 0°.$$

- Determine θ. Because θ is a quadrantal angle and x is negative while y is 0, the terminal side of the angle must coincide with the *negative* x-axis, so $\theta = 180°$.
- Find r:

$$r = \sqrt{x^2 + y^2} = \sqrt{\left(-1\right)^2 + 0^2} = 1.$$

The equivalent polar coordinates are $(r, \theta) = (\mathbf{1, 180°})$.

d. To convert (–5,–12) into polar coordinates, find θ and r when $x = –5$ and $y = –12$.

- Find the reference angle:

$$\theta_{\text{Ref}} = \tan^{-1}\left(\left|\frac{-12}{-5}\right|\right) = \tan^{-1}(2.4) \approx 67.4°.$$

- Determine θ. Because x and y are both negative in Quadrant III, θ is a Quadrant III angle. Hence:

$$\theta = 180° + \theta_{\text{Ref}} = 180° = 67.4° – 247.4°.$$

- Find r:

$$r = \sqrt{5^2 + 12^2} = \sqrt{25 + 144} = \sqrt{169} = 13.$$

The equivalent polar coordinates are (r, θ) = **(13, 247.4°)**.

Converting Between Polar and Rectangular Equations

Equations, like individual points, can be converted between the two coordinate systems.

Exercise 3   **Converting an Equations from Polar to Rectangular Form**

Write $r = 10\sin\theta$ in rectangular form.

Solution: If $r = 10\sin\theta$, then $r = 10\left(\dfrac{y}{r}\right)$, so $r^2 = 10y$. Substituting $x^2 + y^2$ for r^2 gives $x^2 + y^2 = 10y$ or, equivalently, $x^2 + y^2 – 10y = 0$. Complete the square for y:

$$x^2 + y^2 – 10y + 25 = +25$$
$$x^2 + \left(y – 5\right)^2 = 25$$

Thus, the graph of $r = 10\sin\theta$ is the circle with center at (0, 5) and radius 5.

Exercise 4   **Converting an Equation from Rectangular to Polar Form**

Convert $xy = 6$ to polar form.

Solution: If $xy = 6$, then $(r\cos\theta)(r\sin\theta) = 6$, so

$$r^2 = \frac{6}{\sin\theta\cos\theta} = \frac{6}{\dfrac{1}{2}\sin 2\theta} = 12\csc 2\theta.$$

The Polar Form of a Complex Number

KEY IDEAS

When complex numbers of the form $a + bi$ are graphed, the rectangular coordinates (a,b) are plotted where a is measured along the horizontal real axis and b is measured along the vertical imaginary axis. The set of all such points forms the **complex plane**.

In Lesson 13-2, we saw that representing the same point in polar and rectangular coordinate systems produced formulas that allowed us to convert from one system to the other. In a similar way, superimposing the complex plane on the plane created by the polar coordinate system produces formulas that allow us to convert a complex number into polar form.

Polar Representation of a Complex Number

Let point (a,b) in the complex plane correspond to the complex number $a + bi$, and let (r,θ) represent the polar coordinates of the same point, as illustrated in Figure 13.9. Using the familiar relationships $a = r \cos \theta$ and $b = r \sin \theta$ that derive from the right triangle in this figure, we can write the complex number $a + bi$ in polar form as

$$a + bi = (r \cos \theta) + (r \sin \theta)i = r(\cos \theta + i \sin \theta).$$

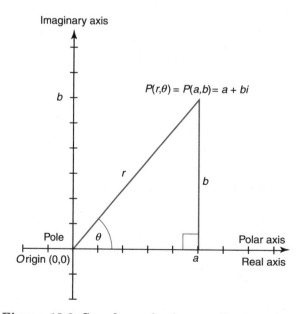

Figure 13.9 Complex and polar coordinate systems

Polar Form of a Complex Number

The **polar form** of the complex number $z = a + bi$ is

$$z = r(\cos\theta + i\sin\theta),$$

where $r^2 = a^2 + b^2$ and $\tan\theta = \dfrac{b}{a}$. The r-value is called the **modulus** of z, and θ is termed an **argument** of z.

The polar form of the complex number, $r(\cos\theta + i\sin\theta)$, is sometimes referred to as the **trigonometric form** or the **"rcis" form** of a complex number. For example, the shorthand notation $3cis\,60°$ means $3(\cos 60° + i\sin 60°)$.

Exercise 1   Converting Complex Numbers from Rectangular to Polar Form

Convert each complex number into polar form.

a. $1 - \sqrt{3}\,i$ b. $5 + 0i$ c. $0 - 4i$

Solutions:

a. To write $z = 1 - \sqrt{3}\,i$ in polar form, find θ and r when $a = 1$ and $b = -\sqrt{3}$.

- Find the reference angle:

$$\theta_{\text{Ref}} = \tan^{-1}\left(\left|\frac{-\sqrt{3}}{1}\right|\right) = \tan^{-1}\left(\sqrt{3}\right) = 60°.$$

- Determine θ. Because a is positive and b is negative in Quadrant IV, θ is a Quadrant IV angle. Hence:

$$\theta = 360° - \theta_{\text{Ref}} = 360° - 60° = 300°.$$

- Find r:

$$r = \sqrt{1^2 + y^2} = \sqrt{1^2 + \left(\sqrt{3}\right)^2} = \sqrt{4} = 2.$$

The polar form of the complex number is $z = r(\cos\theta + i\sin\theta)$, where $r = 2$ and $\theta = 300°$:

$$z = 2(\cos 300° + i\sin 300°).$$

b. To write $z = 5 + 0i$ in polar form, find θ and r when $a = 5$ and $b = 0$.

- Find the reference angle:

$$\theta_{\text{Ref}} = \tan^{-1}\left(\left|\frac{0}{2}\right|\right) = \tan^{-1}(0) = 0°.$$

- Determine θ. Because a is positive, the terminal side of θ coincides with the positive x-axis. Hence:

$$\theta = \theta_{\text{Ref}} = 0°.$$

- Find r:

$$r = \sqrt{5^2 + 0^2} = 5.$$

Because $r = 5$ and $\theta = 0°$, the polar form of the complex number is

$$z = 5(\cos 0° + i \sin 0°)$$

c. To write $z = 0 - 4i$ in polar form, find θ and r when $a = 0$ and $b = -4$.

- Find the reference angle:

$$\theta_{\text{Ref}} = \tan^{-1}\left(\left|\frac{-4}{0}\right|\right) \quad \leftarrow \text{Tangent is undefined}$$

- Determine θ. You know that $\tan \theta$ is undefined for quadrantal angles of $90°$ and $270°$. Because b is negative, the terminal side of $\angle\theta$ coincides with the negative y-axis. Hence:

$$\theta = 270°.$$

- Find r:

$$r = \sqrt{0^2 + (-4)^2} = 4.$$

Because $r = 4$ and $\theta = 270°$, the polar form of the complex number is

$$z = 4(\cos 270° + i\sin 270°).$$

Multiplying and Dividing Complex Numbers in Polar Form

To multiply (divide) two complex numbers in polar form, multiply (divide) their moduli and add (subtract) their amplitudes.

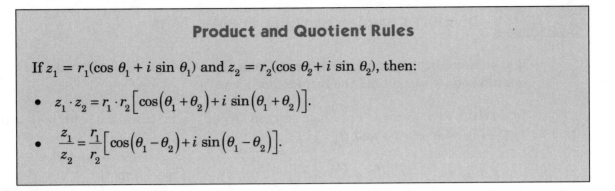

Product and Quotient Rules

If $z_1 = r_1(\cos\theta_1 + i\sin\theta_1)$ and $z_2 = r_2(\cos\theta_2 + i\sin\theta_2)$, then:

- $z_1 \cdot z_2 = r_1 \cdot r_2\left[\cos\left(\theta_1 + \theta_2\right) + i\sin\left(\theta_1 + \theta_2\right)\right]$.

- $\dfrac{z_1}{z_2} = \dfrac{r_1}{r_2}\left[\cos\left(\theta_1 - \theta_2\right) + i\sin\left(\theta_1 - \theta_2\right)\right]$.

Exercise 2 Multiplying Complex Numbers in Polar Form

If $z_1 = 3(\cos 130° + i\sin 130°)$ and $z_2 = 4(\cos 95° + i\sin 95°)$, find $z_1 \cdot z_2$. Express the product in both the polar and the rectangular form.

Solution: To express the product in polar form, use the product rule with $r_1 = 3$, $\theta_1 = 130°$, $r_2 = 4$, and $\theta_2 = 95°$:

$$z_1 \cdot z_2 = 3 \cdot 4\left[\cos\left(130° + 95°\right) + i\sin\left(130° + 95°\right)\right]$$

$$= 12\left(\cos 225° + i\sin 225°\right)$$

To express the product in rectangular form, evaluate the trigonometric functions:

$$z_1 \cdot z_2 = 12\left(-\cos 45° + i\left(-\sin 45°\right)\right)$$

$$= 12\left(\frac{-\sqrt{2}}{2} + \frac{-\sqrt{2}}{2}i\right)$$

$$= -6\sqrt{2} - 6\sqrt{2}\,i$$

Exercise 3 Multiplying Complex Numbers in Polar Form

If $z_1 = 4\left(\cos\dfrac{5\pi}{12} + i\sin\dfrac{5\pi}{12}\right)$ and $z_2 = \dfrac{5}{2}\left(\cos\dfrac{2\pi}{3} + i\sin\dfrac{2\pi}{3}\right)$, find $z_1 \cdot z_2$ in polar form.

Solution: Use the product rule with $r_1 = 4$, $\theta_1 = \dfrac{5\pi}{12}$, $r_2 = \dfrac{5}{2}$, and $\theta_2 = \dfrac{2\pi}{3}$:

$$z_1 \cdot z_2 = \left(4 \cdot \frac{5}{2}\right)\left[\cos\left(\frac{5\pi}{12} + \frac{2\pi}{3}\right) + i\sin\left(\frac{5\pi}{12} + \frac{2\pi}{3}\right)\right]$$

$$= 10\left(\cos\left(\frac{13\pi}{12}\right) + i\sin\left(\frac{13\pi}{12}\right)\right)$$

Dividing Complex Numbers in Polar Form

If $z_1 = 24(\cos 320° + i \sin 320°)$ and $z_2 = 8(\cos 170° + i \sin 170°)$, find $\dfrac{z_1}{z_2}$. Express the quotient in both the polar and the rectangular form.

Solution: To express the quotient in polar form, use the quotient rule with $r_1 = 24, \theta_1 = 320°, r_2 = 8,$ and $\theta_2 = 170°$:

$$\frac{z_1}{z_2} = \frac{24}{8}\left[\cos\left(320° - 170°\right) + i \sin\left(320° - 170°\right)\right]$$

$$= 3\left(\cos 150° + i \sin 150°\right)$$

To express the quotient in rectangular form, evaluate the trigonometric functions:

$$\frac{z_1}{z_2} = 3\left(-\cos 30° + i \sin 30°\right) = -\frac{3\sqrt{3}}{2} + \frac{1}{2}i.$$

Lesson 13-4

Powers and Roots of Complex Numbers

KEY IDEAS

By repeatedly applying the multiplication rule for complex numbers in polar form, we see that

$$\left[r\left(\cos\theta + i\sin\theta\right)\right]^n = r^n\left(\cos n\cdot\theta + i \sin n\cdot\theta\right).$$

This result is known as **DeMoivre's theorem**, named after the French mathematician Abraham DeMoivre (1667–1754).

Powers of Complex Numbers

The power of a complex number can be calculated by converting the complex number into polar form and then applying DeMoivre's theorem.

Evaluating the Power of a Complex Number

Evaluate $(1 + i)^{12}$, and write the result in rectangular form.

Solution: Convert $1 + i$ to polar form. Then apply DeMoivre's theorem.

- Determine θ, where $a = 1$ and $b = 1$.

$$\theta_{\text{Ref}} = \tan^{-1}\left(\left|\frac{1}{1}\right|\right) = \tan^{-1}(1) = 45°.$$

Because both a and b are positive, θ is a Quadrant I angle, so

$$\theta = \theta_{\text{Ref}} = 45°.$$

- Find r:

$$r = \sqrt{1^2 + 1^2} = \sqrt{2}.$$

- Apply DeMoivre's theorem. If $z = \sqrt{2}\left(\cos 45° + i \sin 45°\right)$, then z^{12} is the complex number:

$$\left(\sqrt{2}\right)^{12}\left[\cos\left(12 \cdot 45°\right) + i \sin\left(12 \cdot 45°\right)\right] = 64\left[\cos 540° + i \sin 540°\right]$$
$$= 64\left[\cos 180° + i \sin 180°\right]$$
$$= 64\left[-1 + i \cdot 0\right]$$
$$= \mathbf{-64}$$

Roots of Complex Numbers

DeMoivre's theorem holds also for rational values of n. This fact allow us to find roots of complex numbers. To find an nth root of $z = a + bi$, replace n in DeMoivre's theorem with $\dfrac{1}{n}$:

$$z^{\frac{1}{n}} = \sqrt[n]{r}\left[\cos\left(\frac{\theta}{n}\right) + i \sin\left(\frac{\theta}{n}\right)\right].$$

A complex number, however, has a total of n distinct nth roots: *two* square roots, *three* cube roots, and so on. By adding successive multiples of $360°$ to θ, each of the nth roots of a complex number can be calculated.

Formula for the *n*th Roots of a Complex Number
If $z = r(\cos \theta + i \sin \theta)$, then the n distinct nth roots of z are given by the expression $$\sqrt[n]{r}\left[\cos\left(\frac{\theta + 360° \cdot k}{n}\right) + i \sin\left(\frac{\theta + 360° \cdot k}{n}\right)\right],$$ where $k = 0$, $k = 1$, $k = 2$, ..., and $k = n - 1$. For example, to find the three cube roots of z, evaluate this formula three times, using $k = 0$, $k = 1$, and $k = 2$.

Exercise 2   Finding the Cube Roots of a Complex Number

Determine the cube roots of $z = 125(\cos 210° + i \sin 210°)$.

Solution: Let $n = 3$, $r = 125$, and $\theta = 210°$. The three cube roots of z are the complex numbers

$$\sqrt[3]{125}\left[\cos\left(\frac{210° + 360° \cdot k}{3}\right) + i\ \sin\left(\frac{210° + 360° \cdot k}{3}\right)\right]$$

for $k = 0$, $k = 1$, and $k = 2$.

- Let $k = 0$:

$$z_1 = \sqrt[3]{125}\left[\cos\left(\frac{210° + 360° \cdot 0}{3}\right) + i\ \sin\left(\frac{210° + 360° \cdot 0}{3}\right)\right] = 5\left(\cos 70° + i \sin 70°\right).$$

- Let $k = 1$:

$$z_2 = \sqrt[3]{125}\left[\cos\left(\frac{210° + 360° \cdot 1}{3}\right) + i\ \sin\left(\frac{210° + 360° \cdot 1}{3}\right)\right] = 5\left(\cos 190° + i \sin 190°\right).$$

- Let $k = 2$:

$$z_3 = \sqrt[3]{125}\left[\cos\left(\frac{210° + 360° \cdot 2}{3}\right) + i\ \sin\left(\frac{210° + 360° \cdot 2}{3}\right)\right] = 5\left(\cos 310° + i \sin 310°\right).$$

In rectangular form, with rounding to the *nearest hundredth*, the three cube roots are **1.72 + 4.70i, –4.92 – 0.87i**, and **3.21 – 3.83i**.

Exercise 3   Finding the Fourth Roots of a Complex Number

Determine the fourth roots of $z = -8 - 8\sqrt{3}\,i$ in rectangular form, with rounding to the *nearest hundredth*.

Solution: First convert $-8 - 8\sqrt{3}\,i$ to polar form, where $a = -8$ and $b = -8\sqrt{3}$. The reference angle is

$$\theta_{\text{Ref}} = \tan^{-1}\left(\left|\frac{-8\sqrt{3}}{-8}\right|\right) = \tan^{-1}\left(\sqrt{3}\right) = 60°.$$

Because both a and b are negative in Quadrant III, θ is a Quadrant III angle, so

$$\theta = 180° = \theta_{\text{Ref}} \ 180° + 60° = 240°.$$

Now find r:

$$r = \sqrt{a^2 + b^2} = \sqrt{\left(-8\right)^2 + \left(-8\sqrt{3}\right)^2} = \sqrt{256} = 16.$$

The polar form of the complex number is $z = 16(\cos 240° + i \sin 240°)$. The fourth roots of z are the complex numbers

$$\sqrt[4]{16}\left[\cos\left(\frac{240° + 360° \cdot k}{4}\right) + i \ \sin\left(\frac{240° + 360° \cdot k}{4}\right)\right]$$

for $k = 0$, $k = 1$, $k = 2$, and $k = 3$.

- Let $k = 0$:

$$z_1 = \sqrt[4]{16}\left[\cos\left(\frac{240° + 360° \cdot 0}{4}\right) + i \ \sin\left(\frac{240° + 360° \cdot 0}{4}\right)\right] = 2\left(\cos 60° + i \sin 60°\right).$$

- Let $k = 1$:

$$z_2 = \sqrt[4]{16}\left[\cos\left(\frac{240° + 360° \cdot 1}{4}\right) + i \ \sin\left(\frac{240° + 360° \cdot 1}{4}\right)\right] = 2\left(\cos 150° + i \sin 150°\right).$$

- Let $k = 2$:

$$z_3 = \sqrt[4]{16}\left[\cos\frac{240° + 360° \cdot 2}{4} + i \ \sin\frac{240° + 360° \cdot 2}{4}\right] = 2\left(\cos 240° + i \sin 240°\right).$$

- Let $k = 3$:

$$z_4 = \sqrt[4]{16}\left[\cos\left(\frac{240° + 360° \cdot 3}{4}\right) + i \ \sin\left(\frac{240° + 360° \cdot 3}{4}\right)\right] = 2\left(\cos 330° + i \sin 330°\right).$$

In rectangular form, with rounding to the *nearest hundredth*, the four fourth roots are **$1 + 1.73i$, $-1.73 + i$, $-1 - 1.73i$**, and **$1.73 - i$**.

Plotting the *n*th Roots of a Number

When the roots obtained in Exercise 2 are plotted in the complex plane, as shown in Figure 13.10, the three cube roots are equally spaced around a circle whose center is at the origin and whose radius is 5. The four fourth roots in Exercise 3 are equally spaced around a circle whose center is at the origin and whose radius is 2, as illustrated in Figure 13.11. In each case, the roots are spaced $\dfrac{360°}{n}$ apart on a circle centered at the origin whose radius is the *n*th root of the *r*-value.

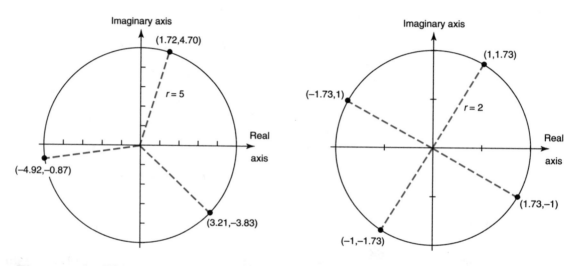

Figure 13.10 Plotting cube roots ($n = 3$)   **Figure 13.11** Plotting fourth roots ($n - 4$)

TIPS

- The *n*th roots of the complex number $r(\cos \theta + i \sin \theta)$ are spaced $\dfrac{360°}{n}$ apart on a circle centered at the origin with radius $\sqrt[n]{r}$. If $n > 2$, the plotted points that represent the *n*th roots of a nonzero complex number are the vertices of a regular *n*-sided polygon centered at the origin.

- The roots of a complex number begin repeating for values of k greater than or equal to n. A root is repeated on a graph when it is a 360° rotation of another root, so that the two roots coincide. For example, when $k = 0$, the angle in the *n*th roots formula is $\dfrac{\theta}{n}$. When $k = n$, the angle is

$$\frac{\theta + 360° \cdot n}{n} = \frac{\theta}{n} + 360°.$$

Hence, the root obtained when $k = 0$ is repeated when $k = n$.

Chapter 13	CHECKUP EXERCISES

1—4. In each case, perform the indicated operation, and express the result in rectangular form.

1. $2\left(\cos 60° + i\sin 60°\right) \cdot 3\left(\cos 30° + i\sin 30°\right)$

2. $0.8(\cos 160° + i\sin 160°) \div 2.4(\cos 10° + i\sin 10°)$

3. $1.5\left(\cos\dfrac{3\pi}{8} + i\sin\dfrac{3\pi}{8}\right) \cdot 4\left(\cos\dfrac{7\pi}{8} + i\sin\dfrac{7\pi}{8}\right)$

4. $\dfrac{12\left(\cos 175° + i\sin 175°\right)}{3\left(\cos 235° + i\sin 235°\right)}$

5. Evaluate the following expression, using the product and quotient rules. Write the result in rectangular form.

$$\left[\dfrac{42\left(\cos\dfrac{2\pi}{3} + i\sin\dfrac{2\pi}{3}\right)}{28\left(\cos\dfrac{\pi}{6} + i\sin\dfrac{\pi}{6}\right)}\right]^2$$

6—9. Write each polar equation in rectangular form.

6. $r = 6\cos\theta$ 7. $r = -4\sin\theta$ 8. $r = 2\sec\theta$ 9. $r = \sin\theta - 2\cos\theta$

10—13. Convert each rectangular equation into polar form.

10. $(x - 4)^2 + y^2 = 16$ 12. $(x + 2)^2 + (y - 5)^2 = 29$

11. $\dfrac{x^2}{3} + \dfrac{y^2}{4} = 1$ 13. $xy - 2x = 0$

14—21. Graph each pair of parametric equations. Then find the rectangular equation for each curve.

14. $x = t - 1, y = t^2 - 1; -2 \le t \le 4$ 18. $x = 2 + 4\cos t, y = 4\sin t - 1; 0 \le t \le \pi$

15. $x = 6t^2, y = 3t; -2 \le t \le 2$ 19. $x = 3\cos t, y = 2\sin t; \dfrac{\pi}{2} \le t \le \dfrac{3\pi}{2}$

16. $x = \sin t, y = 1 - \cos^2 t; 0 \le t \le \pi$ 20. $x = 2\sec\theta - 3, y = 1 + 3\tan\theta$

17. $x = \ln\sqrt{t}, y = \dfrac{1}{t}; e \le t < \infty$ 21. $y = \cos 2t, y = \sin 2t; -\pi \le t \le \pi$

22. Find a set of parametric equations that represents the graph of $y = \sqrt{x+4}$, using the parameter $t = 1 + \frac{1}{4}x$.

23. The path of a projectile launched at a height of h feet above the horizontal ground at an angle, θ, with the horizontal can be modeled by the parametric equations

$$x = \left(v_0 \cos\theta\right)t \quad \text{and} \quad y = -16t^2 + \left(v_0 \sin\theta\right)t + h,$$

where t is measured in seconds and v_0 is the initial velocity of the projectile in feet per second. The centerfield wall in a ballpark is 14 feet in height and 420 feet from home plate. When a baseball player swings his bat at home plate, the bat makes contact with the ball 3 feet above the ground at an angle, θ, with the horizontal. Write a set of parametric equations that describes the path of the ball when it leaves the bat at a speed of 140 miles per hour.

24. Using your answer from Exercise 23, determine the minimum value of $\angle\theta$, correct to the *nearest tenth of a degree*, for which the ball will go over the centerfield wall.

25. *a.* Show that the parametric equations

$$x = x_A + t\left(x_B - x_A\right) \quad \text{and} \quad y = y_A + t\left(y_B - y_A\right)$$

represent the oblique line that contains points (x_A, y_A) and (x_B, y_B).

b. Find the rectangular form of the equation of a line represented by the parametric equations

$$x = 1 + 2t \quad \text{and} \quad y = 2 - t.$$

26 and 27. Evaluate each of the indicated powers, and express the result in polar form.

26. $\left[2\left(\cos\dfrac{2\pi}{5} + i\sin\dfrac{2\pi}{5}\right)\right]^8$

27. $\left[3\left(\cos\dfrac{4\pi}{9} + i\sin\dfrac{4\pi}{9}\right)\right]^5$

28. The radar screen of an air traffic controller shows two airplanes flying at the same altitude with polar coordinates of $(19.0 \text{ mi}, 103°)$ and $(7.0 \text{ mi}, 28°)$. At this moment, what is the distance between the planes to the *nearest tenth of a mile*?

29–34. Evaluate each of the indicated powers, and express the result in rectangular form.

29. $\left[3\left(\cos\dfrac{3\pi}{2} + i\sin\dfrac{3\pi}{2}\right)\right]^5$

31. $\left(-1 - i\right)^{12}$

33. $\left(\sqrt{2} - \sqrt{2}\,i\right)^{10}$

30. $\left[\dfrac{1}{2}\left(\cos\pi + i\sin\pi\right)\right]^6$

32. $\left(-1 + \sqrt{3}\,i\right)^5$

34. $\left(-\sqrt{3} - i\right)^6$

35—40. Find the nth roots of the complex number z for the specified value of n.

35. $z = 4\sqrt{3} - 4i;\ n = 3$

38. $z = -16 + 16i\sqrt{3};\ n = 5$

36. $z = -6 - 6i;\ n = 2$

39. $z = -64i;\ n = 6$

37. $z = -81;\ n = 4$

40. $z = 3 + 4i;\ n = 3$

41 and 42. Find the nth roots of the complex number z for the specified value of n. Express the roots in rectangular form, rounding to the *nearest hundredth*.

41. $z = 243\left(\cos 45° + i \sin 45°\right);\ n = 5$

42. $z = 6 + 2\sqrt{7}\,i;\ n = 3$

43. *a.* Express the cube roots of $z = -8i$ in rectangular form.

   b. Plot these roots in the complex plane, and determine the perimeter of the triangle formed by connecting the plotted points with line segments.

44 and 45. In each case, find the indicated roots and then plot them.

44. The cube roots of $z = -27$ expressed in rectangular form.

45. The fourth roots of unity ($z = 1$) expressed in rectangular form.

Chapter 14

CONIC SECTIONS AND THEIR EQUATIONS

OVERVIEW

ircles, ellipses, parabolas, and hyperbolas are curves in the plane, called **conic sections**, that can be formed by cutting a double cone with a plane at different angles, as shown in the figures below.

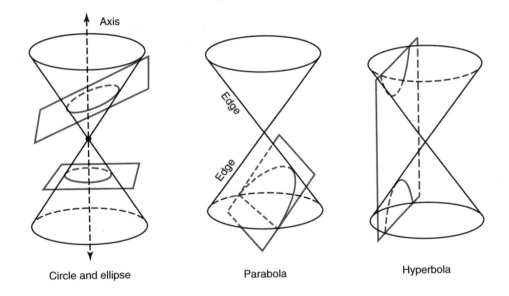

| Circle and ellipse | Parabola | Hyperbola |

- A *circle* is formed when a plane intersects one cone and is perpendicular to the axis; if the cutting plane is not perpendicular to the axis, an *ellipse* is created.
- A *parabola* is formed when a plane intersects one cone and is parallel to the edge of the cone.
- A *hyperbola* is formed when a plane intersects both cones.

Conic sections arise in diverse areas of applied mathematics. They can be used to help describe projectile and planetary motion, orbits of satellites and spaceships, and the shapes and properties of light-reflecting surfaces.

Equations for each of the conic sections can be developed from their geometric definitions. This chapter systematically investigates the relationships between the equations of the conic sections and their graphs, culminating in a unified description of a conic section in both rectangular and polar coordinate systems.

Lesson 14-1   The Parabola

KEY IDEAS

Parabolas were encountered when studying the graphs of quadratic functions. A more useful form of an equation of a parabola can be developed from the geometric definition of a parabola.

Geometric Definition of a Parabola

Figure 14.1 shows a parabola formed by locating all points that are the same distance from point F as they are from line ℓ.

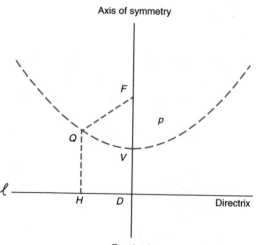

Figure 14.1 Finding all Points Q such that $QF = QH$

Geometric Definition of a Parabola

A **parabola** is the set of all points in a plane that are the same distance from both a fixed point and a given line. The fixed point is the **focus,** and the given line is the **directrix**. The line that contains the focus is the **focal axis**.

Some facts you should know about parabolas are illustrated in Figure 14.1:

- The *focal axis* is a line of symmetry and is perpendicular to the directrix. The *vertex* is the point at which the parabola intersects the focal axis.
- The letter p represents the directed distance from the vertex to the focus.
- The vertex is midway between the focus and the point at which the focal axis intersects the directrix.

The focal axis may be either horizontal or vertical. In either case, the focus always lies in the interior of the parabola, as illustrated in Table 14.1. The sign of p indicates how the parabola opens:

- If the focal axis is a vertical line, the parabola opens *up* when $p > 0$, and opens *down* if $p < 0$.
- If the focal axis is a horizontal line, the parabola opens *to the right* when $p > 0$, and opens *to the left* if $p < 0$.

Table 14.1 Vertical and Horizontal Parabolas

Focal Axis	$p > 0$	$p < 0$
Vertical	Focus is above the vertex. Parabola opens up.	Focus is below the vertex. Parabola opens down.
Horizontal	Focus is to the right of the vertex Parabola opens to the right.	Focus is to the left of the vertex Parabola opens to the left.

Deriving Parabola Equations: Vertex at (0,0)

To derive an equation for a parabola, place the vertex, V, of the parabola at $(0,0)$, as shown in Figure 14.2. The focus is p units above the vertex at $F(0, p)$, and the directrix is p units below the vertex at $y = -p$.

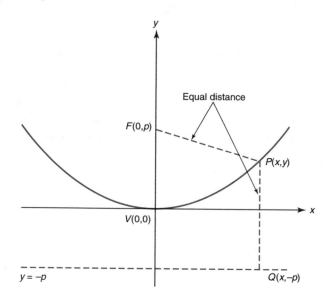

Figure 14.2 Coordinate derivation of an equation for a vertical parabola

According to the definition of a parabola, the distance from any point, $P(x,y)$, on the parabola to the focus, $F(0,p)$, must be equal to the distance from the same point to the directrix, $y = -p$.

- Express this fact using the distance formula:

$$PF = PQ$$
$$\sqrt{(x-0)^2 + (y-p)^2} = \sqrt{(x-x)^2 + (y-(-p))^2}$$
$$\sqrt{x^2 + (y-p)^2} = \sqrt{(y+p)^2}$$

- Square the binomials underneath the radical signs, and then square both sides of the equation to eliminate the radicals:

$$x^2 + y^2 - 2py + p^2 = y^2 + 2py + p^2.$$

- Simplify and collect the y-terms on the right side of the equation:

$$x^2 + \cancel{y^2} - 2py + \cancel{p^2} = \cancel{y^2} + 2py + \cancel{p^2}$$

Equation of a vertical parabola: $x^2 = 4py$

Interchange the roles of x and y to obtain the equation of a horizontal parabola whose vertex is at (0,0):

Equation of a horizontal parabola: $y^2 = 4px$

Deriving Parabola Equations: Vertex at (h,k)

Unfortunately, the vertex of a parabola may not always be on the origin. To find the standard equation of a parabola whose vertex has the general coordinates (h,k), shift $x^2 = 4py$ and $y^2 = 4px$ by $|h|$ units horizontally and $|k|$ units vertically. In other words, replace x with $x - h$ and y with $y - k$ in each equation.

Parabola Equations in Standard Form		
Orientation	Vertex: (0,0)	Vertex: (h,k)
Vertical	$x^2 = 4py$	$(x-h)^2 = 4p(y-k)$
Horizontal	$y^2 = 4px$	$(y-k)^2 = 4p(x-h)$

TIP

To use a graphing calculator to graph an equation of a horizontal parabola such as $(y - 1)^2 = x$, solve for y:

$$y = \pm\sqrt{x} + 1.$$

Then graph the two equations, $Y_1 = \sqrt{x} + 1$ and $Y_2 = -\sqrt{x} + 1$, that comprise the horizontal parabola. Here is the graph of $(y - 1)^2 = x$ in a basic Decimal window:

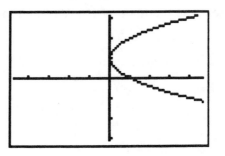

Table 14.2 summarizes the key properties of vertical and horizontal parabolas that you need to know.

Table 14.2 Key Properties of Vertical and Horizontal Parabolas

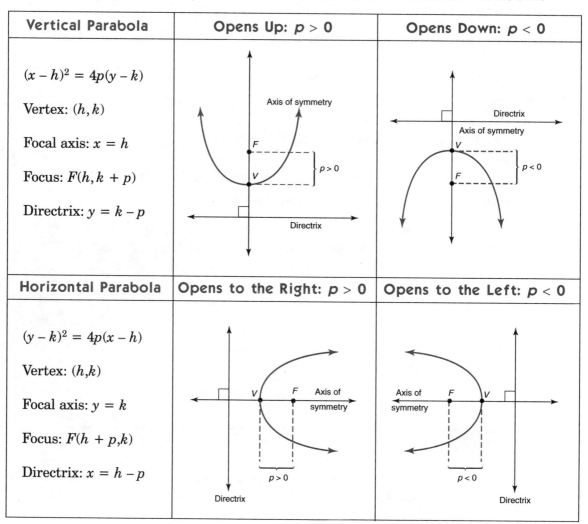

Vertical Parabola	Opens Up: $p > 0$	Opens Down: $p < 0$
$(x - h)^2 = 4p(y - k)$ Vertex: (h, k) Focal axis: $x = h$ Focus: $F(h, k + p)$ Directrix: $y = k - p$		
Horizontal Parabola	**Opens to the Right: $p > 0$**	**Opens to the Left: $p < 0$**
$(y - k)^2 = 4p(x - h)$ Vertex: (h, k) Focal axis: $y = k$ Focus: $F(h + p, k)$ Directrix: $x = h - p$		

Interpreting the Parabola Equation

By inspecting a parabola equation written in standard form, you can easily tell the coordinates of the vertex, the distance of the focus from the vertex, and the orientation of the parabola.

EXAMPLE: If $(x + 1)^2 = -8(y - 4)$, then

$$\left(x - \underbrace{(-1)}_{h}\right)^2 = -8\left(y - \underbrace{4}_{k}\right).$$

Since the equation contains an x^2–term, the parabola has a vertical focal axis and the vertex is at $(-1, 4)$. Because $4p = -8$, $p = -2$. Since p is *negative*, the parabola opens *down* and the focus is located 2 units *below* the vertex at $(-1, 4 - 2) = (-1, 2)$.

EXAMPLE: If $\left(y-2\right)^2 = -\dfrac{1}{2}\left(x+5\right)$, the parabola is a horizontal since the equation contains a y^2–term. The vertex is at (–5, 2). Because $4p=-\dfrac{1}{2}$, $p=-\dfrac{1}{8}$. Since p is *negative*, the parabola opens to the *left* with the focus located $\dfrac{1}{8}$ unit to the *left* of the vertex at $\left(-5\dfrac{1}{8},2\right)$.

Exercise 1   Finding a Parabola Equation, Given Its Vertex and Focus

Write the standard form of the equation of a parabola whose vertex is at $V(0,0)$ and whose focus is at $F(0,3)$, as shown in the accompanying figure. Determine the equations of the axis of symmetry and the directrix.

Solutions: Since the vertex and the focus have the same x-coordinate but different y-coordinates, they are located on the same vertical line. The vertex is at the origin, so use the form $x^2 = 4py$ of the parabola equation.

Since the focus lies 3 units above the vertex, $p = +3$. Hence, the equation of the parabola is $x^2 = 12y$.

The axis of symmetry is the y-axis, so its equation is $x = 0$. Since the axis of symmetry is a vertical line, the directrix is a horizontal line on the opposite side of the vertex and at the same distance from the vertex as the focus. Hence, the equation of the directrix is $y = -3$.

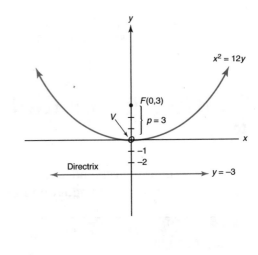

Exercise 2   Finding a Parabola Equation, Given Its Focus and Directrix

Find the standard form of the equation of a parabola with focus $F(-3,3)$ and directrix $x = 1$.

Solution: Since the directrix is the vertical line $x = 1$, the focal axis is the horizontal line $y = 3$ that contains the focus, $F(-3,3)$, as shown in the accompanying figure.

- The directrix intersects the focal axis at (1, 3). Since the vertex is midway between this point and the focus, the coordinates of the vertex are

$$V\left(\frac{-3+1}{2}, \frac{3+3}{2}\right) = V\left(-1, 3\right).$$

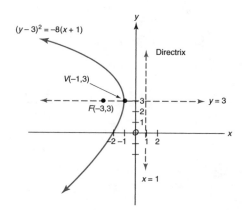

- The focus lies to the left of the vertex and is a distance of 2 units from it, so $p = -2$.

- The equation of this horizontal hyperbola is $(y - k)^2 = 4\,p(x - h)$, where $h = -1$, $k = 3$, and $p = -2$:

$$\left(y - 3\right)^2 = 4\left(-2\right)\left(x - \left(-1\right)\right)$$
$$\left(y - 3\right)^2 = -8\left(x + 1\right)$$

Writing a Parabola Equation in Standard Form

When a parabola equation is given in the form $y = ax^2 + bx + c$ or $x = ay^2 + by + c$, you can rewrite the equation in standard form by completing the square.

Exercise 3   Finding the Focus and Directrix, Given the Parabola Equation

Determine the coordinates of the vertex and of the focus of $y = \dfrac{1}{2}x^2 - 2x + 3$. Find the equations of the focal axis and the directrix.

Solution: Put the given equation into the standard form $\left(x - h\right)^2 = 4p\left(y - k\right)$ by isolating the x-terms and then completing the square:

$$\frac{1}{2}\left(x^2 - 4x\right) = y - 3$$
$$\frac{1}{2}\left(x^2 - 4x + \boxed{4}\right) = y - 3 + \boxed{\frac{1}{2} \times 4}$$
$$\frac{1}{2}\left(x - 2\right)^2 = y - 1$$
$$\left(x - 2\right)^2 = 2\left(y - 1\right)$$

- Since $h = 2$ and $k = 1$, the vertex is at $V(2,1)$
- From the parabola equation, $4p = 2$, so $p = \dfrac{1}{2}$. Since $p = \dfrac{1}{2}$, the parabola opens upward and the focus is located $\dfrac{1}{2}$ unit *above* the vertex at $F\left(2, 1 + \dfrac{1}{2}\right) = F\left(2, \dfrac{3}{2}\right)$.

Sketch the parabola, as shown in the accompanying figure.

- Since the vertical focal axis contains $V(2,1)$, its equation is $x = 2$.

- The directrix is a horizontal line that is located $\dfrac{1}{2}$ unit *below* the vertex, so its equation is $y = 1 - \dfrac{1}{2} = \dfrac{1}{2}$.

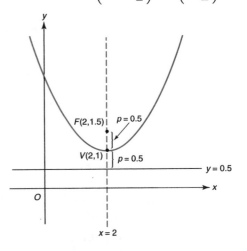

The Ellipse

KEY IDEAS

In the equation of the circle $x^2 + y^2 = 16$, the coefficients of the x^2- and y^2-terms are equal. If one coefficient is made larger than the other, as in $x^2 + 8y^2 = 16$, the curve it describes becomes more oval and is called an *ellipse*. The equation of an ellipse is usually written so that 1 appears alone on the right side of the equation, as in $\dfrac{x^2}{16} + \dfrac{y^2}{2} = 1$.

Definition of an Ellipse

A circle is defined in terms of a single point and a fixed distance, while an ellipse is defined in terms of the sum of the distances from *two* fixed points.

Geometric Definition of an Ellipse

An **ellipse** is the set of all points in the plane the sum of whose distances from two fixed points, called **foci**, is constant. In Figure 14.3, the foci are labeled F and F'. The sum $PF + PF'$ is the same for any point P on the ellipse. The line that contains F and F' is the **focal axis**.

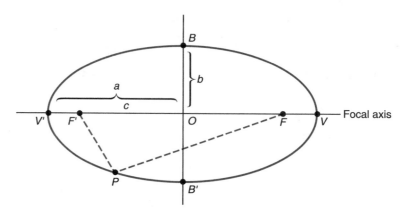

Figure 14.3 Terms related to an ellipse

Some facts you should know about an ellipse are illustrated in Figure 14.3.

- Point O, the midpoint of $\overline{FF'}$, is the *center* of the ellipse. The center of an ellipse is the midpoint of the segment whose endpoints are the foci. Each of the foci is c units from the center.
- Points V and V' are the *vertices* of the ellipse. The vertices are the two points at which the focal axis intersects the ellipse.
- Line segment VV' is the *major axis*. The major axis coincides with the focal axis and is an axis of symmetry. Since each of the vertices is a units from the center, the length of the major axis is $2a$.
- Points B and B' are the endpoints of the *minor axis*. The minor axis is a line segment through the center of the ellipse that is perpendicular to the major axis and whose endpoints are points on the ellipse. The length of the minor axis is $2b$. The major axis of an ellipse is always longer than the minor axis.

Deriving the Pythagorean Relation for Ellipses: $a^2 = b^2 + c^2$

The distances a, b, and c are related by the equation $a^2 = b^2 + c^2$. To understand why this relationship is true, consider Figure 14.4 and reason as follows:

- The distance from F to V is $a - c$, and the distance from F' to V is $a + c$. The sum of the distances from the two foci to V is $(a - c) + (a + c) = 2a$.
- Drawing $\overline{BF'}$ and $\overline{BF}$ forms two right triangles. When two triangles agree in the measures of two sides and their included angle, as do triangles BOF and BOF', every other pair of dimensions of the two triangles must also match. Hence, the lengths of $\overline{BF'}$ and $\overline{BF}$ must be the same.
- Because the sum of the distances from the foci to any point on the ellipse is the same, the sum of the distances from the foci to B $(0,b)$ must be equal to $2a$. Hence, $BF' + BF = 2a$, so $BF' = BF = a$.

- Applying the Pythagorean theorem to either right triangle gives $a^2 = b^2 + c^2$, where a is the length of the hypotenuse.

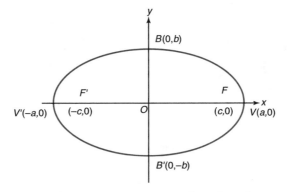

Figure 14.4 Deriving $a^2 = b^2 + c^2$

Deriving an Ellipse Equation: Center at (0,0)

It is now possible to derive an equation for an ellipse that is centered at the origin, as shown in Figures 14.5 and 14.6.

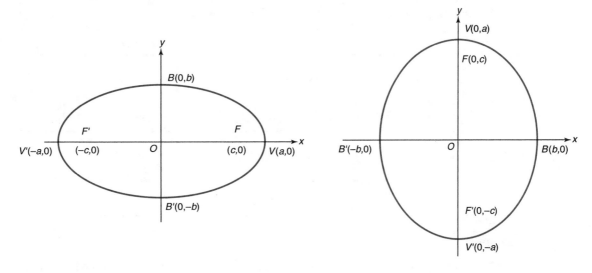

Figure 14.5 Horizontal ellipse   **Figure 14.6** Vertical ellipse

Choose any point $P(x,y)$ on the ellipse in Figure 14.5. Because the sum of the distances from the two foci to any point on the ellipse is $2a$, $PF' + PF = 2a$.

- Rewrite $PF' + PF = 2a$, using coordinates by applying the distance formula:

$$\overbrace{\sqrt{(x-(-c))^2+(y-0)^2}}^{PF'} + \overbrace{\sqrt{(x-c)^2+(y-0)^2}}^{PF} = 2a$$

$$\sqrt{(x+c)^2+y^2} + \sqrt{(x-c)^2+y^2} = 2a$$

- Isolate one of the radicals, square both sides of the equation, and then repeat the procedure to eliminate the remaining radical. The result is

$$\frac{x^2}{a^2} + \frac{y^2}{a^2 - c^2} = 1.$$

- Because $a^2 = b^2 + c^2$, $a^2 - c^2 = b^2$. Then:

$$\frac{x^2}{a^2} + \frac{y^2}{b^2} = 1.$$

Use a similar approach to develop the standard equation for the vertical ellipse in Figure 14.6.

Ellipses: Center at (0,0) $(a > b > 0)$	
Horizontal Ellipse	**Vertical Ellipse**
$\dfrac{x^2}{a^2} + \dfrac{y^2}{b^2} = 1$	$\dfrac{x^2}{b^2} + \dfrac{y^2}{a^2} = 1$
Foci: $\left(\pm c, 0\right)$	Foci: $\left(0, \pm c\right)$
Vertices: $\left(\pm a, 0\right)$ Major axis: $2a$	Vertices: $\left(0, \pm a\right)$ Major axis: $2a$
The numbers a, b, and c are related by the equation $a^2 = b^2 + c^2$.	

TIPS

- When the equation of an ellipse is in standard form, you can tell whether the ellipse is horizontal or vertical by comparing the denominators of the variable terms.
  - If the denominator of the x^2-term is greater than the denominator of the y^2-term, the major axis is *horizontal*.
  - If the denominator of the y^2-term is greater than the denominator of the x^2-term, the major axis is *vertical*.
- When the center of an ellipse is at the origin, the x-intercepts are the vertices for a horizontal ellipse and the y-intercepts are the vertices for a vertical ellipse.
- To use your graphing calculator to graph an ellipse equation, $\dfrac{x^2}{a^2} + \dfrac{y^2}{b^2} = 1$, first solve the equation for y. Then graph the two equations that comprise the ellipse:

$$Y_1 = \frac{b}{a}\sqrt{a^2 - x^2} \quad \text{and} \quad Y_2 = -\frac{b}{a}\sqrt{a^2 - x^2}.$$

Exercise 1   Describing the Properties of an Ellipse from Its Equation

Describe the key properties of the graph of $\dfrac{x^2}{9} + \dfrac{y^2}{4} = 1$.

Solution:

- **Center and orientation of the ellipse:** The ellipse equation has the form $\dfrac{x^2}{a^2} + \dfrac{y^2}{b^2} = 1$, where $a^2 = 9$ and $b^2 = 4$. Because the larger denominator is associated with the x^2-term, the ellipse equation describes a horizontal ellipse whose center is at $(0,0)$.

- **Lengths of major and minor axes:** Because $a^2 = 9$ and $b^2 = 4$, $a = 3$ and $b = 2$. The length of the major axis is $2a = 2 \times 3 = 6$, and the length of the minor axis is $2b = 2 \times 2 = 4$.

- **x-intercepts and vertices:** To find the x-intercepts, set $y = 0$ and solve for x, making $\dfrac{x^2}{9} = 1$, so $x^2 = 9$ and $x = \pm 3$. Hence, the x-intercepts of the ellipse are at $(-3,0)$ and $(3,0)$. Because this horizontal ellipse is centered at the origin, the x-intercepts are also the vertices of the ellipse.

- **y-intercepts:** To find the y-intercepts, set $x = 0$ and solve for y, making $\dfrac{y^2}{4} = 1$, so $y^2 = 4$ and $y = \pm 2$. Hence, the y-intercepts of the ellipse are at $(0,-2)$ and $(0,2)$.

- **Location of the foci:** Because $c^2 = a^2 - b^2 = 9 - 4 = 5$, $c = \sqrt{5}$. The foci are on the x-axis, $c = \sqrt{5}$ units on either side of the origin at $\left(\sqrt{5},0\right)$ and $\left(-\sqrt{5},0\right)$.

Exercise 2   Finding an Equation of an Ellipse

Determine the standard form of an equation of an ellipse whose foci are at $(-5,0)$ and $(5,0)$ and whose vertices are at $(-13,0)$ and $(13,0)$.

Solution: The foci are on the x-axis, so this is a horizontal ellipse.

- The center of the ellipse is on the same line as the foci and midway between them:

$$\text{Center} = \left(\frac{-5+5}{2}, \frac{0+0}{2}\right) = (0,0).$$

- Because the foci are 5 units from the center, $c = 5$ and $c^2 = 25$. The vertices are 13 units from the center, so $a = 13$ and $a^2 = 169$. For an ellipse:

$$b^2 = a^2 - c^2 = 169 - 25 = 144.$$

- The equation of this horizontal ellipse is $\dfrac{x^2}{a^2} + \dfrac{y^2}{b^2} = 1$, where $a^2 = 169$ and $b^2 = 144$:

$$\boldsymbol{\dfrac{x^2}{169} + \dfrac{y^2}{144} = 1.}$$

Deriving an Ellipse Equation: Center at (h,k)

An ellipse may not be centered at the origin. The equation of an ellipse with its center shifted from $(0,0)$ to (h,k) can be obtained by replacing x with $x - h$ and y with $y - k$. This translation also shifts the vertices and foci of the ellipse without changing the lengths of the major and minor axes, as shown in Figures 14.7 and 14.8.

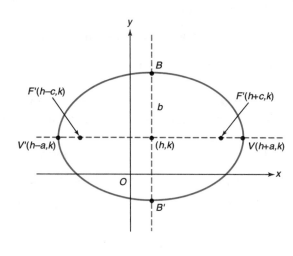

Figure 14.7 Horizontal ellipse centered at (h,k) with focal axis $y = k$

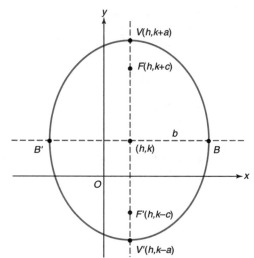

Figure 14.8 Vertical ellipse centered at (h,k) with focal axis $x = h$

Ellipses: Center at (h,k) $(a > b > 0)$	
Horizontal Ellipse	**Vertical Ellipse**
$$\dfrac{(x-h)^2}{a^2} + \dfrac{(y-k)^2}{b^2} = 1$$	$$\dfrac{(x-h)^2}{b^2} + \dfrac{(y-k)^2}{a^2} = 1$$
Foci: $(h \pm c,k)$	Foci: $(h,k \pm c)$
Focal axis: $y = k$	Focal Axis: $x = h$
Vertices: $(h \pm a,k)$	Vertices: $(h,k \pm a)$
The numbers a, b, and c are related by the equation $a^2 = b^2 + c^2$.	

Exercise 3   Finding the Center and Foci, Given an Ellipse Equation

Determine the coordinates of the center and the foci of the ellipse whose equation is

$$\frac{(x+1)^2}{3} + \frac{(y-7)^2}{19} = 1.$$

Solution: Because the denominator of the y^2-term is greater than the denominator of the x^2-term, the ellipse is vertical with $a^2 = 19$ and $b^2 = 3$.

- The numerator of the first term of the given equation can be written as $(x -(-1))^2$, so $h = -1$. The center of this ellipse is at **(–1,7)**.
- To locate the foci, first calculate c:

$$c^2 = a^2 - b^2 = 19 - 3 = 16, \text{ so } c = \sqrt{16} = 4.$$

The foci are on the vertical major axis, $c = 4$ units on either side of the center at $(h, k \pm c) : (-1, 7 \pm 4) = $ **(–1,11)** and **(–1,3)**.

Exercise 4   Finding the Center, Vertices, and Equation of an Ellipse

The foci of an ellipse are located at (–1,2) and (7,2).

   a. If the length of the minor axis is 6 units, find the coordinates of the center and the vertices.

   b. Write the standard form of the equation of the ellipse.

Solutions: *a.* First determine whether the ellipse is horizontal or vertical. Since the foci have the same y-coordinate but different x-coordinates, the foci lie on a horizontal line, so this is a horizontal ellipse.

- The center (h,k) is on the same line as the foci and midway between them. According to the midpoint formula, the coordinates of the center are

$$\left(h, k\right) = \left(\frac{-1+7}{2}, \frac{2+2}{2}\right) = \left(\mathbf{3, 2}\right).$$

- To locate the vertices, you need to know a. Since the distance from the center (3,2) to the focus (7,2) is $7 - 3 = 4$ units, $c = 4$. The length of the vertical minor axis is given as 6, so $2b = 6$ and $b = 3$. Thus, $a = \sqrt{b^2 + c^2} = \sqrt{3^2 + 4^2} = 5$.

The vertices are on the vertical major axis, $a = 5$ units on either side of the center at $(h \pm a, k) : (3 \pm 5, 2) = \textbf{(–2,2)}$ and **(8,2)**, as shown in the figure at the right.

b. The equation of this horizontal ellipse is $\dfrac{(x-h)^2}{a^2} + \dfrac{(y-k)^2}{b^2} = 1$, where $h = 3$, $k = 2$, $a^2 = 25$, and $b^2 = 9$:

$$\frac{(x-3)^2}{25} + \frac{(y-2)^2}{9} = 1.$$

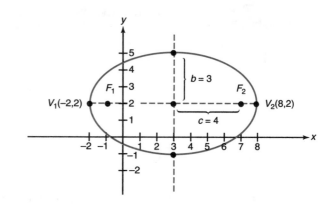

Eccentricity of an Ellipse

Some ellipses are more elongated and less circular than other ellipses. The closer the foci are to the vertices, the closer a and c are in value, and the more elongated the ellipse. Therefore, the closer the ratio $\dfrac{c}{a}$ is to 1, the more elongated the ellipse.

Definition of Eccentricity of an Ellipse

The **eccentricity** of an ellipse, denoted as e, is the ratio $\dfrac{c}{a}$, where a and c have their usual meanings for ellipses.

- Because $e = \dfrac{c}{a}$ and $0 < c < a$, e is between 0 and 1; that is, $0 < e < 1$.
- The closer e is to 0, the closer the foci are to the center and the more circular the ellipse.
- The closer e is to 1, the closer the foci are to the vertices and the more elongated the ellipse.

The concept of eccentricity is useful when describing elliptical paths and orbits. For example, the eccentricity of Earth's elliptical orbit is approximately 0.0167, indicating that Earth's orbit is almost circular.

| Exercise 5 | Using Eccentricity to Help Write an Ellipse Equation |

The vertices of an ellipse with eccentricity 0.875 are located at (3,1) and (3,–15). Determine the standard form of the equation of the ellipse.

Solution: First determine whether the ellipse is horizontal or vertical. Since the vertices have the same x-coordinate but different y-coordinates, the vertices lie on a vertical line, so this is a vertical ellipse.

- The center (h,k) of the ellipse is on the line that contains the vertices and is midway between them. Hence:

$$(h,k) = \left(\frac{3+3}{2}, \frac{-15+1}{2} \right) = (3,-7).$$

- Find a and a^2. Because the distance from $(3,-7)$ to $(3,1)$ is $1 - (-7) = 8$ units, $a = 8$ and $a^2 = 64$.

- Find c and b^2. Since $e = 0.875$, $\dfrac{c}{a} = \dfrac{c}{8} = 0.875$. Hence:

$$c = 0.875 \times 8 = 7 \quad \text{and} \quad b^2 = a^2 - c^2 = 8^2 - 7^2 = 64 - 49 = 15.$$

- The equation of this vertical ellipse is $\dfrac{(x-h)^2}{b^2} + \dfrac{(y-k)^2}{a^2} = 1$, where $h = 3$, $k = -7$, $a^2 = 64$, and $b^2 = 15$:

$$\frac{(x-3)^2}{15} + \frac{(y+7)^2}{64} = 1.$$

Lesson 14-3 — The Hyperbola

KEY IDEAS

A *hyperbola* consists of two disconnected branches that are mirror images of each other, as shown in the accompanying figure. A hyperbola, like an ellipse, can be defined in terms of two fixed points called *foci*:

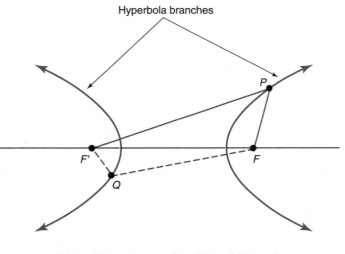

$$PF' - PF = k \quad \text{and} \quad QF - QF' = k.$$

Geometric Definition of a Hyperbola

For any point on a hyperbola, the *difference* of the distances from that point to the foci is the same.

> ### Geometric Definition of a Hyperbola
>
> A **hyperbola** is the set of all points in the plane the difference of whose distances from two fixed points, the **foci**, is the same. The **focal axis** is the line that contains the foci. The foci are always located in the interior of the branches of the hyperbola.

Terms Related to a Hyperbola

The terms *foci*, *focal axis*, *center*, and *vertices* used for ellipses have the same meanings for hyperbolas, as do the distances a and c. In Figure 14.9:

- The *center* of the hyperbola is the midpoint of $\overline{FF'}$. The letter c represents the distance from the center to either of the two foci.
- The *vertices* are labeled as V and V'. Unlike the case in an ellipse, the vertices are closer to the center than are the foci. The letter a represents the distance from the center to either of the two vertices. Unlike an ellipse, $c > a$ for a hyperbola.

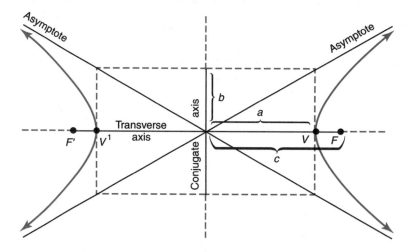

Figure 14.9 Axes and asymptotes of a hyperbola

- The *transverse axis* is the line segment whose endpoints are V and V'. The length of the transverse axis is $2a$.
- The *conjugate axis* is the line segment through the center that is perpendicular to the transverse axis and whose endpoints are b units from the center, where $b^2 = c^2 - a^2$. The length of the conjugate axis is $2b$. The conjugate axis, like the transverse axis, is an axis of symmetry.

- The *Pythagorean relation* for a hyperbola is $c^2 = a^2 + b^2$.
- The *asymptotes* are the two lines through the center of the hyperbola that the branches of the hyperbola approach as the curve extends away from the center. The asymptotes contain the diagonals of the rectangle that has the same center as the hyperbola and whose adjacent sides measure $2a$ and $2b$.

Deriving Hyperbola Equations: Center at (0,0)

The equations for horizontal and vertical hyperbolas, shown in Figures 14.10 and 14.11, can be developed in much the same way as the equations for an ellipse.

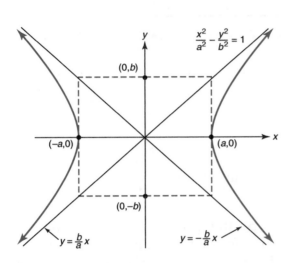

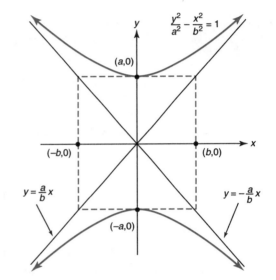

Figure 14.10 Equation of a horizontal hyperbola with center at (0,0)

Figure 14.11 Equation of a vertical hyperbola with center at (0,0)

Hyperbolas: Center at (0,0)	
Horizontal Hyperbola	**Vertical Hyperbola**
$\dfrac{x^2}{a^2} - \dfrac{y^2}{b^2} = 1$	$\dfrac{y^2}{a^2} - \dfrac{x^2}{b^2} = 1$
Foci: $(\pm c, 0)$	Foci: $(0, \pm c)$
Vertices: $(\pm a, 0)$	Vertices: $(0, \pm a)$
Asymptote: $y = \pm \dfrac{b}{a} x$	Asymptote: $y = \pm \dfrac{a}{b} x$
The numbers a, b, and c are related by the equation $c^2 = a^2 + b^2$.	

TIPS

For a hyperbola:

- The Pythagorean relation is $c^2 = a^2 + b^2$, so c is always the greater of the three distances.
- The order in which the x^2- and y^2-terms are written matters. In the equation for a horizontal hyperbola, the x^2-term comes before the y^2-term; in the equation for a vertical hyperbola, the y^2-term comes first.
- There is no requirement that a be greater than b or that a be not equal to b. Either of the two axes of a hyperbola may be longer than the other axis, or both axes may have the same length.

Exercise 1   Finding the Key Features of a Horizontal Hyperbola

Discuss the key features of the hyperbola whose equation is $9x^2 - y^2 = 81$.

Solution: First rewrite the equation in standard form, $\dfrac{x^2}{a^2} - \dfrac{y^2}{b^2} = 1$, by dividing each member of the equation by 81:

$$\frac{9x^2}{81} - \frac{y^2}{81} = \frac{81}{81} \quad \text{or, equivalently,} \quad \frac{x^2}{9} - \frac{y^2}{81} = 1.$$

- **Center and orientation of the hyperbola:** Because the x^2-term comes before the y^2-term, the hyperbola has a horizontal focal axis with its center at $(0,0)$.
- **Lengths of transverse and conjugate axes:** From the equation you know that a^2, the denominator of the first term of the equation, is 9 and $b^2 = 81$, so $a = 3$ and $b = 9$. Hence, the length of the transverse axis is $2 \times 3 = 6$, and the length of the conjugate axis is $2 \times 9 = 18$.
- **Vertices:** The vertices of this horizontal hyperbola are on the x-axis, $a = 3$ units on either side of the origin at $V(3,0)$ and $V'(-3,0)$.
- **Location of the foci:** Because $c^2 = a^2 + b^2 = 9 + 81 = 90$, $c = \sqrt{90} = 3\sqrt{10}$. The foci of this horizontal hyperbola are located $3\sqrt{10}$ units on either side of the origin along the x-axis at $F\left(3\sqrt{10},0\right)$ and $F'\left(-3\sqrt{10},0\right)$.
- **Equations of the asymptotes:** For a horizontal hyperbola, the equations of the asymptotes are $y = \pm \dfrac{b}{a}x$. Hence, the two asymptotes are $y = \pm \dfrac{9}{3}x = \pm 3x$.

| Exercise 2 | Finding the Key Features of a Vertical Hyperbola |

Discuss the key features of the hyperbola whose equation is $25y^2 - 4x^2 = 100$.

Solution: First rewrite the equation in standard form, $\dfrac{y^2}{a^2} - \dfrac{x^2}{b^2} = 1$, by dividing each member of the equation by 100:

$$\frac{25y^2}{100} - \frac{4x^2}{100} = \frac{100}{100} \quad \text{or, equivalently,} \quad \frac{y^2}{4} - \frac{x^2}{25} = 1.$$

- **Center and orientation of the hyperbola:** Because the y^2-term comes before the x^2-term, the hyperbola has a vertical focal axis with its center at $(0,0)$.
- **Lengths of transverse and conjugate axes:** From the equation you know that a^2, the denominator of the first term of the equation, is 4 and $b^2 = 25$, so $a = 2$ and $b = 5$. Hence, the length of the transverse axis is $2 \times 4 = 8$, and the length of the conjugate axis is $2 \times 5 = 10$.
- **Vertices:** The vertices of this vertical hyperbola are on the y-axis, $a = 2$ units on either side of the origin at $V(0,2)$ and $V'(0,2)$.
- **Location of the foci:** Because $c^2 = a^2 + b^2 = 4 + 25 = 29$, $c = \sqrt{29}$. The foci of this vertical hyperbola are located $\sqrt{29}$ units on either side of the origin along the y-axis at $F\left(0, \sqrt{29}\right)$ and $F'\left(0, -\sqrt{29}\right)$.
- **Equations of the asymptotes:** For a vertical hyperbola, the equations of the asymptotes are $y = \pm \dfrac{a}{b}x$. Hence, the two asymptotes are $y = \dfrac{2}{5}x$ and $y = -\dfrac{2}{5}x$.

Sketching a Hyperbola and Its Asymptotes

To sketch a horizontal or vertical hyperbola whose center is at $(0,0)$:

1. Locate the vertices, and mark off points along the other coordinate axis that are $\pm b$ units from the center. Draw vertical or horizontal segments through these four points so that a rectangle is formed, as shown in Figures 14.10 and 14.11.
2. Draw and then extend the diagonals of the rectangle. These lines are the asymptotes of the hyperbola.
3. Using the asymptotes as guides, sketch the two branches of the hyperbola.

| Exercise 3 | Finding an Equation of a Hyperbola and Its Asymptotes |

The foci of a hyperbola are at $(0, \pm\sqrt{13})$, and the vertices are at $(0, \pm 3)$. Find the equation of the hyperbola, and sketch its graph.

Solution: The foci are located on the y-axis the same distance on either side of the origin, so this is a vertical hyperbola centered at the origin.

- For a hyperbola, $c^2 = a^2 + b^2$. Since $c = \sqrt{13}$ and $a = 3$;

$$b = \sqrt{c^2 - a^2} = \sqrt{13 - 9} = \sqrt{4} = 2.$$

- The equation of this vertical hyperbola is $\dfrac{y^2}{a^2} - \dfrac{x^2}{b^2} = 1$, where $a^2 = 9$ and $b^2 = 4$:

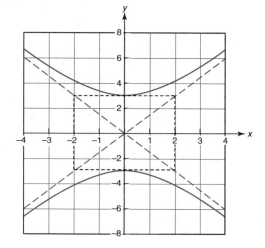

$$\frac{y^2}{9} - \frac{x^2}{4} = 1.$$

- The equations of the asymptotes of this vertical hyperbola are $y = \pm\dfrac{a}{b}x$, where $a = 3$ and $b = 2$:

$$y = \frac{3}{2}x \text{ and } y = -\frac{3}{2}x.$$

The graph of the equation is shown in the accompanying figure.

Deriving Hyperbola Equations: Center at (h,k)

When a hyperbola is shifted horizontally $\left|h\right|$ units and vertically $\left|k\right|$ units, its center moves from $(0, 0)$ to (h,k). The translation also shifts the vertices, foci, and asymptotes of the hyperbola without changing the lengths of the transverse and conjugate axes, as illustrated in Figures 14.12 and 14.13.

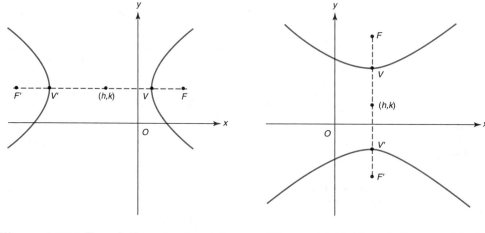

Figure 14.12 Translating a horizontal hyperbola

Figure 14.13 Translating a vertical hyperbola

Hyperbolas: Center at (h,k)	
Horizontal Hyperbola	**Vertical Hyperbola**
$\dfrac{(x-h)^2}{a^2}-\dfrac{(y-k)^2}{b^2}=1$	$\dfrac{(y-k)^2}{a^2}-\dfrac{(x-h)^2}{b^2}=1$
Foci: $(h\pm c,k)$	Foci: $(h,k\pm c)$
Focal axis: $y=k$	Focal axis: $x=h$
Vertices: $(h\pm a,k)$	Vertices: $(h,k\pm a)$
Asymptotes: $y=\pm\dfrac{b}{a}(x-h)+k$	Asymptotes: $y=\pm\dfrac{a}{b}(x-h)+k$
The numbers a, b, and c are related by the equation $c^2=a^2+b^2$.	

Exercise 4   Discussing a Hyperbola, Given Its Equation

Discuss the key features of the hyperbola whose equation is $\dfrac{(x-1)^2}{36}-\dfrac{(y+4)^2}{13}=1$.

Solution: Find h and k:

$$(x-1)^2=\left(x-(\underline{1})\right)^2 \text{ and } (y+4)^2=\left(y-(\underline{-4})\right)^2 \text{, so } h=1 \text{ and } k=-4.$$

- **Center and orientation of the hyperbola:** In the hyperbola equation, the x^2-term comes before the y^2-term, so the hyperbola has a horizontal focal axis with its center at $(1,-4)$.
- **Lengths of transverse and conjugate axes:** From the given equation you know that a^2, the denominator of the first term of the equation, is 36 and $b^2=13$, so $a=6$ and $b=\sqrt{13}$. Hence, the length of the transverse axis is $2\times6=12$, and the length of the conjugate axis is $2\sqrt{13}$.
- **Vertices:** The vertices are on the horizontal focal axis, $a=6$ units on either side of $(1,-4)$ at $V(h\pm a,k)=(1\pm6,-4)=V(7,-4)$ and $V'(-5,-4)$.
- **Location of the foci:** Because $c^2=a^2+b^2=36+13=49$, $c=\sqrt{49}=7$. The foci are 7 units on either side of $(1,-4)$ along the horizontal focal axis at $F(h\pm c,k)=(1\pm7,-4)=V(8,-4)$ and $V'(-6,-4)$.
- **Equations of the asymptotes:** For a horizontal hyperbola, the equations of the asymptotes are $y=\pm\dfrac{b}{a}(x-h)+k$. Hence, the two asymptotes are

$$y=\frac{\sqrt{13}}{6}(x-1)-4 \quad \text{and} \quad y=-\frac{\sqrt{13}}{6}(x-1)-4.$$

Exercise 5 | Writing the Equation of a Hyperbola in Standard Form

Determine the coordinates of the center, vertices, and foci of the hyperbola whose equation is $x^2 - 9y^2 - 10x - 54y - 47 = 0$.

Solution: To locate the center, vertices, and foci, put the equation into standard form by completing the squares for x and for y:

$$x^2 - 9y^2 - 10x - 54y - 47 = 0$$

$$\left(x^2 - 10x\right) + \left(-9y^2 - 54y\right) = 47$$

$$\left(x^2 - 10x + \boxed{25}\right) - 9\left(y^2 + 6y + \boxed{9}\right) = 47 + \boxed{25 - 81}$$

$$\left(x - 5\right)^2 - 9\left(y + 3\right)^2 = -9$$

$$\frac{\left(x - 5\right)^2}{-9} - \frac{9\left(y + 3\right)^2}{-9} = \frac{-9}{-9}$$

$$\frac{\left(y + 3\right)^2}{1} - \frac{\left(x - 5\right)^2}{9} = 1$$

- The hyperbola equation describes a vertical hyperbola for which $h = 5$ and $k = -3$, so the center is at **(5,–3)**.
- Because $a^2 = 1$, $a = 1$. The vertices are located on the vertical focal axis at

$$(h, k \pm a) = (5, -3 \pm 1) = \textbf{(5,–4)} \text{ and } \textbf{(5,–2)}.$$

- To find the foci, you need to know the value of c. Because $a^2 = 1$ and $b^2 = 9$,

$$c = \sqrt{a^2 + b^2} = \sqrt{1 + 9} = \sqrt{10}.$$

The foci are located on the vertical focal axis at

$$\left(h, k \pm c\right) = \left(5, -3 \pm \sqrt{10}\right) = \left(\textbf{5, –3} - \sqrt{\textbf{10}}\right) \text{ and } \left(\textbf{5, –3} + \sqrt{\textbf{10}}\right).$$

Lesson 14-4	General Equations of Conics

KEY IDEAS

It should not be surprising that, since a parabola, a circle, an ellipse, and a hyperbola are all cross sections of a right circular cone, they can be described by a single equation. The equation

$$Ax^2 + Bxy + Cy^2 + Dx + Ey + F = 0$$

is called the **general equation of a conic section**, provided that at least one of the constants A, B, and C is not 0.

So far, we have considered only conics whose focal axis is parallel to a coordinate axis, but this is not always the case.

Degenerate Conic Sections

For certain values of the numerical constants A, B, C, D, E, and F, the equation $Ax^2 + Bxy + Cy^2 + Dx + Ey + F = 0$ may describe a **degenerate conic section**, such as a point, line, or graph with no points.

EXAMPLE: The equation $9x^2 - 4y^2 = 0$ is a pair of intersecting lines whose equations are $y = \pm\dfrac{3}{2}x$.

EXAMPLE: The graph of the equation $3x^2 + 2y^2 + 12 = 0$ contains no points since there are no real-valued ordered pairs that satisfy the equation.

EXAMPLE: The equation $x^2 + 2x + y^2 - 6y + 5 = 0$ looks like the equation for a circle since the coefficients of the x^2- and y^2-terms are the same. However, an attempt to write the equation in standard form by completing the squares for both x and y gives

$$(x^2 + 2x)(y^2 - 6y) = -10$$
$$(x^2 + 2x + \boxed{1}) + (y^2 - 6y + \boxed{9}) = -10 + \boxed{1+9}$$
$$(x + 1)^2 + (y - 3)^2 = -10 + 10$$
$$(x + 1)^2 + (y - 3)^2 = 0$$

The graph of the equation $(x + 1)^2 + (y - 3)^2 = 0$ is point $(-1,3)$. Thus, the equation $x^2 + 2x + y^2 - 6y + 5 = 0$ describes a degenerate conic section.

General Equation of Conics: $B = 0$

The xy-term of an equation is referred to as the **cross-product term**. If, in the general equation of a conic, $B = 0$, the cross-product term vanishes and the general equation simplifies to

$$Ax^2 + Cy^2 + Dx + Ey + F = 0.$$

This equation describes a conic whose focal axis is parallel to a coordinate axis. The values of A and C determine the type of conic.

Classifying a Conic When $B = 0$

For the general equation $Ax^2 + Cy^2 + Dx + Ey + F = 0$, where A and C are not both 0, and degenerate conics are excluded:

- if $A \times C < 0$, the graph is a hyperbola.
- if $A \times C = 0$, the graph is a parabola.
- if $A \times C > 0$, the graph is an ellipse or a circle; if $A = C$, the graph is a circle.

Exercise 1   Identifying a Conic from Its General Equation

Identify the conic section whose general equation is $x^2 = 9y^2 - 10x - 54y - 47 = 0$.

Solution: The given equation has the form $Ax^2 + Cy^2 + Dx + Ey + F = 0$, where $A = 1$, and $C = -9$, so $A \times C = -9 < 0$. Therefore, the equation defines a **hyperbola**. The standard form of the equation of this hyperbola was determined in Exercise 5 of Lesson 14-3.

Rotation of Axes

The coordinate axes need not be treated as immovable objects in the coordinate plane. A **rotation of axes** is a transformation in which the origin of the xy-plane remains fixed as the coordinate axes are rotated about the origin to a new position with axes denoted by x' and y'. The **angle of a rotation** is the angle through which the x-axis must be rotated so that it coincides with the x'-axis. Figure 14.14 shows a counterclockwise rotation of the coordinate axes through an angle, θ. The coordinates of point P in the xy-plane are different from the coordinates of the same point in the new $x'y'$-plane. The size and shape of a graph are not affected by a rotation of the coordinate axes.

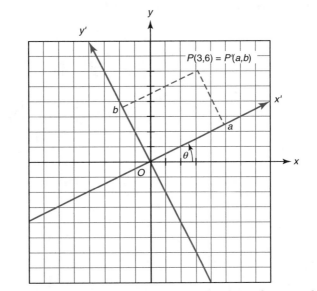

Figure 14.14 Rotating the coordinate axes through an angle, θ

Formulas for the Rotation of Axes

If the coordinate axes of the xy-plane are rotated counterclockwise through an acute angle, θ, the coordinates (x,y) of point P in the xy-plane and the coordinates (x',y') of the same point in the new $x'y'$-plane are related by the formulas

$$x = x'\cos\theta - y'\sin\theta \quad \text{and} \quad y = x'\sin\theta + y'\cos\theta.$$

Exercise 2   Finding an Equation of a Conic After a Rotation of Axes

Write an equation of the conic section $xy - 1 = 0$ after a rotation of $45°$.

Solution: Using the rotation formulas, substitute, into the given equation, $x'\cos\theta - y'\sin\theta$ for x and $x'\sin\theta + y'\cos\theta$ for y, where $\theta = 45°$:

- $x = x'\cos 45° - y'\sin 45° = x'\left(\dfrac{\sqrt{2}}{2}\right) - y'\left(\dfrac{\sqrt{2}}{2}\right) = \dfrac{\sqrt{2}}{2}\left(x' - y'\right).$

- $y = x'\sin 45° + y'\cos 45° = x'\left(\dfrac{\sqrt{2}}{2}\right) + y'\left(\dfrac{\sqrt{2}}{2}\right) = \dfrac{\sqrt{2}}{2}\left(x' + y'\right).$

Then do the substitutions for x and y in the given equation:

$$\overbrace{\left[\frac{\sqrt{2}}{2}\left(x'-y'\right)\right]}^{x} \cdot \overbrace{\left[\frac{\sqrt{2}}{2}\left(x'+y'\right)\right]}^{y} - 1 = 0$$

$$\frac{2}{4}\left[\left(x'\right)^2 - \left(y'\right)^2\right] = 1$$

$$\frac{\left(x'\right)^2}{2} - \frac{\left(y'\right)^2}{2} = 1$$

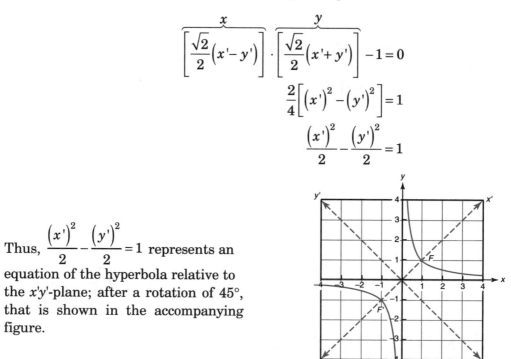

Thus, $\dfrac{\left(x'\right)^2}{2} - \dfrac{\left(y'\right)^2}{2} = 1$ represents an equation of the hyperbola relative to the $x'y'$-plane; after a rotation of $45°$, that is shown in the accompanying figure.

General Equation of Conics: $B \ne 0$

If you are to make use of much of what you learned in preceding lessons, the focal axis of the conic must be parallel to a coordinate axis. This not the case when $B \ne 0$, and, as a result, the general conic equation includes a cross-product term. For instance, the equation $x^2 - 2\sqrt{5}xy + 5y^2 + x - 1 = 0$ contains the cross-product term $-2\sqrt{5}xy$, so its focal axis will not be parallel to a coordinate axis.

In Example 2 you found that a suitable rotation of the coordinate axes could eliminate the xy-term from the equation of the conic, thereby aligning the rotated x-axis with the focal axis of the conic. This approach can be applied also to the equations of other conics provided that you choose an appropriate angle of rotation.

Formula for Angle of Rotation

If the equation of a conic contains a cross-product term, this term can be eliminated by rotating the coordinate axes through an angle, θ, where

$$\tan 2\theta = \frac{B}{A-C} \quad \text{or, equivalently,} \quad \theta = \frac{1}{2}\tan^{-1}\left(\frac{B}{A-C}\right).$$

Exercise 3   Finding an Angle of Rotation Formula

Determine to the *nearest tenth of a degree* the angle through which the coordinate axes must be rotated so that the new coordinate axes are aligned with the focal axis of the graph of $x^2 - 2\sqrt{5}xy + 5y^2 + x - 1 = 0$.

Solution: Find the value of θ, the angle of rotation, using the formula

$$\theta = \frac{1}{2}\tan^{-1}\left(\frac{B}{A-C}\right), \text{ where } A = 1, B = -2\sqrt{5}, \text{ and } C = 5:$$

$$\theta = \frac{1}{2}\tan^{-1}\left(\frac{-2\sqrt{5}}{1-4}\right) = \frac{1}{2}\tan^{-1}\left(\frac{\sqrt{5}}{2}\right) \approx \frac{1}{2}\left(48.2°\right) = \mathbf{24.1°}.$$

Graphing $Ax^2 + Bxy + Cy^2 + Dx + Ey + F = 0$ $(B \neq 0)$

You can use your graphing calculator to display the graph of an equation that has the form

$$Ax^2 + Bxy + Cy^2 + Dx + Ey + F = 0$$

by first solving for y in terms of x.

Exercise 4   Graphing an Equation of a Conic When $B \neq 0$

Graph $x^2 - 2\sqrt{5}xy + 5y^2 + x - 1 = 0$.

Solution:

- Rewrite the given equation as a quadratic equation in y:

$$\overset{a}{\overbrace{5}}\,y^2 + \overset{b}{\overbrace{\left(-2\sqrt{5}x\right)}}y + \overset{c}{\overbrace{\left(x^2 + x - 1\right)}} = 0.$$

- Use the quadratic formula to solve for y, where $a = 5, b = -2\sqrt{5}\,x$, and $c = x^2 + x - 1$:

$$y = \frac{-b \pm \sqrt{b^2 - 4ac}}{2a}$$

$$= \frac{2\sqrt{5}x \pm \sqrt{\left(-2\sqrt{5}\,x\right)^2 - 4\left(5\right)\left(x^2 + x - 1\right)}}{2\left(5\right)}$$

$$= \frac{2\sqrt{5}x \pm \sqrt{20x^2 - 20x^2 - 20x + 20}}{10}$$

$$= \frac{2\sqrt{5}x \pm \sqrt{20\left(1 - x\right)}}{10}$$

- Graph *Y*1 and *Y*2 in a Decimal window, where

$$Y1 = \frac{2\sqrt{5}x + \sqrt{20(1-x)}}{10} \quad \text{and} \quad Y2 = \frac{2\sqrt{5}x - \sqrt{20(1-x)}}{10}.$$

This graph is shown on the left.

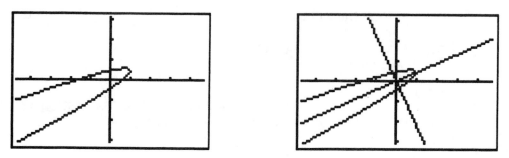

From Exercise 3 you know that rotating the coordinate axes 24.1° brings the new coordinate axes into approximate alignment with the focal axis of the parabola. The graph on the right shows the rotation of the coordinate axes through an angle of 24.1°.

The Equation of a Conic After a Rotation of Axes

When the coordinate axes are rotated through an angle, *θ*, the general equation of a conic in the *xy*-plane:

$$Ax^2 + Bxy + Cy^2 + Dx + Ey + F = 0,$$

is transformed into an equation in the *x'y'*-plane that has the form

$$A'x'^2 + B'x'y' + C'y'^2 + D'x' + E'y' + F' = 0.$$

The coefficients of the two equations are related by this set of equations:

$$A' = A\cos^2\theta + \frac{1}{2}B\sin 2\theta + C\sin^2\theta$$
$$B' = B\cos 2\theta + (C - A)\sin 2\theta$$
$$C' = C\cos^2\theta - \frac{1}{2}B\sin 2\theta + A\sin^2\theta$$
$$D' = D\cos\theta + E\sin\theta$$
$$E' = E\cos\theta - D\sin\theta$$
$$F' = F$$

| **Exercise 5** | Finding an Equation of a Conic in the $x'y'$-Plane |

Graph $4x^2 + 24xy + 11y^2 - 8x + 6y - 25 = 0$. Rotate the coordinate axes to eliminate the xy-term, and then write the equation of the conic relative to the $x'y'$-plane.

Solution: This problem requires a lot of work, so divide it into manageable parts.

Part I: Graph the original equation.

- Before you can graph the equation, solve for y in terms of x. Rearrange the terms of the equation so it has the form of a quadratic equation in y:

$$11y^2 + (24x + 6)y + (4x^2 - 8x - 25) = 0.$$

- Solve for y using the quadratic formula, where $a = 11$, $b = (24x + 6)$, and $c = 4x^2 - 8x - 25$:

$$y = \frac{-b \pm \sqrt{b^2 - 4ac}}{2a}$$

$$= \frac{-(24x+6) \pm \sqrt{(24x+6)^2 - 4(11)(4x^2 - 8x - 25)}}{2(11)}$$

$$= \frac{-(24x+6) \pm \sqrt{(576x^2 + 288x + 36) - (176x^2 - 352x - 1100)}}{22}$$

$$= \frac{-(24x+6) \pm \sqrt{400x^2 + 640x + 1136}}{22}$$

- Graph Y1 and Y2 in a Decimal window where

$$Y1 = \frac{-(24x+6) + \sqrt{400x^2 + 640x + 1136}}{22}$$

and

$$Y2 = \frac{-(24x+6) - \sqrt{400x^2 + 640x + 1136}}{22}$$

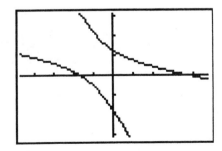

From the accompanying graph, it appears that the given equation describes a hyperbola.

Part II: Find the angle of rotation and other needed angles.

To find the angle of rotation, θ, needed to eliminate the cross-product term from the original equation, use the formula $\tan 2\theta = \dfrac{B}{A-C}$, where $A = 4$, $B = 24$, and $C = 11$:

$$\tan 2\theta = \frac{24}{4-11} = -\frac{24}{7}.$$

Referring to the coefficient relationships on page 390, you will also need to know the *exact* values of $\sin\theta$ and $\cos\theta$. The value of $\tan 2\theta$ can be used to determine the value of $\cos 2\theta$, which in turn can be used to find the exact values of $\sin\theta$ and $\cos\theta$:

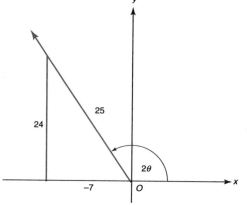

- Because $\tan 2\theta < 0$, $90° < 2\theta < 180°$, locate the reference triangle in Quadrant II such that

$$\tan 2\theta = \frac{24}{-7} = \frac{y}{x},$$

as shown in the accompanying figure. Hence,

$$\cos 2\theta = \frac{-7}{25} \quad \text{and} \quad \sin 2\theta = \frac{24}{25}.$$

- Use the half-angle identities to find the values of $\sin\theta$ and $\cos\theta$:

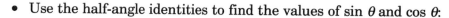

$$\cos\theta = \sqrt{\frac{1+\cos 2\theta}{2}} = \sqrt{\frac{1+\left(-\dfrac{7}{25}\right)}{2}} = \frac{3}{5} \quad \text{and} \quad \sin\theta = \sqrt{\frac{1-\cos 2\theta}{2}} = \sqrt{\frac{1-\left(-\dfrac{7}{25}\right)}{2}} = \frac{4}{5}.$$

Part III: Find the coefficients of the equation in the $x'y'$-plane.

Make the appropriate substitutions into the coefficient relationships given on page 390:

$$A' = A\cos^2\theta + \frac{1}{2}B\sin 2\theta + C\sin^2\theta$$
$$= 4\left(\frac{3}{5}\right)^2 + \frac{1}{2}(24)\left(\frac{24}{25}\right) + 11\left(\frac{4}{5}\right)^2$$
$$= 20$$

$$B' = B\cos 2\theta + (C-A)\sin 2\theta$$
$$= 24\left(-\frac{7}{25}\right) + (11-4)\left(\frac{24}{25}\right)$$
$$= 0$$

$$C' = C\cos^2\theta - \frac{1}{2}B\sin 2\theta + A\sin^2\theta$$
$$= 11\left(\frac{3}{5}\right)^2 - \frac{1}{2}(24)\left(\frac{24}{25}\right) + 4\left(\frac{4}{5}\right)^2$$
$$= 5$$

$$D' = D\cos\theta + E\sin\theta$$
$$= -8\left(\frac{3}{5}\right) + 6\left(\frac{4}{5}\right)$$
$$= 0$$

$$E' = E\cos\theta - D\sin\theta$$
$$= 6\left(\frac{3}{5}\right) - (-8)\left(\frac{4}{5}\right)$$
$$= 10$$

$$F' = F = -25$$

Part IV: Write the Standard Form of the Equation in the $x'y'$-plane.

Because $A' = 20$, $B' = 0$, $C' = 5$, $D' = 0$, $E' = 10$, and $F' = -25$, an equation of the hyperbola relative to the $x'y'$-plane is

$$20x'^2 - 5y'^2 + 10y' - 25 = 0.$$

Complete the square in y':

$$4x'^2 - y'^2 + 2y' = 5$$
$$4x'^2 - \left(y'^2 - 2y' + 1\right) = 5 - 1$$
$$\frac{4x'^2}{4} - \frac{\left(y' - 1\right)^2}{4} = \frac{4}{4}$$
$$x'^2 - \frac{\left(y' - 1\right)^2}{4} = 1$$

The equation $\left(x'\right)^2 - \dfrac{\left(y' - 1\right)^2}{4} = 1$ defines a hyperbola in the $x'y'$-plane with center at $(0,1)$ and a horizontal major axis.

Classifying a Conic Using the Discriminant

When the equation of a conic contains an xy-term, you can use the discriminant test to determine the type of conic that the equation represents.

The Discriminant Test for Classifying Conics

The **discriminant of the equation**

$$Ax^2 + Bxy + Cy^2 + Dx + Ey + F = 0,$$

is the quantity $B^2 - 4AC$. Assume that A, B, and C are not all 0, and that degenerate cases are excluded. Then:

- if $B^2 - 4AC < 0$, the graph is an ellipse or a circle.
- if $B^2 - 4AC = 0$, the graph is a parabola.
- if $B^2 - 4AC > 0$, the graph is a hyperbola.

Exercise 6   Looking Back at Exercise 5

Determine the type of conic described by the equation

$$4x^2 + 24xy + 11y^2 - 8x + 6y - 25 = 0.$$

Solution: Since you are required only to determine the type of conic represented by the given equation, you can simply apply the discriminant test, where $A = 4$, $B = 24$, and $C = 11$. Because

$$B^2 - 4AC = (24)^2 - 4(4)(11) = 400 > 0,$$

the equation describes a **hyperbola**. This fact was discovered in Exercise 5 only after the equation was graphed and then confirmed by obtaining the standard form of the equation after a rotation of axes.

Solving a System of Quadratic Equations Algebraically

The solution set of a system of two quadratic equations contains, at most, four ordered pairs of numbers. Real-valued ordered pairs represent the coordinates of the points at which the graphs of the two equations intersect. To solve a quadratic-quadratic system of equations algebraically, eliminate one of the variables either by substitution or by adding the two equations.

Exercise 7   Solving a Quadratic-Quadratic System by Substitution

Solve the system:

$$x^2 + y^2 = 13$$
$$xy - 6 = 0$$

Solution: Start with the simpler equation, $xy - 6 = 0$, and solve for y in terms of x: $y = \dfrac{6}{x}$.

Eliminate y in the first equation by replacing it with $\dfrac{6}{x}$:

$$x^2 + \left(\frac{6}{x}\right)^2 = 13$$

$$x^2 + \frac{36}{x^2} = 13$$

$$x^2\left(x^2 + \frac{36}{x^2}\right) = 13x^2$$

$$x^4 + 36 = 13x^2$$

$$x^4 - 13x^2 + 36 = 0$$

Since $x^4 - 13x^2 + 36$ has the form of a quadratic trinomial, try factoring it as the product of two binomials:

$$\left(x^2 - 4\right)\left(x^2 - 9\right) = 0$$
$$x^2 = 4 \quad or \quad x^2 = 9$$
$$x = \pm\sqrt{4} = \pm 2 \quad or \quad x = \pm\sqrt{9} = \pm 3$$

Substitute each value of x into the equation $y = \dfrac{6}{x}$ to find the corresponding value of y. The solutions are $(-2,-3)$, $(2,3)$, $(-3,-2)$, and $(3,2)$.

Since all of the solutions are real, they could have been obtained graphically, as shown in the accompanying figure.

<table>
<tr><td>

Lesson 14-5

</td><td>

Polar Equations of Conics

</td></tr>
</table>

KEY IDEAS

Using the concepts of eccentricity, focus, and directrix, we can give a single definition for parabolas, ellipses, and hyperbolas. This alternative definition of conic sections is particularly well suited to polar coordinates.

The Focus-Directrix Definition of a Conic Section

The focus-directrix definition of a parabola given in Lesson 14-1 can be generalized to include parabolas, ellipses, and hyperbolas. In Figure 14.15, P is any point on the conic section, F is the focus, and H is the point on the directrix closest to P. The eccentricity, e, for each conic is $\dfrac{PF}{PH}$. Conics can be classified according to how their eccentricity values compare to 1.

In this unified approach to conics, it is assumed that:

- the focus does not lie on the directrix meaning that $e \neq 0$;
- e is always positive;
- circles and degenerate conics are not included.

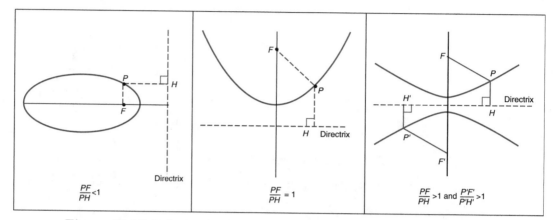

Figure 14.15 Relationship between eccentricity value and type of conic

Unified Focus-Directrix Definition of a Conic

A **conic** is the set of all points in the plane whose distances from a given point, called the **focus**, to a fixed line, called the **directrix**, have a constant ratio. The constant ratio is called the **eccentricity** of the conic and is denoted as e. The value of e determines whether the conic section is an ellipse, a parabola, or a hyperbola:

- if $e < 1$, the conic section is an ellipse.
- If $e = 1$, the conic section is a parabola.
- If $e > 1$, the conic section is a hyperbola.

Polar Form of an Equation of a Conic

In rectangular coordinates, the equations of conic sections become simpler when the centers of the conic sections are located at the origin. Similarly, positioning a conic section in polar coordinates, so that a focus is at the pole, also simplifies matters.

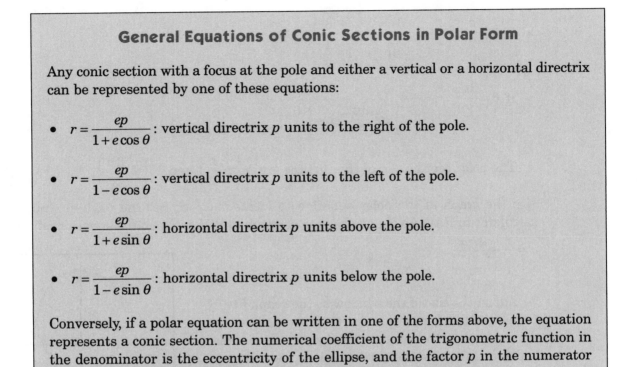

General Equations of Conic Sections in Polar Form

Any conic section with a focus at the pole and either a vertical or a horizontal directrix can be represented by one of these equations:

- $r = \dfrac{ep}{1 + e\cos\theta}$: vertical directrix p units to the right of the pole.

- $r = \dfrac{ep}{1 - e\cos\theta}$: vertical directrix p units to the left of the pole.

- $r = \dfrac{ep}{1 + e\sin\theta}$: horizontal directrix p units above the pole.

- $r = \dfrac{ep}{1 - e\sin\theta}$: horizontal directrix p units below the pole.

Conversely, if a polar equation can be written in one of the forms above, the equation represents a conic section. The numerical coefficient of the trigonometric function in the denominator is the eccentricity of the ellipse, and the factor p in the numerator represents the distance between the focus (pole) and the directrix.

Exercise 1   Determining the Type of Conic from Its Polar Equation

Classify the conic section whose polar equation is $r = \dfrac{18}{6 - 3\sin\theta}$.

Determine the polar coordinates of the vertices and the foci.

Solutions: Divide the numerator and the denominator of the given equation by 6 so that the first term of the denominator of the new fraction will be 1:

$$r = \frac{18 \div 6}{\left(6 - 3\sin\theta\right) \div 6} = \frac{3}{1 - \dfrac{1}{2}\sin\theta}.$$

The polar equation has the form $r = \dfrac{ep}{1 - e\sin\theta}$, where $e = \dfrac{1}{2}$, $p = 6$, and the directrix is horizontal. Since $e < 1$, the polar equation represents an **ellipse**. Because the ellipse has a horizontal directrix, the major axis is vertical. The values of θ that determine a vertical major axis are $\dfrac{\pi}{2}$ and $\dfrac{3\pi}{2}$. The vertices are, therefore, located on the vertical major axis at

$V\left(r, \dfrac{\pi}{2}\right)$ and $V'\left(r, \dfrac{3\pi}{2}\right)$, where the values of r can be calculated by substituting the corresponding values of θ into the polar equation.

- If $\theta = \dfrac{\pi}{2}$, $r = \dfrac{3}{1 - \dfrac{1}{2}\sin\dfrac{\pi}{2}} = \dfrac{3}{1 - \dfrac{1}{2}(1)} = 6$.

- If $\theta = \dfrac{3\pi}{2}$, $r = \dfrac{3}{1 - \dfrac{1}{2}\sin\dfrac{3\pi}{2}} = \dfrac{3}{1 - \dfrac{1}{2}(-1)} = 2$.

- The polar coordinates of the vertices are $V\left(6, \dfrac{\pi}{2}\right)$ and $V'\left(2, \dfrac{3\pi}{2}\right)$.

Use the graph of the polar equation to locate the foci. Set the mode of your graphing calculator to RADIANS, and the type of graph to POLar. Open the equation editor, and set

$$r = \dfrac{3}{1 - \dfrac{1}{2}\sin\theta}.$$

Adjust the size of the window so the graph fits. If the ellipse looks more circular than oval, adjust the dimensions of the window so that the horizontal width is 1.5 times the vertical width, as shown in the accompanying figure. You can also press

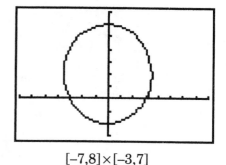

[–7,8]×[–3,7]

$\boxed{\text{ZOOM}}$ $\boxed{5}$ to view the ellipse in a square window.

From the figure you can see that the pole is 2 units above V'. The pole is also one of the foci. Hence, the other focus must be the same 2 units from V, or at $\left(6 - 2, \dfrac{\pi}{2}\right) = \left(4, \dfrac{\pi}{2}\right)$.

Thus, the polar coordinates of the two foci are **(0,0)** and $\left(4, \dfrac{\pi}{2}\right)$.

Exercise 2   **Changing from Polar to Rectangular Form**

Find the equation, in rectangular coordinates, for the ellipse discussed in Exercise 1.

Solution: For the ellipse discussed in Exercise 1:

- The center is at (0,4), so $(h,k) = (0,4)$.
- The length of the major axis is 8. Hence, $2a = 8$, so $a = 4$ and $a^2 = 16$. Each focus is 2 units from a vertex, so $c = 2$. You can also calculate c using the relationship $e = \dfrac{c}{a}$, where $e = \dfrac{1}{2}$ and $a = 4$.
- The value of b^2 is given by the Pythagorean relation for an ellipse:

$$b^2 = a^2 - c^2 = 4^2 - 2^2 = 12.$$

- The equation of this vertical ellipse is $\dfrac{(x-h)^2}{b^2} + \dfrac{(y-k)^2}{a^2} = 1$, where $a^2 = 16$, $b^2 = 12$, and $(h,k) = (0, 4)$:

$$\dfrac{x^2}{12} + \dfrac{(y-4)^2}{16} = 1.$$

| Chapter 14 | **CHECKUP EXERCISES** |

1—12. Determine the coordinates of the center, the vertices, and the foci for each conic section.

1. $\dfrac{x^2}{4} + \dfrac{y^2}{13} = 1$

2. $\dfrac{x^2}{49} - \dfrac{y^2}{25} = 1$

3. $9x^2 + 16y^2 = 144$

4. $x^2 = -\dfrac{1}{2}y$

5. $y^2 + 2 = 6x - 1$

6. $\dfrac{(x-2)^2}{16} + \dfrac{(y+1)^2}{36} = 1$

7. $x^2 - 2x + y^2 - 8x + 1 = 0$

8. $(x-4)^2 = 2(y+3)$

9. $\dfrac{(x+2)^2}{16} - \dfrac{(y-5)^2}{81} = 1$

10. $8x + 14 - y^2 + 6y + 10 = 0$

11. $2x^2 - 16x + y^2 + 4y - 5 = 0$

12. $\dfrac{(y-2)^2}{36} - \dfrac{(x-4)^2}{100} = 1$

13—17. Write an equation of the ellipse with the indicated properties.

13. Foci at $\left(\pm\sqrt{7}, 0\right)$; minor axis length is 4.

14. Center at $(-4,1)$; vertex at $(-4,7)$; focus at $\left(-4, 3\sqrt{3}\right)$.

15. Vertices at $(-14,2)$ and $(12,2)$; focus at $(-13,2)$.

16. Center at $\left(2, \dfrac{\sqrt{7}}{4}\right)$; horizontal major axis length is 5; passes through point $\left(5, \sqrt{7}\right)$.

17. Foci at $(-1,3)$ and $(-1,-5)$; vertex at $(-1,-8)$.

18—22. Write an equation of the hyperbola with the indicated properties.

18. Center at $(-1,3)$; focus at $(-14,3)$; vertex at $(11,3)$.

19. Vertices at $(-2,0)$ and $(-2,8)$; focus at $(-2,-1)$.

20. Foci at (−6,4) and (14,4); vertex at (−1,4).

21. Vertex at (0,9); an equation of an asymptote is $y = \dfrac{6}{5}x$.

22. Vertex at (−4,−2); an equation of an asymptote is $y = \dfrac{2}{3}(x+1) - 2$.

23−28. Write an equation of the parabola with the indicated properties.

23. Vertex at (−1,3); focus at (−1,1).

24. Vertex at (4,−2); focus at (8,−2)

25. Focus at (−3,0); directrix is the line $y = 4$.

26. Vertex at (4,5); directrix is the line $x = -2$.

27. Focus at (−4,1); directrix is the line $y = -3$.

28. Tangent to the line $y = 12$; x-intercepts at (−3,0) and (9,0).

29 and 30. Find the coordinates of the foci of a hyperbola with the indicated properties.

29. Vertex at (6,−1); an equation of an asymptote is $y = \dfrac{1}{2}(x+2) - 1$.

30. Equations of the asymptotes are $y = 3x - 8$ and $y = -3x + 10$; transverse axis length is 6.

31. When radio signals that are all parallel to the focal axis of a parabola-shaped satellite dish strike the dish, the signals are reflected along the axis of symmetry onto the focus point, where a receiving device that captures the signals is located, as illustrated in the accompanying figure.

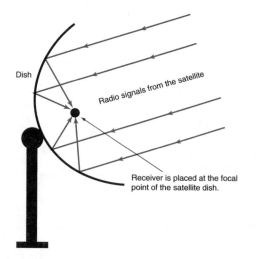

If the opening of the satellite dish has a maximum width of 6 feet and the maximum depth from the center of the base of the dish is 2.25 feet, describe where the receiver should be located.

32–37. Without graphing or completing the square, determine the type of conic that each equation represents.

32. $-x^2 + 3y^2 + 4x + 12y - 10 = 0$

33. $4x^2 + y^2 - 12x - 6y + 9 = 0$

34. $x^2 - 6xy + 9y^2 + 5x - y - 3 = 0$

35. $2x^2 - 3\sqrt{2}xy - y^2 + 7x + 2y - 5 = 0$

36. $-x^2 + 2\sqrt{3}xy - 4y^2 + 6x - y - \sqrt{3} = 0$

37. $-7x^2 + 4\sqrt{5}xy - 3y^2 - 10x + y + \dfrac{2}{\sqrt{5}} = 0$

38 and 39. For each equation use the quadratic formula to solve for y in terms of x. Then use a graphing calculator to display the graph of the equation.

38. $x^2 - \sqrt{3}xy + y^2 - 2x + y - 2 = 0.$

39. $4x^2 - 8xy - 2y^2 + \sqrt{5}x - \dfrac{2\sqrt{5}y}{5} + 4 = 0$

40–45. Rotate the axes to eliminate the xy-term in each equation. Then write the standard form of the equation of the conic relative to $x'y'$-plane.

40. $3x^2 - 4xy + 1 = 0$

41. $7x^2 - 6\sqrt{7}xy + 9y^2 - 36x - 12\sqrt{7}y - 48 = 0$

42. $13x^2 + 6\sqrt{3}\,xy + 7y^2 + 4x - 4\sqrt{3}y - 12 = 0$

43. $x^2 + 2xy + y^2 - \dfrac{7}{\sqrt{2}}x - \dfrac{9}{\sqrt{2}}y + 5 = 0$

44. $14x^2 + 13y^2 + 4\sqrt{5}\,xy - 48\sqrt{5}\,x - 96y + 270 = 0$

45. $181x^2 - y^2 - 130\sqrt{11}\,xy + 36 = 0$

46–49. Solve each system of equations.

46. $x^2 + y^2 - 4y = 5$
    $x - y - 1 = 0$

47. $x^2 - xy = -12$
    $5x^2 - 2xy = 12$

48. $\left(x + 5\right)^2 + \left(y + 2\right)^2 = 169$
    $xy - 21 = 0$

49. $2y^2 - 11x^2 = -1$
    $5x^2 + 2xy = 3$

50–52. *a.* Classify the conic, and determine the polar coordinates of the vertices and foci.

      b. Write an equation of the conic in the rectangular coordinate system.

50. $r = \dfrac{18}{6 + 9\cos\theta}$

51. $r = \dfrac{30}{5 - 5\sin\theta}$

52. $r = \dfrac{42}{6 + 4\cos\theta}$

STUDY UNIT V:
NUMBER PATTERNS
AND COUNTING

Chapter 15

SEQUENCES, SERIES, AND COUNTING

OVERVIEW

Lists of numbers sometimes follow predictable patterns. Arithmetic and geometric sequences have the property that each term in a sequence after the first is obtained by adding or multiplying the preceding term by the same number.

Some number patterns that involve only nonnegative integers can be generalized by statements that look like algebraic formulas. These formulas are proved using a special method called *mathematical induction*.

Expanding a binomial such as $(a + b)^2$ is not difficult. However, when the exponent of the binomial is greater than 2, a good deal of effort is required to expand the binomial using repeated multiplication. The binomial theorem provides an algebraic formula that tells how to find each term in the expansion of a binomial of the form $(a + b)^n$, where n is a positive integer.

Lessons in Chapter 15
Lesson 15-1: Arithmetic Sequences and Series
Lesson 15-2: Geometric Sequences and Series
Lesson 15-3: Generalized Sequences
Lesson 15-4: Mathematical Induction
Lesson 15-5: Permutations and Combinations
Lesson 15-6: The Binomial Theorem

Lesson 15-1	Arithmetic Sequences and Series

KEY IDEAS

A **sequence** is a list of numbers, called *terms*, written in a specific order. A list of numbers such as 2, 5, 8, 11, 14, . . . is called an **arithmetic sequence** since each term after the first is obtained by adding a constant, 3 in this case, to the term that precedes it. A sequence of numbers may be *finite* or *infinite*. A **finite sequence** has a definite number of terms. An **infinite sequence** is nonending. An infinite sequence uses three trailing periods to indicate that the pattern never ends, as in 2, 5, 8, 11, 14, ...

Common Difference

In an arithmetic sequence, subtracting any term from the term that follows it always results in the same number. This number is called the **common difference** and is denoted as d. The common difference for the arithmetic sequence 2, 5, 8, 11, 14, . . . is 3 since

$$5 - 2 = 8 - 5 = 11 - 8 = 14 - 11 = \cdots = 3.$$

A Generalized Arithmetic Sequence

Consider a finite arithmetic sequence with n terms in which a_1 is the first term, a_n is the last term, and d is the common difference. Then:

> **Last Term of an Arithmetic Sequence**
>
> $$a_n = a_1 + (n - 1)d$$

For example, to find the 36th term of 2, 5, 8, 11, 14, . . . , set $a_1 = 2$, $n = 36$, and $d = 3$:

$$a_n = a_1 + (n - 1)d$$
$$a_6 = 2 + (36 - 1)3$$
$$= 2 + 35 \cdot 3$$
$$= 107$$

The 36th term is 107.

Arithmetic Series: Sum of Terms

The indicated sum of the terms of a sequence is called a **series**. An **arithmetic series** is the sum of the terms in the corresponding arithmetic sequence. For example,

$$2 + 5 + 8 + 11 + 14 + \cdots,$$

is an arithmetic series. To find the sum, S_n, of the first n terms of an arithmetic series, use this formula:

> ### Sum of First n Terms of an Arithmetic Sequence
>
> $$S_n = \frac{n}{2}\left(a_1 + a_n\right)$$

For the arithmetic sequence 2, 5, 8, 11, 14, . . ., you can find the sum

$$2 + 5 + 8 + 11 + 14 + \cdots + 107$$

by using the formula $S_n = \dfrac{n}{2}\left(a_1 + a_n\right)$, where $a_1 = 2$, $a_{36} = 107$, and $n = 36$:

$$S_{36} = \frac{36}{2}\left(2 + 107\right)$$
$$= 18 \cdot 109$$
$$= 1962$$

Replacing a_n in the formula $S_n = \dfrac{n}{2}\left(a_1 + a_n\right)$ by $a_1 + (n-1)d$ produces another useful formula that allows you to find the sum of the first n terms of an arithmetic sequence without first finding the last term:

> ### Sum of First n Terms of an Arithmetic Sequence: Alternative Formula
>
> $$S_n = \frac{n}{2}\left(2a_1 + (n-1)d\right)$$

To find the sum of the first 27 terms of the sequence 2, 5, 8, 11, 14, . . ., use this formula with $a_1 = 2$, $d = 3$, and $n = 27$:

$$S_n = \frac{n}{2}\left(2a_1 + (n-1)d\right)$$
$$S_{27} = \frac{27}{2}\left(2(2) + (23-1)2\right)$$
$$= \frac{27}{2}\left(4 + 44\right)$$
$$= 648$$

Exercise 1   Finding the Terms of an Arithmetic Sequence

An arithmetic sequence whose terms decrease in value includes 12.0 and –2.0. If 12.0 and –2.0 are separated by three terms of this sequence, what are the values of these three terms?

Solution: You need to determine the three numbers that make the sequence of five numbers from 12.0 to –2.0 an arithmetic sequence:

$$\ldots, \underbrace{12.0, \underline{}, \underline{}, \underline{}, -2.0}_{d\,=\,?}, \ldots$$

To find the common difference of this sequence, use the formula $a_n = a_1 + (n-1)d$, where $a_1 = 12$, $a_5 = -2.0$, and $n = 5$:

$$a_5 = a_1 + \left(5-1\right)d$$
$$-2.0 = 12.0 + 4d$$
$$4d = -14.0$$
$$d = \frac{-14.0}{4} = -3.5$$

Since $d = -3.5$, the three terms are:

$$a_2 = 12.0 - 3.5 = \mathbf{8.5}$$
$$a_3 = 8.5 - 3.5 = \mathbf{5.0}$$
$$a_4 = 5.0 - 3.5 = \mathbf{1.5}$$

Exercise 2   Finding the Sum of the Terms of an Arithmetic Sequence

If the eighth term of an arithmetic sequence is –18 and the third term is 7, find the sum of the first 30 terms of this sequence.

Solution: First find the common difference. Since $a_3 = 7$ and $a_8 = -18$, there are four terms between 7 and –18:

$$\underline{}, \underline{}, \underbrace{7, \underline{}, \underline{}, \underline{}, \underline{}, -18}_{\text{six terms}}, \ldots$$

- Find the common difference of the six terms of the arithmetic sequence whose first term is 7 and whose last term is –18 by using the formula $a_n = a_1 + (n-1)d$, where $a_1 = 7$, $a_6 = -18$ and $n = 6$:

$$-18 = 7 + (6-1)d$$
$$5d = -25$$
$$d = \frac{-25}{5} = -5$$

- Consider the original sequence, in which $a_3 = 7$. Because $d = -5$,
$$a_2 + (-5) = a_3 = 7, \quad \text{so} \quad a_2 = 12,$$
$$and$$
$$a_1 + (-5) = a_2 = 12, \quad \text{so} \quad a_1 = 17.$$

- Find the sum of the first 30 terms using the formula $S_n = \frac{n}{2}\left(2a_1 + (n-1)d\right)$, where $a_1 = 17$, $n = 30$, and $d = -5$:

$$S_{30} = \frac{30}{2}\left(2(17) + (30-1)(-5)\right)$$
$$= 15(34 - 145)$$
$$= 15 \cdot (-111)$$
$$= \mathbf{-1665}$$

Exercise 3 **Counting the Number of Terms in a Sequence That Are Divisible by a Given Number**

How many numbers from 15 to 633 are divisible by 3?

Solution: Since 15 and 633 are both divisible by 3, the numbers from 15 to 633 that are divisible by 3 form an arithmetic sequence in which the first term is 15, the nth term is 633, and the common difference is 3: $a_1 = 15$, $a_n = 633$, and $d = 3$. Find n using the formula $a_n = a_1 + (n-1)d$, where $a_1 = 15$, $a_n = 633$, and $d = 3$:

$$633 = 15 + (n-1)3$$
$$\frac{618}{3} = \frac{(n-1)\cancel{3}}{\cancel{3}}$$
$$206 = n - 1$$
$$207 = n$$

There are **207** numbers from 15 to 633 that are divisible by 3.

KEY IDEAS

A list of numbers such as 2, 6, 18, 54, . . . is called a **geometric sequence** because each term after the first is obtained by multiplying the preceding term by the same number, 3.

Common Ratio

In a geometric sequence, dividing any term after the first by the term that precedes it always results in the same nonzero number. This number is called the **common ratio** and is denoted as r. The common ratio for the geometric sequence 2, 6, 18, 54, . . . is 3 since

$$\frac{6}{2} = \frac{18}{6} = \frac{54}{18} = \cdots = 3.$$

Formulas for Finite Geometric Sequences

For a geometric sequence with n terms in which a_1 is the first term, a_n is the last term, and r is the common ratio ($r \neq 1$), the following formulas can be used to find the last term and the sum of the terms:

Geometric Sequence Formulas	
Last-Term Formula	**Sum-of-Terms Formula**
$a_n = a_1 r^{n-1}$	$S_n = \dfrac{a_1\left(1 - r^n\right)}{1 - r}$

Exercise 1   Finding the Sum of the Terms of a Geometric Sequence

Find the sum of the first nine terms of this geometric sequence:

$$8,\ 4,\ 2,\ 1, \ldots .$$

Solution: The common ratio is $\dfrac{4}{8} = \dfrac{2}{4} = \cdots = \dfrac{1}{2}$. Find the sum of the first nine terms by

using the formula $S_n = \dfrac{a_1\left(1 - r^n\right)}{1 - r}$, where $a_1 = 8$, $r = \dfrac{1}{2}$, and $n = 9$:

$$S_9 = \frac{8\left(1 - \left(\dfrac{1}{2}\right)^9\right)}{1 - \dfrac{1}{2}} = 16\left(1 - \frac{1}{2^9}\right) = 2^4\left(\frac{2^9 - 1}{2^9}\right) = \frac{511}{2^5} = \mathbf{\frac{511}{32}}.$$

Exercise 2 **Finding the Terms of a Geometric Sequence and Their Sum**

a.   If the first term of a geometric sequence is 36 and the fifth term is $\dfrac{64}{9}$, what are the
second, third, and fourth terms of this sequence?

b.   Find the sum of the first six terms of this sequence.

Solutions: *a.* Find the common ratio using the formula $a_n = a_1 r^{n-1}$, where

$a_1 = 36$, $a_5 = \dfrac{64}{9}$, and $n = 5$:

$$\frac{64}{9} = 36 r^{5-1}$$

$$r^4 = \frac{64}{36 \cdot 9} = \frac{\cancel{4} \cdot 16}{\cancel{4} \cdot 9 \cdot 9} = \frac{16}{81}$$

$$r = \sqrt[4]{\frac{16}{81}} = \frac{2}{3}$$

Hence:

$$a_2 = \frac{2}{3} \times a_1 = \frac{2}{3} \times 36 = \mathbf{24}$$

$$a_3 = \frac{2}{3} \times a_2 = \frac{2}{3} \times 24 = \mathbf{16}$$

$$a_4 = \frac{2}{3} \times a_3 = \frac{2}{3} \times 16 = \mathbf{\frac{32}{3}}$$

b. To find the sum of the first six terms of this sequence, use the formula $S_n = \dfrac{a_1\left(1-r^n\right)}{1-r}$,

where $a_1 = 36$, $r = \dfrac{2}{3}$, and $n = 6$:

$$S_7 = \frac{36\left(1-\left(\dfrac{2}{3}\right)^6\right)}{1-\dfrac{2}{3}} = 36 \cdot 3\left(1-\frac{2^6}{3^6}\right) = 36\left(\frac{3^6-2^6}{3^5}\right) = 36\left(\frac{665}{243}\right) = \frac{2660}{27}$$

Infinite Geometric Series

If the common ratio, r, of a nonending geometric sequence is between 0 and 1 or between −1 and 0, the sum of its terms converges to a real number that can be determined using the following formula:

Sum of Terms of an Infinite Geometric Series

$$S_\infty = \frac{a_1}{1-r}, \text{ provided that } |r| < 1.$$

Exercise 3   **Representing a Repeating Decimal as an Infinite Geometric Series**

Express the repeating decimal 0.131313. . . as the ratio of two integers.

Solution: Because   0.131313. . . = 0.13 + 0.13(.01) + 0.13(.01)^2 + 0.13(.01)^3 + ⋯, the repeating decimal 0.131313. . . can be written as the sum of the terms of an infinite geometric sequence in which the common ratio is 0.01. To find the sum of the terms, use the formula $S_\infty = \dfrac{a_1}{1-r}$, where $a_1 = 0.13$ and $r = 0.01$:

$$S_\infty = \frac{0.13}{1-0.01} = \frac{0.13}{0.99} = \frac{13}{99}.$$

Hence, the repeating decimal 0.131313. . . can be written in ratio form as $\dfrac{13}{99}$.

Generalized Sequences

KEY IDEAS

Sequences can be described by formula-type expressions. The sum of any number of consecutive terms of a generalized sequence can be indicated by using the symbol Σ, the capital Greek letter sigma.

Viewing a Sequence as a Function

A **sequence** is a function whose *domain* is a set of consecutive whole numbers that represent the position numbers of the terms in the sequence and whose *range* is the corresponding terms of the sequence. Instead of using standard function notation such as $a(1)$ to indicate the first term of the sequence, $a(2)$ to indicate the second term, and so forth, the terms or function values of a sequence are represented by the subscripted variables a_1, a_2, a_3, ..., a_n, ..., where the position of a term is indicated by its subscript. The notation $\{a_n\}$ refers to the entire sequence in which a_n is a general term.

For the sequence 10, 15, 20, 25, and 30, the domain is the set of position numbers, $\{1, 2, 3, 4, 5\}$, and the range is the corresponding set of terms, $\{10, 15, 20, 25, 30\}$. If a represents this sequence function, then $a_1 = 10$, $a_2 = 15$, $a_3 = 20$, $a_4 = 25$, and $a_5 = 30$.

Explicit Versus Recursive Formulas

A sequence may be defined *explicitly* or *recursively*.

- A sequence is defined **explicitly** when a formula is given that tells how to obtain all of the terms of the sequence without knowing the identity of any specific term. The formula $a_n = 2n + 1$ defines a sequence explicitly since it tells how to obtain each term without knowing any other term:

$$\text{First term:} \quad a_1 = 2(1) + 1 = 3$$
$$\text{Second term:} \quad a_2 = 2(2) + 1 = 5$$
$$\text{Third term:} \quad a_3 = 2(3) + 1 = 7$$
$$\vdots \qquad\qquad \vdots$$
$$n\text{th term:} \quad a_n = 2n + 1$$

- A sequence is defined **recursively** when all of its terms can be obtained by a statement that relates a general term of the sequence to one or more terms of the sequence that preceded it. Suppose that a_1 represents the initial balance in year 1 of a savings account in which interest is compounded yearly at the rate of 5%. If the initial deposit is $3000, the recursion formula

$$a_n = 1.05(a_{n-1}); \ a_1 = 3000$$

gives the account balance in each of the successive years $n = 2, 3, \ldots$.

Exercise 1   Finding a Recursion Formula

Write a recursion formula that tells how to obtain the terms of the sequence

$$2, 4, 10, 28, 82, \ldots,$$

assuming that the number pattern continues to hold.

Solution: In the sequence 2, 4, 10, 28, 82, . . . , each term after the first is obtained by multiplying the preceding term by 3 and then subtracting 2 from the product. The recursion formula that expresses this number pattern is

$$a_n = 3(a_{n-1}) - 2.$$

This recursion formula, however, does not provide enough information by itself to obtain each term of the sequence it describes. If the fact that $a_1 = 2$ accompanies the recursion formula, then each term of the sequence can be obtained by direct substitution into the formula:

$$a_2 = 3\left(a_1\right) - 2 = 3\left(2\right) - 2 = 4$$
$$a_3 = 3\left(a_2\right) - 2 = 3\left(4\right) - 2 = 10$$
$$a_4 = 3\left(a_3\right) - 2 = 3\left(10\right) - 2 = 28$$
$$a_5 = 3\left(a_4\right) - 2 = 3\left(28\right) - 2 = 82$$

$$\begin{array}{ccc} \cdot & \cdot & \cdot \\ \cdot & \cdot & \cdot \\ \cdot & \cdot & \cdot \end{array}$$

Thus, the sequence 2, 4, 10, 28, 82, . . ., is defined recursively by the statements

$$a_n = 3(a_{n-1}) - 2 \quad \text{and} \quad a_1 = 2.$$

Summation Notation

The notation $\sum_{i=1}^{n} a_i$ represents the sum of the numbers a_i as the index variable i takes on consecutive integer values from 1 to n. For example:

$$\sum_{i=1}^{5} a_i = a_1 + a_2 + a_3 + a_4 + a_5.$$

EXAMPLE: The arithmetic series $3 + 6 + 9 + 12 + 15 + 18$ can be represented as $\sum_{i=1}^{6} 3i$.

EXAMPLE: The geometric series $2 + 4 + 8 + 16 + 32$ can be represented as $\sum_{k=1}^{5} 2^k$.

EXAMPLE: A constant can be passed through the summation sign:

$$\sum_{k=1}^{3} 4(2k-1)^2 = 4\sum_{k=1}^{3}(2k-1)^2$$

$$= 4\left[(2(1)-1)^2 + (2(2)-1)^2 + (2(3)-1)^2\right]$$

$$= 4\left[\quad 1 \quad + \quad 9 \quad + \quad 25 \quad\right]$$

$$= 140$$

Exercise 2   Representing a Series Using Sigma Notation

Assuming that the same pattern continues to hold, represent the following series using sigma notation:

$$(-1) + 4 + (-9) + 16 + (-25) + \cdots.$$

Solution: The terms of the series are the squares of consecutive whole numbers that alternate in sign. The odd-numbered terms of the series are negative, and the even-numbered terms are positive. The sign of each term can be represented by $(-1)^k$, where k is the position number of the term. Hence, the general term of the series is $(-1)^k \cdot k^2$, where k is a positive integer. Thus:

$$(-1) + 4 + (-9) + 16 + (-25) + \cdots + (-1)^n \cdot n^2 = \sum_{k=1}^{n}(-1)^k \cdot k^2.$$

Exercise 3	Evaluating an Arithmetic Series

Evaluate: $\displaystyle\sum_{k=1}^{25} (3k-5)$.

Solution: The given sum represents an arithmetic series that can be seen by writing the first few terms and the last term:

$$\sum_{k=1}^{25} (3k-5) = (3(1)-5) + (3(2)-5) + (3(3)-5) + \cdots + (3(25)-5)$$
$$= (-2) + 1 + 4 + \cdots + 70$$

To find the sum of this arithmetic series, use the formula $S_n = \dfrac{n}{2}(a_1 + a_n)$, where $a_1 = -2$, $a_{25} = 70$, and $n = 25$:

$$S_{25} = \frac{25}{2}(-2+70) = \frac{25}{2}(68) = 25\cdot 34 = \mathbf{850}.$$

Exercise 4	Evaluating an Infinite Geometric Series

Evaluate: $\displaystyle\sum_{k=1}^{\infty} \left(\frac{4}{7}\right)^k$.

Solution: The given sum represents an infinite geometric series whose first term is $\dfrac{4}{7}$ and whose common ratio is $\dfrac{4}{7}$:

$$\sum_{k=1}^{\infty} \left(\frac{4}{7}\right)^k = \frac{4}{7} + \frac{4}{7}\left(\frac{4}{7}\right) + \frac{4}{7}\left(\frac{4}{7}\right)^2 + \frac{4}{7}\left(\frac{4}{7}\right)^3 + \cdots.$$

Since the common ratio is between 0 and 1, the sum can be evaluated using the formula $S_\infty = \dfrac{a_1}{1-r}$, where $a_1 = r = \dfrac{4}{7}$:

$$S_\infty = \frac{\dfrac{4}{7}}{1-\dfrac{4}{7}} = \frac{\dfrac{4}{7}}{\dfrac{3}{7}} = \frac{4}{3}.$$

Exercise 5   Evaluating a Finite Geometric Series

Evaluate $\displaystyle\sum_{k=2}^{6}\left(\frac{3}{2}\right)^{k}$.

Solution: The given sum represents a geometric series with five terms whose first term

is $\left(\dfrac{3}{2}\right)^{k=2} = \dfrac{9}{4}$, and whose common ratio is $\dfrac{3}{2}$. To find the sum of this series, use the for-

mula $S_n = \dfrac{a_1\left(1-r^n\right)}{1-r}$, where $a_1 = \dfrac{9}{4}$, $r = \dfrac{3}{2}$, and $n = 5$:

$$S_5 = \frac{\dfrac{9}{4}\left(1-\left(\dfrac{3}{2}\right)^5\right)}{1-\dfrac{3}{2}} = \frac{\dfrac{9}{4}\left(\dfrac{2^5-3^5}{2^5}\right)}{-\dfrac{1}{2}} = -\frac{9}{2}\left(-\frac{211}{32}\right) = \frac{\textbf{1899}}{\textbf{64}}.$$

Mathematical Induction

KEY IDEAS

Mathematical induction is a special method of mathematical proof that is particularly useful when it is necessary to prove formula-type statements that depend on a positive integer n.

The Need for Induction

Let P_n represent a statement expressed in terms of a positive integer n, such as

$$P_n:\ 1+3+5+7+\cdots+\left(2n-1\right)=n^2,$$

where n is the number of terms in the series. You can easily verify that statements $P_1, P_2,$ $P_3,$ and P_4 are true:

$$P_1: 1 = 1^2 \qquad\qquad\qquad P_3: 1+3+5 = 9 = 3^2$$
$$P_2: 1+3 = 4 = 2^2 \qquad\qquad P_4: 1+3+5+7 = 16 = 4^2$$

Although statements $P_1, P_2, P_3,$ and P_4 are true, can you be certain that P_n will continue to hold true for each successive integer replacement for n? Since it is not practical or possible to test P_n for every possible integer value of n, a method for proving (or disproving) that this statement is always true is needed.

The Domino Principle of Induction

Imagine an endless row of standing dominoes in which P_1 written on the face of the first domino, P_2 on the face of the second domino, and so forth. All of the dominoes falling in succession, one pushing over the domino that follows it, corresponds to the situation in which statements $P_1, P_2, P_3, \ldots, P_n$ are all true. For all of the dominoes to fall in succession, two conditions are required:

- The first domino falls.
- If another domino after the first falls, it pushes over the domino that follows it.

Mathematical Principle of Induction

To prove by mathematical induction that the statement P_n is true for all possible positive-integer replacements for n, complete these two steps:

- VERIFICATION STEP: Show that $P_{n=\text{initial value}}$ is true. This corresponds to verifying that the first domino falls. Completing this step involves replacing n with the starting value of n, typically 1, and then confirming that the resulting statement is true. The starting value for n is called the **anchor value**.
- INDUCTION STEP: Show that $P_{n=k+1}$ is true assuming that $P_{n=k}$, where $k > 1$, is true. This step corresponds to showing that, when a domino after the first falls, the next consecutive domino also falls. The assumption that P_k is true is called the **induction hypothesis**.

Mathematical Induction: an Example

To prove that

$$P_n : 1 + 3 + 5 + 7 + \cdots + (2n - 1) = n^2$$

is true for all positive-integer values of n, perform the verification and induction steps.

- VERIFICATION STEP. The anchor value for n is 1. The statement P_1 is true because replacing n with 1 leads to $1 = 1^2$, which is true.
- INDUCTION STEP. Assume that P_k is true, which means that

$$P_k : 1 + 3 + 5 + 7 + \cdots + (2k - 1) = k^2. \quad \leftarrow \text{Induction hypothesis}$$

Now show that P_{k+1} is also true. Since the next term on the left side of the equation would be $(2k - 1 + 2) = (2k + 1)$, add $(2k + 1)$ to both sides of P_k:

$$1 + 3 + 5 + 7 + \cdots + (2k - 1) + \boxed{2k + 1} = k^2 + \boxed{2k + 1}.$$

Simplify the right side of the equation by factoring $k^2 + 2k + 1$ as $\left(k + 1\right)^2$:

$$1 + 3 + 5 + 7 + \cdots + \left(2k - 1\right) + \left(2k + 1\right) = \left(k + 1\right)^2.$$

The last equation is the statement P_{k+1} since it is the same statement that is obtained by replacing n with $k+1$ in P_n. Because P_{k+1} is true when P_k is true, the induction step is finished. Hence, by mathematical induction, P_n is true for all positive integers.

Exercise 1   **Proof by Mathematical Induction**

Prove: $1 + 2 + 4 + 8 + \cdots + 2^{n-1} = 2^n - 1$ for all positive-integer values of n.

Solution: Complete the two steps required for an induction proof.

- VERIFICATION STEP: The anchor value for n is 1. The statement P_1 is true because replacing n with 1 leads to $1 = 2^1 - 1 = 1$, which is true.
- INDUCTION STEP: Write the induction hypothesis by replacing n with k:

$$P_k : 1 + 2 + 4 + 8 + \cdots + 2^{k-1} = 2^k - 1.$$

Since the next consecutive term on the left side of the equation would be 2^k, add 2^k to both sides of the preceding equation:

$$1 + 2 + 4 + 8 + \cdots + 2^{k-1} + 2^k = 2^k + 2^k - 1.$$

On the right side of the equation, rewrite $2^k + 2^k$ as $2 \cdot 2^k = 2^1 \cdot 2^k = 2^{k+1}$, giving

$$1 + 2 + 4 + 8 + \cdots + 2^{k-1} + 2^k = 2^{k+1} - 1.$$

The last equation is the statement P_{k+1} since it is the same statement as is obtained by letting $n = k + 1$. Because both the verification and the induction steps have been completed, the proof by mathematical induction is complete.

Proving Divisibility

Mathematical induction is used also in proofs other than those that establish general sum formulas. Some algebraic expressions that depend on a positive integer n are always divisible by the same constant. For example:

- $4^2 - 1$ is divisible by 3 since $\dfrac{4^2 - 1}{3} = \dfrac{15}{3} = 5$.

- $4^3 - 1$ is divisible by 3 since $\dfrac{4^3 - 1}{3} = \dfrac{63}{3} = 21$.

- $4^4 - 1$ is divisible by 3 since $\dfrac{4^4 - 1}{3} = \dfrac{255}{3} = 85$.

If P_n represents the statement " $4^n - 1$ is divisible by 3 for all positive-integer values of n," then mathematical induction can be used to prove that statement P_n is always true. Here is the proof.

- VERIFICATION STEP: The anchor value is 1. Statement P_1 is true because $4^1 - 1 = 3$ and 3 is divisible by 3.
- INDUCTION STEP: Assume that the induction hypothesis, " $4^k - 1$ is divisible by 3," is true. If $4^k - 1$ is divisible by 3, then $4^k - 1$ must be a whole-number multiple of 3. Thus, $4^k - 1 = 3p$, where p is a whole number.

1. Multiply both sides of $4^k - 1 = 3p$ by 4: $4\left(4^k - 1\right) = 4\left(3p\right)$, so $4^{k+1} - 4 = 12p$.

2. Add 3 to each side of $4^{k+1} - 4 = 12p$: $4^{k+1} - 1 = 12p + 3$ or, equivalently, $4^{k+1} - 1 = 3\left(4p + 1\right)$.

3. Because $4p + 1$ is a positive integer, $3\left(4p + 1\right)$ is an integer multiple of 3, meaning that $4^{k+1} - 1$ is divisible by 3. The last statement corresponds to the statement P_{k+1}. Because P_{k+1} is true, the proof by mathematical induction is complete.

Proving a General Inequality

Some pairs of algebraic expressions that depend on a positive-integer value of n maintain the same size relationship. For example, compare $n!$ with 2^n for integer values of n greater than or equal to 4:

$n!$	2^n	Size Relationship
$4! = 4 \times 3 \times 2 \times 1 = 24$	$2^4 = 16$	$4! > 2^4$
$5! = 5 \times 4 \times 3 \times 2 \times 1 = 120$	$2^5 = 32$	$5! > 2^5$
.	.	.
.	.	.
.	.	.
$n!$	2^n	$n! > 2^n$

The inequality statement $n! > 2^n$ for $n \geq 4$ can be proved using mathematical induction:

- VERIFICATION STEP: Since the anchor value is 4, you need to show that P_4 is true. Because $4! = 4 \times 3 \times 2 \times 1 = 24$ and $2^4 = 16$, $4! > 2^4$ is true.
- INDUCTION STEP: Assume that the induction hypothesis, " $P_k : k! > 2^k$," is true.

1.   Multiply each side of $k! > 2^k$ by $(k+1)$: $(k + 1)(k!) > (k + 1)(2^k)$. Recognizing that $(k + 1)(k!) = (k + 1)!$, write the last inequality as $(k + 1)! > (k + 1)(2^k)$.

2.   Because the anchor value is 4, $k > 4$:

$$2^k = 2^k$$
$$\underline{\underline{(k+1)}} \cdot 2^k > \underline{\underline{(2)}} \cdot 2^k \qquad \leftarrow \text{Because } k+1 > 2$$
$$\underline{\underline{(k+1)}} \cdot 2^k > 2^{k+1} \qquad \leftarrow \text{Rewrite } 2 \cdot 2^k \text{ as } 2^{k+1}.$$

3.   You now know that

$$(k + 1)! > (k + 1)(2^k) \quad \text{and} \quad (k + 1) \cdot 2^k > 2^{k + 1}.$$

Hence, $(k + 1)! > 2^{k + 1}$. The last inequality is the same inequality that is obtained by

replacing n with $k + 1$ in P_n. Hence, P_{k+1} is true, so the proof by mathematical induction is complete.

Lesson 15-5 — Permutations and Combinations

KEY IDEAS

The letters A, B, and C may be arranged in six different ways:

(1) A, B, C	(3) B, A, C	(5) C, A, B
(2) A, C, B	(4) B, C, A	(6) C, B, A

Although there are six different *permutations* of the letters A, B, and C, there is exactly one *combination* of the three letters: $\{A, B, C\}$. A **combination** is an unordered selection of objects, while a **permutation** is any ordered arrangement of those objects.

Permutation Notation

The product of consecutive positive integers from n down to 1, inclusive, is called **n factorial** and is written as $n!$, where $0!$ is defined to be equal to 1. For example, $4! = 4 \times 3 \times 2 \times 1 = 24$. The number of different ways in which n objects can be arranged in a line is $n!$. Each different arrangement is called a *permutation*.

EXAMPLE: The number of ways in which five children of different heights can be arranged in a line in height order is $5! = 5 \times 4 \times 3 \times 2 \times 1 = 120$.

Arranging n Objects in Fewer than n Slots

The number of objects or people to be arranged may be greater than the available number of slots. For example, if seven students run in a race in which there are no ties, the number of different arrangements of first, second, and third place that are possible is

$$\boxed{7} \times \boxed{6} \times \boxed{5} = 210.$$

Thus, the total number of possible arrangements of seven students taken three at a time is the product of the three greatest factors of 7. This product can be represented using the shorthand notation $_7P_3$; that is, $_7P_3 = 7 \times 6 \times 5 = 210$. Since the number of students is greater than the number of finishing positions, not every student will appear in each possible arrangement.

Counting Arrangements Using Permutation Notation

- $_nP_n$ is read as "the permutation of n things taken n at time," and represents the total number of arrangements of n objects in n slots. To calculate $_nP_n$, use:

$$_nP_n = n!.$$

- $_nP_r$ is read as "the permutation of n objects taken r at a time," and represents the total number of arrangements of n objects in r slots, where $1 \leq r \leq n$. To calculate $_nP_r$, find the product of the r greatest factors of $n!$:

$$_nP_r = \underbrace{n \times (n-1) \times (n-2) \times \cdots \times (n-r+1)}_{r \text{ greatest factors of } n!}.$$

Exercise 1   Counting Arrangements of Objects When There Are Fewer Positions Than Objects

Shari remembers that the last four digits of a seven-digit telephone number but knows only that each of the first three digits of the telephone number is a different odd number. Find the maximum number of telephone calls she must make until she dials the correct number.

Solution: There are five odd digits: 1, 3, 5, 7, and 9. These five digits can fill the three available slots in $_nP_r$ ways, where $n = 5$ and $r = 3$. Hence, the maximum number of telephone calls Shari must make until she dials the correct number is $_5P_3 = 5 \times 4 \times 3 = \mathbf{60}$.

| Exercise 2 | Counting Ordered Arrangements Subject to Conditions |

Seven students with unequal heights are arranged in a line. In how many ways can the students be arranged in the line so that the shortest student is first and the tallest student is last?

Solution: The shortest student must fill the first of the seven positions, and the tallest student must fill the last position. The remaining five positions can be filled in 5! ways:

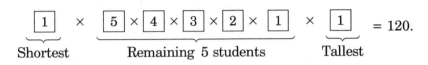

$$\underbrace{\boxed{1}}_{\text{Shortest}} \times \underbrace{\boxed{5} \times \boxed{4} \times \boxed{3} \times \boxed{2} \times \boxed{1}}_{\text{Remaining 5 students}} \times \underbrace{\boxed{1}}_{\text{Tallest}} = 120.$$

Thus, the students can be arranged in **120** different ways.

Combination Notation

A *combination* is a selection of people or objects in which the identity, rather than the order, of the people or objects is important. If exactly two letters are selected from $\{A, B, C, D\}$, each of the six possible two-letter subsets is called a *combination of the four letters taken two at a time*:

$$\{A,B\}, \{A,C\}, \{A,D\}, \{B,C\}, \{B,D\}, \text{ and } \{C,D\}.$$

When sets have larger numbers of elements, making lists can be time consuming and error prone. In such cases, a convenient formula can be used to figure out the number of combinations of n objects taken r at a time, which is denoted by $_nC_r$.

Combinations Formula

$$_nC_r = \frac{n!}{r!\,(n-r)!}, \text{ where } 1 \le r \le n.$$

EXAMPLE: To find the number of different subcommittees consisting of three teachers that can be formed from a group of seven teachers, evaluate $_nC_r$ for $n = 7$ and $r = 3$:

$$_7C_3 = \frac{7!}{3!(7-3)!} = \frac{7 \times 6 \times 5 \times 4 \times 3 \times 2 \times 1}{(3!)(4!)}$$

$$= \frac{7 \times 6 \times 5 \times \overset{1}{\cancel{4 \times 3 \times 2 \times 1}}}{(3 \times 2 \times 1)(\cancel{4 \times 3 \times 2 \times 1})}$$

$$= \frac{7 \times \overset{1}{\cancel{6}} \times 5}{(\cancel{3 \times 2 \times 1})}$$

$$= 35$$

You may be able to save some time and effort by remembering that:

- $_nC_0 = 1.$ **EXAMPLE:** $_7C_0 = 1.$

- $_nC_1 = n.$ **EXAMPLE:** $_7C_1 = 7.$

- $_nC_n = 1.$ **EXAMPLE:** $_7C_7 = 1.$

Evaluating $_nP_r$ and $_nC_r$ Using a Calculator

To evaluate $_nP_r$ or $_nC_r$ using your calculator:

1. Press $\boxed{\text{2nd}}$ $\boxed{\text{QUIT}}$ to return to the home screen. Then enter the value of n.

2. Press $\boxed{\text{MATH}}$ $\boxed{\triangleright}$ $\boxed{\triangleright}$ $\boxed{\triangleright}$ $\boxed{3}$ to select $_nP_r$ or $_nC_r$ from the MATH PRB menu.

3. Enter the value of r. Then press $\boxed{\text{ENTER}}$.

Not all calculators work in the same way. If this procedure does not work with your calculator, read the manual that came with the calculator.

Multiplying Combinations

To figure out the total number of selections that can be made from two or more different groups, multiply together the number of possible selections for each of the groups.

Exercise 3   Counting Using Combinations

There are six pens and seven books on a desk. In how many different ways can four pens *and* three books be selected from the desk?

Solution: Figure out the number of possible selections for four from six pens and the number of possible selections for three from seven books. Then multiply the results together.

- Four pens can be selected from six pens in $_6C_4$ ways. Use your calculator to find that $_6C_4 = 15$.
- Three books can be selected from seven books in $_7C_3$ ways. Use your calculator to find that $_7C_3 = 35$.
- The number of different ways in which four pens *and* three books can be selected is $15 \times 35 = \mathbf{525}$.

Solving Combination Problems with Conditions

At least means "is equal to or is greater than," and *at most* means "is equal to or is less than." Example 4 illustrates how to solve counting problems with combinations involving these two types of conditions.

Exercise 4   Forming Combinations Subject to Conditions

A four-member subcommittee is selected at random from a U.S. Senate committee consisting of five Republicans and three Democrats.

a. How many four-member subcommittees will include *at least* two Democrats?
b. How many four-member subcommittees will include *at most* three Republicans?

Solutions: *a.* Since the committee includes three Democrats, any subcommittee cannot include more than three Democrats. The four-member subcommittee will include *at least* two Democrats if it consists of 2 Democrats and 2 Republicans *or* 3 Democrats and 1 Republican.

Subcommittee Make-up	Number of Subcommittees
• 2 Democrats *and* 2 Republicans	$_3C_2 \times {_5}C_2 = 3 \times 10 = 30$
• 3 Democrats *and* 1 Republican	$_3C_3 \times {_5}C_1 = 1 \times 5 = 5$

Thus, $30 + 5 = $ **35** four-member subcommittees will include *at least* two Democrats.

b.   The four member subcommittee will include *at most* three Republicans if it consists of 3 Republicans and 1 Democrat, 2 Republicans and 2 Democrats, 1 Republican and 3 Democrats, or no Republican and 4 Democrats, which is impossible since there are only three Democrats.

Subcommittee Make-up	Number of Subcommittees
• 3 Republicans *and* 1 Democrat	$_5C_3 \times {_3}C_1 = 10 \times 3 = 30$
• 2 Republicans *and* 2 Democrats	$_5C_2 \times {_3}C_2 = 10 \times 3 = 30$
• 1 Republican *and* 3 Democrats	$_5C_1 \times {_3}C_3 = 5 \times 1 = 5$

Thus, $30 + 30 + 5 = $ **65** four-member subcommittees will include *at most* three Republicans.

Lesson 15-6   The Binomial Theorem

KEY IDEAS

A binomial such as $(a + b)^5$ may be expanded by repeated multiplication using $(a + b)$ as a factor five times, or by using an algebraic formula called the *binomial theorem*.

Expanding $(a + b)^n$

Using repeated multiplication, you can verify that

$$(a+b)^5 = \underbrace{1 \cdot a^5}_{\text{Term 1}} + \underbrace{5a^4b^1}_{\text{Term 2}} + \underbrace{10a^3b^2}_{\text{Term 3}} + \underbrace{10a^2b^3}_{\text{Term 4}} + \underbrace{5a^1b^4}_{\text{Term 5}} + \underbrace{1 \cdot b^5}_{\text{Term 6}}.$$

- The expansion of $(a + b)^5$ has $5 + 1$ or 6 terms. The first term is a^5, and the last term is b^5.
- The expansion consists of descending powers of a and ascending powers of b such that the sum of the exponents of a and b in each term is 5.

- The numerical coefficients of the terms of the expansion can be represented using combinations, where the numerical coefficient of the kth term of $(a + b)^n$ is $_nC_{k-1}$.

The **binomial theorem** tells how to expand a binomial to a given power without performing repeated multiplication.

> ### The Binomial Theorem
>
> If n is a positive whole number, then
> $$\left(a+b\right)^n = {_nC_0}\, a^n + {_nC_1}\, a^{n-1}\, b^1 + {_nC_2}\, a^{n-2}\, b^2 + \cdots + {_nC_n}\, a^0 b^n = \sum_{r=0}^{n} {_nC_r}\, a^{n-r} b^r .$$

Binomial Coefficients and Pascal's Triangle

When the values of $_nC_r$ are arranged in a triangular pattern in which the rows correspond to successive values of n, starting with $n = 0$, the result is an interesting array of numbers called **Pascal's triangle,** shown in the accompanying figure. Pascal's triangle is named for the French mathematician Blaise Pascal (1623–1662).

row 0:	$_0C_0$	1
row 1:	$_1C_0 \quad _1C_1$	1 1
row 2:	$_2C_0 \quad _2C_1 \quad _2C_2$	1 2 1
row 3:	$_3C_0 \quad _3C_1 \quad _3C_2 \quad _3C_3$	1 3 3 1
row 4:	$_4C_0 \quad _4C_1 \quad _4C_2 \quad _4C_3 \quad _4C_4$	1 4 6 4 1
row 5:	$_5C_0 \quad _5C_1 \quad _5C_2 \quad _5C_3 \quad _5C_4 \quad _5C_5$	1 5 10 10 5 1
. . .	$\cdots$	$\cdots$

Figure 15.1 Pascal's triangle

In Pascal's triangle, each number after 1 is the sum of the two numbers directly above it. When this fact and the symmetry pattern that results from it are known, it is easy to create Pascal's triangle.

Pascal's triangle allows you to quickly find the value of $_nC_r$ for different values of n and r, where n is the row number and $r = 0$ corresponds to the first entry on each row.

Exercise 1   Expanding a Binomial Using the Binomial Theorem

Write the expansion of $\left(x - 3y\right)^4$.

Solution: Use the binomial theorem to expand $\left(a + b\right)^n$, where $a = x$, $b = -3y$, and $n = 4$:

$$\left(x - 3y\right)^4 = {}_4C_0\, x^4 + {}_4C_1\, x^3\left(-3y\right)^1 + {}_4C_2\, x^2\left(-3y\right)^2 + {}_4C_3\, x^1\left(-3y\right)^3 + {}_4C_4\, x^0\left(-3y\right)^4.$$

Evaluate the combinations either by using a calculator or by copying the values from row 4 of Pascal's triangle in the preceding figure:

$$= 1x^4 + 4x^3\left(-3y\right) + 6x^2\left(-3y\right)^2 + 4x\left(-3y\right)^3 + 1\left(-3y\right)^4$$
$$= x^4 - 12x^3 y + 54x^2 y^2 - 108xy^3 + 81y^4$$

Exercise 2   Applying Pascal's Triangle

Write the expansion of $\left(x + y\right)^6$.

Solution: According to the binomial theorem:

$$\left(x + y\right)^6 = {}_6C_0\, x^6 + {}_6C_1\, x^5\, y + {}_6C_2\, x^4\, y^2 + {}_6C_3\, x^3 y^3 + {}_6C_4\, x^2\, y^4 + {}_6C_5\, x\, y^5 + {}_6C_6\, y^6$$

To evaluate the binomial coefficients, use row 6 of Pascal's triangle. To obtain row 6 from row 5 of Pascal's triangle in the preceding figure, write 1 as the first and the last member of row 6. Then obtain each of the remaining numbers in row 6 by adding the two numbers directly above from row 5:

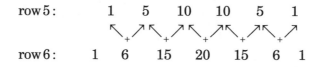

Hence,

$$\left(x + y\right)^6 = 1x^6 + 6x^5\, y + 15x^4\, y^2 + 20\, x^3 y^3 + 15\, x^2\, y^4 + 6\, x\, y^5 + 1y^6.$$

Writing the kth Term of $(a + b)^n$

Sometimes the only thing you want to know about the power of a binomial is what a particular term looks like.

The kth Term of a Binomial Expansion

The kth term of $(a+b)^n$ is ${}_nC_{k-1}\,a^{n-(k-1)}\,b^{k-1}$.

From Example 1, you know that the third term of the expansion of $(x-3y)^4$ is $54x^2y^2$.

This particular term can also be obtained using the expression ${}_nC_{k-1}\,a^{n-(k-1)}\,b^{k-1}$, where $n = 4$, $k = 3$, $a = x$, and $b = -3y$:

$$
\begin{aligned}
{}_4C_{3-1}\,x^{4-(3-1)}\left(-3y\right)^{3-1} &= {}_4C_2\,x^{4-2}\left(-3y\right)^2 \\
&= 6\left(x^2\right)\left(9y^2\right) \\
&= 54x^2y^2
\end{aligned}
$$

| **Exercise 3** | **Finding a Specific Term in a Binomial Expansion** |

What is the middle term of the expansion of $\left(\dfrac{x}{3}+y^2\right)^4$?

Solution: Since the expansion of $\left(\dfrac{x}{3}+y^2\right)^4$ consists of $4 + 1$ or 5 terms, the middle term

of the expansion is the third term. To find the third term of the expansion of $\left(\dfrac{x}{3}+y^2\right)^4$,

evaluate ${}_nC_{k-1}\,a^{n-(k-1)}\,b^{k-1}$, where $n = 4$, $k = 3$, $a = \dfrac{x}{3}$, and $b = y^2$:

$$
\begin{aligned}
{}_4C_{3-1}\left(\frac{x}{3}\right)^{4-(3-1)}\left(y^2\right)^{3-1} &= {}_4C_2\left(\frac{x}{3}\right)^2\left(y^2\right)^2 \\
&= 6\left(\frac{x^2}{9}\right)\left(y^4\right) \\
&= \frac{2x^2y^4}{3}
\end{aligned}
$$

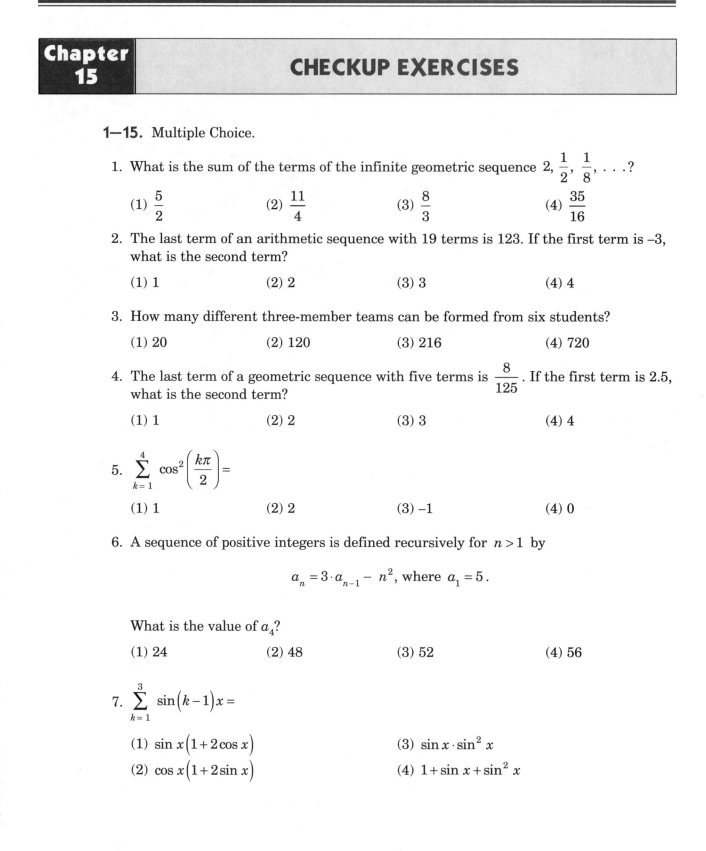

Chapter 15 — CHECKUP EXERCISES

1–15. Multiple Choice.

1. What is the sum of the terms of the infinite geometric sequence $2, \dfrac{1}{2}, \dfrac{1}{8}, \ldots$?

   (1) $\dfrac{5}{2}$ (2) $\dfrac{11}{4}$ (3) $\dfrac{8}{3}$ (4) $\dfrac{35}{16}$

2. The last term of an arithmetic sequence with 19 terms is 123. If the first term is –3, what is the second term?

   (1) 1   (2) 2   (3) 3   (4) 4

3. How many different three-member teams can be formed from six students?

   (1) 20   (2) 120   (3) 216   (4) 720

4. The last term of a geometric sequence with five terms is $\dfrac{8}{125}$. If the first term is 2.5, what is the second term?

   (1) 1   (2) 2   (3) 3   (4) 4

5. $\displaystyle\sum_{k=1}^{4} \cos^2\left(\dfrac{k\pi}{2}\right) =$

   (1) 1   (2) 2   (3) –1   (4) 0

6. A sequence of positive integers is defined recursively for $n > 1$ by

   $$a_n = 3 \cdot a_{n-1} - n^2, \text{ where } a_1 = 5.$$

   What is the value of a_4?

   (1) 24   (2) 48   (3) 52   (4) 56

7. $\displaystyle\sum_{k=1}^{3} \sin(k-1)x =$

   (1) $\sin x\left(1 + 2\cos x\right)$ (3) $\sin x \cdot \sin^2 x$

   (2) $\cos x\left(1 + 2\sin x\right)$ (4) $1 + \sin x + \sin^2 x$

8. The Fibonacci sequence of numbers is defined recursively for $n > 2$ by

$$a_n = a_{n-1} + a_{n-2}, \text{ where } a_1 = a_2 = 1.$$

What is the eighth term in the Fibonacci sequence?

(1) 13   (2) 17   (3) 21   (4) 38

9. In the sequence shown below, $a_1 = \dfrac{1}{2}$ and a_n represents the nth term.

$$\frac{1}{2}, -\frac{1}{4}, \frac{1}{8}, -\frac{1}{16}, \ldots, a_n.$$

Which equation expresses a_n in terms of n?

(1) $a_n = \left(-1\right)^n \cdot \dfrac{1}{2^n}$ (3) $a_n = \left(-1\right)^n \cdot \dfrac{1}{2^{n+1}}$

(2) $a_n = \left(-1\right)^{n+1} \cdot \dfrac{1}{2^n}$ (4) $a_n = \left(-\dfrac{1}{2}\right)^n$

10. The exact sum of the first 14 terms in the geometric sequence $\sqrt{2}, \ 2, \ 2\sqrt{2}, \ 4, \ldots$ is

(1) $254 + 127\sqrt{2}$ (3) $256\sqrt{2}$

(2) $\dfrac{64 + 512\sqrt{2}}{1 - \sqrt{2}}$ (4) $\dfrac{255\sqrt{2}}{1 + \sqrt{2}}$

11. What is the numerical coefficient of the term containing $x^3 y^2$ in the expansion of $\left(x + 2y\right)^5$?

(1) 10   (2) 20   (3) 40   (4) 80

12. What is the third term in the expansion of $\left(x + 2y\right)^5$?

(1) $10x^3 y^2$ (2) $40x^3 y^2$ (3) $80x^2 y^3$ (4) $20x^2 y^3$

13. What is the middle term in the expansion of $\left(3x - 2y\right)^4$?

(1) $-6x^2 y^2$ (2) $36x^2 y^2$ (3) $-216x^2 y^2$ (4) $216x^2 y^2$

14. The fifth term in the expansion of $\left(3a - b\right)^6$ is

(1) $135a^2 b^4$ (2) $540a^3 b^3$ (3) $-18ab^5$ (4) $-135a^2 b^4$

15. What is the middle term in the expansion of $\left(x^2 + \dfrac{1}{x} \right)^6$?

    (1) $10x^3$ (2) $30x^2$ (3) $20x^3$ (4) $\dfrac{10}{x^2}$

16–21. In each case, evaluate and express the result in simplest form.

16. $\displaystyle\sum_{k=1}^{4} \left(k^3 - 1 \right)^2$ 18. $\displaystyle\sum_{k=1}^{4} (-1)^k \cdot \left(k^2 + 1 \right)$ 20. $\displaystyle\sum_{k=0}^{3} \log\left(10^{k+1} \right)$

17. $\displaystyle\sum_{j=2}^{4} \frac{1}{3} \left(\frac{6j}{j-1} \right)^j$ 19. $\displaystyle\sqrt{1 + \sum_{n=1}^{3} \left(n^2 - n \right)}$ 21. $\displaystyle\sum_{k=2}^{5} {}_7C_{k+1}$

22–24. In each case, evaluate and express the result in simplest form.

22. $\displaystyle\sum_{k=1}^{40} \left(3k + 2 \right)$ 23. $\displaystyle\sum_{k=1}^{5} \left(\frac{3}{4} \right)^k$ 24. $\displaystyle\sum_{k=1}^{\infty} \left(\frac{3}{4} \right)^k$

25. How many numbers from 13 to 894 are divisible by 6?

26. Use an infinite geometric series to express 0.407407407. . . as the ratio of two integers.

27. What is the value of $\displaystyle\sum_{k=1}^{\infty} 4\left(\frac{1}{3} \right)^k$?

28. What is the value of $\displaystyle\sum_{k=1}^{8} (-1)^{k+1}\left(\frac{1}{3} \right)^k$?

29. How many four-digit even numbers can be formed if no digit is repeated?

30. The telephone company has run out of seven-digit telephone numbers for an area code. To fix this problem, the company will introduce a new area code. Find the number of new seven-digit telephone numbers that can be generated for the new area code if both of the following conditions must be met:

CONDITION 1: The first digit cannot be 0 or 1.

CONDITION 2: The first three digits cannot be the emergency number (911) or the number used for information (411).

31. Four girls and three boys are arranged in a line so that all of the girls stand in front of the boys. In how many more ways can these boys and girls be arranged if the girls and boys can appear in any order?

32. Find the sum of the first 37 terms of an arithmetic sequence in which the second term is –12 and the ninth term is 65.

33. The length of a side of a square is 8. The midpoints of the adjacent sides of the square are connected by line segments to form another square inside the first. In a similar fashion, a third square is formed inside the second square by connecting the midpoints of its adjacent sides. If this process is continued indefinitely, what is the sum of the perimeters of all such squares?

34 and 35. A ball is dropped from a height of 8 feet and allowed to bounce. Each time the ball bounces, it rebounds back to half its previous height. The vertical distance, d, that the ball travels is given by the formula

$$d = 8 + 16\sum_{k=1}^{n}\left(\frac{1}{2}\right)^{k},$$

where n is the number of bounces.

34. What is the total vertical distance that the ball has traveled after 12 bounces?

35. What is the total vertical distance that the ball travels if it rebounds indefinitely?

36 and 37. The first term of a geometric sequence is 27, and the sixth term is $\dfrac{512}{9}$.

36. Find the seventh term.

37. Find the sum of the first five terms of this sequence.

38 and 39. The second term of an infinite geometric sequence is $\dfrac{4}{9}$, and the sixth term is $\dfrac{9}{64}$.

38. What is the sum of the first five terms of this sequence?

39. What is the sum of all of the terms of this sequence?

40. Forrest receives a box of chocolates containing ten candies: 4 nut clusters, 1 peppermint, 2 jellies, and 3 caramels. What is the number of possible five-candy selections that contain 2 nut clusters, 2 jellies, and 1 caramel?

41. On a math test a student is to select any four out of ten problems of equal difficulty. The test contains 2 geometry, 3 algebra, 1 statistics, and 4 probability problems.

   a. How many different four-problem selections are possible if each selection includes a different type of problem?

   b. How many different four-problem selections are possible if each selection includes the same number of algebra problems as probability problems?

42. A bookshelf contains six mysteries and three biographies. In how many different ways can two books be selected so that *at least* one of the books is a mystery book?

43. The student government at Central High School consists of 4 seniors, 3 juniors, 3 sophomores, and 2 freshmen. How many different nine-student committees can be formed that include, *at most*, one freshman?

44. Christine plans to rent six videos for the weekend. She has narrowed her selection to four comedies, eight mysteries, and three musicals. In how many different ways can Christine select the videos so that her selection includes at *least one* musical? At *most one* mystery?

45. Let $z_1 = 2 - i$. Function f is defined recursively over the set of complex numbers as follows:

$$f(n) = z_n = \begin{cases} z_1; \ n = 1 \\ \left(z_{n-1}\right)^2 + z_1 \ ; \ n > 1 \end{cases}.$$

Determine $f(3)$.

46–51. In each case, use mathematical induction to prove that P_n is true for all positive-integer values of n.

46. $P_n : 2 + 4 + 6 + \cdots + 2n = n^2 + n$

47. $P_n : 1 + 5 + 9 + \cdots + \left(4n - 3\right) = n\left(2n - 1\right)$

48. $P_n : 2 + 6 + 18 + \cdots + 2 \cdot 3^{n-1} = 3^n - 1$

49. $P_n : 1^2 + 3^2 + 5^2 + \cdots + \left(2n - 1\right)^2 = \dfrac{n\left(4n^2 - 1\right)}{3}$

50. $P_n : 3 + 3^2 + 3^3 + \cdots + 3^n = \dfrac{3\left(3^n - 1\right)}{2}$

51. $P_n : \dfrac{1}{1 \cdot 2} + \dfrac{1}{2 \cdot 3} + \dfrac{1}{3 \cdot 4} + \cdots + \dfrac{1}{n\left(n + 1\right)} = \dfrac{n}{n + 1}$

52. Prove that $5^n - 1$ is divisible by 4.

53. Prove that $3^n - 1$ is divisible by 2.

54. Prove that $n^3 + 2n$ is divisible by 3.

55. Prove that $2n + 1 < 2^n$ for $n \geq 3$.

56. Prove $2^n > n^2$, for $n \geq 5$.

Chapter 1

ANSWERS TO CHECKUP EXERCISES

This section contains the answers to each of the exercises found throughout the book. The answers that provide thorough, worked-out solutions and show all work are marked with asterisks, and can be found after the list of answers for each chapter.

1. $x = 3$

2. $n = 9$

3. $x = 7$

4. $y < 3$

5. $p = -13$

6. $q = 18$

7. $w = -6$

8. $y = 2$

9. $x \geq -4$

10. $x = 4$

11. $x = -5$

12. $x < 24$

13. $x \leq \dfrac{2}{3}$

14. $x \leq 3$

15. $x = 5$

16. $\dfrac{-11}{2} < x \leq \dfrac{21}{2}$

17. $z > 10$

18. $x = 1.3$

19. $x^2 - 6xy + 9y^2$

20. $2x^2 + 13x - 7$

21. $16y^2 - 9$

22. $36x^3 - 31x^2 + 23x - 15$

23. $5x^3 - 16x^2 + 7x + 2$

24. $-6y^4 - y + 2$

25. $h^2k + hk^2 - 7$

26. $0.2a^2 - 0.15ab$

27. $(y - 0.3)(y + 0.3)$

28. $(x - 2)(x + 2)(x^2 + 4)$

29. $(x - 7)(x + 3)$

30. $n(n - 4)(n + 3)$

31. $4m^3n^2(2m^2 - 3n^2)$

32. $-(y - 11)(y + 1)$

33. $t(r - s^2)(r + s^2)$

34. $(w - 2v)(w^2 + 2vw + 4v^2)$

35. $(4h + k)(16h^2 - 4hk + k^2)$

36. $(x - 2)(x + 2)(y - 1)(y + 1)$

37. $(3x - 1)(x + 2)$

38. $(2y + 5)(y - 9)$

39. $(a - 3)(a + 3)(a - b)(a + b)$

40. $(2x - 5)(2x + 3)$

41. $(x^2 + 2)(x - 1)(x^2 + x + 1)$

42. $w = \dfrac{p - 2l}{2}$

43. $t = \dfrac{x}{x + s}$

44. $x = \dfrac{c - ab}{a}$

45. $F = \dfrac{9}{5}C + 32$

46. b^n

47. $\dfrac{9a^{2n}}{b^n}$

48. $y = -\dfrac{1}{2}$

49. 5

50. 113

51. $\dfrac{11}{18}$

52. 31

53. 30

54. 8

55. 135

56. 11 years

57. 2 inches

Chapter 2

ANSWERS TO CHECKUP EXERCISES

1. (3)

2. (3)

3. (2)

4. (3)

5. (4)

6. $\dfrac{3}{2 - x}$

7. $\dfrac{5a}{2a + 3b}$

8. $\dfrac{-b}{5a}$

9. $\dfrac{2x - 10}{2x - 5}$

10. $\dfrac{x - 3}{4}$

11. $\dfrac{x - 2y}{x + 2y}$

12. $24\sqrt{3}$

13. $-6x\sqrt{3}$

14. 44

15. $-4 - 4\sqrt{3}$

16. $30 - 12\sqrt{6}$

17. 3

18. $\dfrac{1}{a-1}$

19. $a+b$

20. $3y\sqrt[3]{2}$

21. $y\sqrt[3]{y}$

22. $\dfrac{-3\sqrt{5}}{2}$

23. $5\sqrt{2}$

24. $-\dfrac{1}{2}$

25. $\dfrac{-2(2x+1)}{5}$

26. $\dfrac{x^2+2x-10}{(x-2)(x-3)(x+3)}$

27. $\dfrac{-(2\sqrt{5}+20)}{19}$

28. $\dfrac{\sqrt{x}+1}{x-1}$

29. $\sqrt{3}+1$

30. $-\dfrac{23+6\sqrt{10}}{13}$

31. $\dfrac{1}{x-3}$

32. x

33. $-w$

34. $\dfrac{x}{y}$

35. $\dfrac{m+1}{m-1}$

36. $\dfrac{b+6}{b}$

37. $\dfrac{5x+5}{(x-2)(x+2)(x+3)}$

38. $\dfrac{x^2(x+3)}{y(x-3)}$

39. $\dfrac{x-4}{2(x+4)(x-1)}$

40. $\dfrac{-4}{9(t+2)}$

41. $\dfrac{(x-2)(4x+1)}{x+1}$

Chapter 3 — ANSWERS TO CHECKUP EXERCISES

1. (2)
2. (1)
3. (4)
4. (4)

5. (2)
6. (3)
7. (3)
8. (3)

9. (2)
10. (1)
11. $y=2x-4$
12. -2

13. $-\dfrac{9}{2}$
14. $y=-2x+3$

15.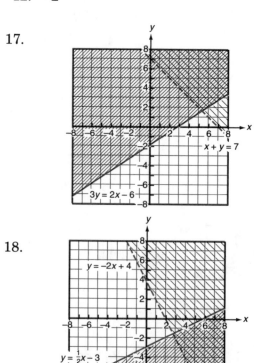

16.

17.

18.

19. $(2,-10)$

20. $\left(\dfrac{7}{5}, \dfrac{28}{5}\right)$

21. $(9,20)$

22. $(3,-4)$

23. $(3,9)$

24. $(5,-2)$

25. $0.70

26. 7

27. $-40°$

28. 90

29. $w = 8.22L - 77.11$; 17.7

30. $y = -3.35x + 99.78$; 87.5

31. $y = 0.72x - 0.13 >$; 2.0

32. $a.$ $y = -6.2x + 12,451.2$

   $b.$ 20.2 thousand $c.$ 2008

Chapter 4 — ANSWERS TO CHECKUP EXERCISES

1. (3)

2. (3)

3. (2)

4. (2)

5. (2)

6. (3)

7. (3)

8. (3)

9. (1)

10. (4)

11. (1)

12. (3)

13. (1)

14. (2)

15. (1)

16. (2)

17. (1)

18. (4)

19. 20.1

20. $4x^{-\frac{1}{3}}$

21. $\dfrac{16}{3}$

22. 2

23. $b.$ $(1,-4)$; $x = 1$

24. $b.$ $(3,2)$; $x = 3$

25. $b.$ $\left(-\dfrac{1}{2}, \dfrac{15}{4}\right)$; $x = -\dfrac{1}{2}$

26. $a.$ 1.5, 7.5

   $b.$ 9

27. 9.13, 25.96

*28. 7 inches

29. -0.16, 6.16

30. 0.54, 2.46

31. 0.56, 4.44

32. $x = \pm\dfrac{3\sqrt{3}}{2}$

33. $y = -1, -2$

34. $x = 0, 6$

35. $h = -5, 0, 8$

36. $p = -5, \dfrac{1}{3}$

37. $x = \dfrac{2}{3}$

38. $r = -\dfrac{1}{2}, 3$

39. $n = -\dfrac{1}{3}, \dfrac{3}{2}$

40. $n = 7, -3$

41. $w = 1, 6$

42. $y = \pm 1, \pm\dfrac{1}{2}$

43. $k = -64, \dfrac{1}{8}$

44. $20 < x < 100$

44. $20 < x < 100$

45. 5600

46. $x \le -5$ or $x \ge 1$

47. $0 < m < 7$

48. $-\dfrac{1}{2} < y < 3$

49. $-9 \le x \le 7$

50. $t \le -2$ or $t \ge 3$

51. $-2 < p < 0$ or $p > 7$

52. $(1,6)$; $(4,9)$

53. $(2,-2)$; $(5,1)$

54. $(2,-3)$; $(6,5)$

55. $a.$ 11.75; 2209

   $b.$ $8.5 < x < 15$

*56. 96,800 square yards

*57. 3

*58. 9.75 inches by 10.25 inches

*59. 17

*60. 34

SELECTED SOLUTIONS

28. If x = width of the box in inches, then the volume (V) is

$$V = l \times w \times h = (3x - 4) \times (x - 2) \times 2 = 102 \text{ cubic inches.}$$

Solve the equation for x: $x = 7$ inches.

56. If x = width, then length = $880 - 2x$, both in yards, so

$$A = x(880 - 2x) = -2x^2 + 800x \text{ square yards.}$$

Then the x-coordinate of the vertex of the parabola $= \dfrac{-b}{2a} = \dfrac{-800}{-4} = 220$, and

$$A_{\text{max}} = 220(880 - 440) = 96,800 \text{ square yards.}$$

57. If x = number of inches that each of the two parallel (12-in.) sides is turned up, then the number of cubic inches in the volume (V) is

$$V = 300(12 - 2x)x = -50x^2 + 300x.$$

This is a maximum at $x = \dfrac{-b}{2a} = \dfrac{-300}{-100} = 3$.

58. Represent the dimensions of the page by x (= length) and $20 - x$ (= width), both in inches. The area (A) of the printed region of the page is

$$A = (x - 2)((20 - x) - 2.5) = -x^2 + 19.5x - 35 \text{ square inches.}$$

This is a maximum at $x = \dfrac{-b}{2a} = \dfrac{-19.5}{-2} = 9.75$ inches = length.

Width = $20 - x = 20 - 9.75 = 10.25$ inches.

59. If y is the total number of apples grown in the orchard when x additional trees are planted, then

$$y = (14 + x)(480 - 10x) = -10x^2 + 340x + 6720.$$

This is a maximum at $x = \dfrac{-b}{2a} = \dfrac{-340}{-20} = 17$.

60. Inscribe the rectangle so that two of its vertices are at $(-x, 0)$ and $(x, 0)$. The remaining vertices of the rectangle must be at $(-x, 16 - x^2)$ and $(x, 16 - x^2)$. The perimeter, P, of the rectangle is

$$P = 2(2x) + 2(16 - x^2) = -2x^2 + 4x + 32.$$

The maximum value of P is at $x = \dfrac{-b}{2a} = \dfrac{-4}{-4} = 1$, so

$$P_{\text{max}} = -2(1^2) + 4(1) + 32 = 34.$$

Chapter 5	ANSWERS TO CHECKUP EXERCISES

1. (4)

2. (3)

3. (4)

4. (2)

5. (1)

6. (3)

7. (2)

8. (4)

9. (3)

10. (2)

11. (4)

12. (1

13. (2)

14. (2)

15. (4)

16. 4

17. 4

18. $21i$

19. $-5i$

20. -28

21. 1

22. $-5i\sqrt{3}$

23. $-i$

24. $9i$

25. 0

26. $-i$

27. $a = 3$, $b = 2$

28. $\dfrac{1 \pm \sqrt{61}}{6}$

29. $\dfrac{5 \pm \sqrt{23}}{2}$

30. $3 \pm i\sqrt{3}$

31. $\dfrac{1}{3} \pm \dfrac{1}{3}i$

32. $\dfrac{7}{4} \pm \dfrac{3}{4}i$

33. $-6 \pm 2\sqrt{10}$

34. $\dfrac{4}{5} \pm \dfrac{3}{5}i$

35. $\dfrac{3}{2} \pm \dfrac{1}{2}i$

36. $-\dfrac{1}{3} \pm \dfrac{1}{3}i$

37. $-4 \pm 3i$

*38. *a.* See "Selected Solutions."
  b. $x^2 + 3x - 10 = 0$

39. $x^2 - 10x + 29 = 0$

40. $\dfrac{-3 + \sqrt{45}}{2}$

41. $1 \pm \dfrac{1}{3}\sqrt{3}i$

42. *a.* 0.1, 3.9

*43. 12.6

*44. 11.8 centimeters by 23.6 centimeters

45. *a.* 23.5
  b. $6.7 < t < 16.8$

SELECTED SOLUTIONS

38.   *a.* If $ax^2 + bx + c = 0$ with $a \ne 0$, then

- Sum of roots $= \dfrac{\left(-b + \sqrt{b^2 - 4ac}\right) + \left(-b - \sqrt{b^2 - 4ac}\right)}{2a} = -\dfrac{b}{a} = -\dfrac{b}{a}$

- Product of roots $= \dfrac{\left(-b + \sqrt{b^2 - 4ac}\right)\left(-b - \sqrt{b^2 - 4ac}\right)}{2a}$

$$= \dfrac{b^2 - \left(b^2 - 4ac\right)}{4a^2}$$

$$= \dfrac{c}{a}$$

Therefore, $\dfrac{ax^2}{a} + \dfrac{b}{a}x + \dfrac{c}{a} = 0$ or, equivalently, $x^2 - \left(-\dfrac{b}{a}\right)x + \left(\dfrac{c}{a}\right) = 0$, so

$$x^2 - \left(\text{sum of roots}\right)x + \left(\text{product of roots}\right) = 0.$$

43. If the length (20 ft) and width (15 ft) of the rectangular deck are each increased by x feet, then the area is $(20+x)(15+x) \leq 900$ square feet. Rewrite the inequality as $x^2 + 35x - 600 \leq 0$, and solve it for x: $x \approx 12.6$.

44. If x = width of the rectangular sheet, then its length is $2x$, both in centimeters. The volume is $(2x-4)(x-4)2 \geq 300$ cubic centimeters. Rewrite the inequality as $x^2 - 6x - 77 \geq 0$, and solve it for x: $x \approx 11.8$ centimeters, and $2x = 23.6$ centimeters.

Chapter 6 — ANSWERS TO CHECKUP EXERCISES

1. (4)
2. (1)
3. (1)
4. (4)
5. (2)
6. (2)
7. (1)
8. (4)
9. (3)
10. (3)
11. (4)
12. (2)
13. (3)
14. (2)
15. (3)
16. $n = 40$
17. $x = 4$
18. $x = 7$

19. $x = 125$
20. $y = 3$
21. $n = 7$
22. $x = \dfrac{16}{81}$
23. $x = -4$
24. $p = -5, -1$
25. $x = 5$
26. $x = -7, 5$
27. $x = -1, \dfrac{19}{16}$
28. -17
29. 3.2
30. $x < -1$ or $x > \dfrac{3}{2}$
31. $-3 < x < 7$
32. $x < -\dfrac{7}{3}$ or $x > 5$

33. $x \geq -\dfrac{1}{2}$
34. $x = -1, 5$
35. $b = 5$
36. $y = -5, 2$
37. $-2 < x < \dfrac{8}{3}$
38. $0 < x < \dfrac{2}{7}$ or $x > 0.4$
39. $t = \dfrac{3}{2}$
40. $x = -\dfrac{4}{3}, 6$
41. $x = -2, 1$
42. $m = 2, 6$
43. $r = -2$
44. No solution
45. $x = -1.9$

46. $\dfrac{5}{2} < t < \dfrac{11}{2}$
47. 3.84
48. 3.5
*49. $-\dfrac{7}{8}, 7$
*50. 8
*51. 50
*52. 35 miles per hour
53. -2
54. 6
55. 3 miles per hour
56. 200
57. 2 miles per hour
58. 12
*59. 28
60. 161

SELECTED SOLUTIONS

49. If $x = (y+1)^{\frac{1}{3}}$, then $2x^2 - 5x + 2 = 0$, so

$$(2x-1)(x-2) = 0 \text{ and } x = \frac{1}{2} = (y+1)^{\frac{1}{3}} \text{ or } x = 2 = (y+1)^{\frac{1}{3}}.$$

Solve each equation for y: $y = -\dfrac{7}{8}$ or $y = 7$.

50. If Antoine takes x minutes to solve the problem working alone, then Mary takes $x + 16$ minutes to solve the same problem. Solve the equation $\dfrac{6}{x} + \dfrac{6}{x+16} = 1$ for x: $x = 8$.

51. If Dave can mow the lawn with a hand mower in x minutes, then he can mow the lawn with a power mower in $x - 20$ minutes. Solve the equation $\dfrac{15}{x-20} + \dfrac{25}{x} = 1$ for x: $x = 50$.

52. If the teacher drove to the conference at an average rate of x miles per hour, then $\dfrac{280}{x}$ represents the time, in hours, of the trip. Hence, $\dfrac{280}{x+5}$ is the number of hours the trip would have taken if the teacher had increased her average speed by 5 miles per hour. Since she would have arrived on time traveling at the faster speed, $\dfrac{280}{x+5} = \dfrac{280}{x} - 1$. Solve the equation for x: $x = 35$ miles per hour.

59. If x members contributed equally to raise \$896, then each of the original members contributed $\dfrac{\$896}{x}$. When x is decreased by 4, the amount contributed per member increases by \$13 in order to raise \$1080. Because each of the remaining $x - 4$ members must now contribute $\left(\dfrac{\$896}{x} + \$13\right)$,

$$\left(\frac{\$896}{x} + \$13\right)(x - 4) = \$1080.$$

Solve the equation for x: $x = 28$.

Chapter 7 — ANSWERS TO CHECKUP EXERCISES

1. (1)

2. (3)

3. (2)

4. (3)

5. (2)

6. 14

7. −52

8. 11

9. −3, 0, 1

10. $0, \pm\sqrt{5}\,i$

11. $\pm 2, \pm 2i$

12. $2, \pm\dfrac{1}{2}, -1 \pm \sqrt{3}\,i$

13. $4x^2 + 7x + 22 + \dfrac{44}{x-2}$

14. $x^3 - 3x^2 + 3x + 1 + \dfrac{10}{x+1}$

15. $3x^2 + x + 2 + \dfrac{-4}{2x-1}$

16. $x^2 + 6x - 3 + \dfrac{x-5}{x^2+2}$

17. $b.\ (x-3)(2x+1)(x-1)$

18. $f(x) = x^4 - 3x^3 + 5x^2 - x - 10$

19. $f(x) = x^4 + 4x^3 - 4x^2 - 24x - 9$

20. $a.\ -2 \quad b.\ 1 \pm \dfrac{\sqrt{2}}{2}\,i$

21. $-2 \pm \sqrt{2}$

22. $2 + i, -2 \pm \dfrac{3\sqrt{2}}{2}$

23. Show that there is a change in sign between −2 and −1; 0 and 1; 3 and 4.

24. − 4.22636

25. $(0,-1); y = x$

26. $(-2,0); y = x + 2$

27. $(2,0); (0,2); y = x + 3$

28. $b.\ (x-1)^2(3x+4)(2x-1)$

29. $-2, 3 \pm \sqrt{10}$

30. $\dfrac{5}{2}, 1 \pm \sqrt{3}\,i$

31. $-\dfrac{3}{2}, 1, 1 \pm \sqrt{5}$

32. $-3, \dfrac{1}{2}, 2 \pm 3i$

33. Asymptotes: x-axis; $x = \pm 4$
     Intercept: $(0,0)$

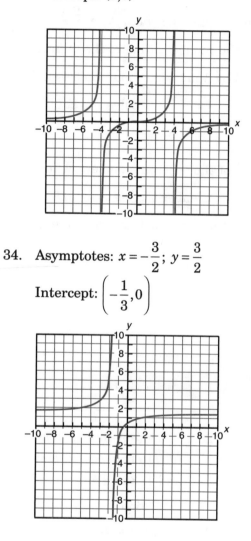

34. Asymptotes: $x = -\dfrac{3}{2}$; $y = \dfrac{3}{2}$

     Intercept: $\left(-\dfrac{1}{3}, 0\right)$

35. Asymptotes: $x = \pm 2$; $y = 1$
     Intercepts: $(-5,0)$, $(5,0)$; $(0, 6.25)$

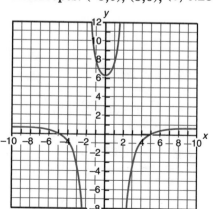

36. Asymptotes: $x = 1$; $y = x + 3$
     Intercepts: $(-2,0)$, $(0,0)$

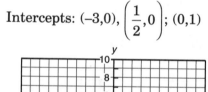
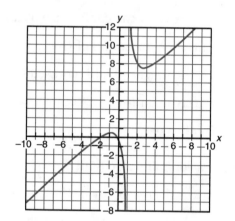

37. Asymptotes: $x = 3$; $y = 2x + 11$

     Intercepts: $(-3,0)$, $\left(\dfrac{1}{2}, 0\right)$; $(0,1)$

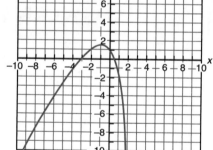

38. Asymptotes: $x = \pm 1$
     Intercepts: $(-1,0)$, $(3,0)$; $(0,-4)$

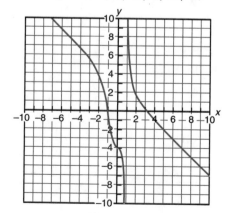

39. $\dfrac{5}{2x} + \dfrac{7}{2(x+2)}$

40. $\dfrac{13}{7(x-6)} + \dfrac{1}{7(x+1)}$

41. $\dfrac{-1}{9x} + \dfrac{1}{9(x+3)} + \dfrac{4}{3(x+3)^2}$

42. $\dfrac{5}{4(x-1)} + \dfrac{7x-3}{4(x^2+2x+5)}$

43. $\dfrac{-3}{4(x+2)} + \dfrac{3}{4(x-2)} + \dfrac{3}{(x-2)^2}$

44. $\dfrac{1}{3(x-2)} + \dfrac{2(x+1)}{3(x^2+2x+4)}$

45. $\dfrac{11}{2(x-1)} - \dfrac{11x+9}{2(x^2+x+1)}$

46. $\dfrac{2}{x} - \dfrac{2}{x-2} + \dfrac{x+6}{(x-2)^2}$

47. $\dfrac{-1}{x} + \dfrac{-5}{21(2x-1)} + \dfrac{20}{21(x+3)}$

48. $3 + \dfrac{1}{4x} - \dfrac{49}{4(x+4)}$

49. $2 - \dfrac{1}{x-2} + \dfrac{2}{x-1}$

50. $x+3 - \dfrac{2}{3x} + \dfrac{32}{3(x-3)}$

Chapter 8	**ANSWERS TO CHECKUP EXERCISES**

1. (2)
2. (3)
3. (2)
4. (4)
5. (3)
6. (4)
7. (3)
8. (2)
9. (2)
10. (1)
11. (2)
12. (1)
13. (1)
14. (3)
15. (3)
16. (4)
17. (3)
18. (3)
19. (3)
20. (1)
21. $-2, 1$
22. $e^4 - 2$
23. $-\dfrac{1}{2}, 4$
24. $\dfrac{5}{3}$
25. 2
26. $-4, 5$
27. $0, \ln 2$
28. $-3, 11$
29. $x = 4, y = -7$

30. $2p - \dfrac{1}{2}q;\ \dfrac{1}{3}(p-q)$

31. -2.42

32. 2.29

33. $-4.16, 2.16$

34. 1500

35. 2.93

36. *a.* 552
    b. 2008

37. *a.* 9140
    b. 9.1

38. *a.* 61.6
    b. 164.2

39. 10 years, 9 months

40. 10 years, 8 months

41. $(0,0),\ \left(\dfrac{1}{2}, \dfrac{1}{2}\right)$

42. *a.* 21%   *b.* 2007

43. 13

44. *a.* $y = 276.67(1.21)^x$
    b. 3297

45. If $y = \log_b x$, then $x = b^y$, so $\log x = \log b^y = y \log b$. Solve for y: $y = \dfrac{\log x}{\log b} = \log_b x$.

46. 20.6

47. 28

48. *a.* $y = 61.8(1.2)^x$; 62
    b. 8

49. $20.0(0.7)^x$;
    10 hours, 21 minutes

Chapter 9

ANSWERS TO CHECKUP EXERCISES

1. (2)

2. (2)

3. (3)

4. (1)

5. (2)

6. (4)

7. (4)

8. (3)

9. (2)

10. (3)

11. (4)

12. (3)

13. (2)

14. (3)

15. (4)

16. (4)

17. (2)

18. (4)

19. (4)

20. (1)

21. a. 13.8
    b. 688

22. 4.1

23. 32.3

24. 102.6

25. a. 11
    b. 22

26. 2240π

27. $\dfrac{12}{13}$

28. $\dfrac{1}{2}$

29. $-\dfrac{17}{15}$

30. $-\dfrac{15}{17}$

31. $-\tan 30°$

32. $-\cos\left(\dfrac{\pi}{3}\right)$

33. $-\csc 50°$

34. $-\cos 20°$

35. $-\cot 20°$

36. $-\sin 15°$

37. $\big(3\tan x\big)\big(\tan x - 8\big)\big(\tan x + 8\big)$

38. $2\big(\sin\theta - 3\cos\theta\big)\big(\sin\theta + 3\cos\theta\big)$

39. $\big(6\sin x - 3\big)\big(\sin x + 1\big)$

40. $\dfrac{1}{\sin y - 2}$

41. $\dfrac{-1}{\tan x + 1}$

42. $\dfrac{1}{\cos x - 3}$

43. $\sin x + \cos x$

44. $-\dfrac{1}{2};\ 2\,(\text{reject})$

45. $-\dfrac{1}{4}, -1$

46. $-\dfrac{5}{4}, 6$

Chapter 10

ANSWERS TO CHECKUP EXERCISES

1. (2)

2. (4)

3. (4)

4. (2)

5. (1)

6. (2)

7. (2)

8. (4)

9. (2)

10. (1)

11. (2)

12. (3)

13. (1)

14. (3)

15. (2)

16. (2)

17. (3)

18. (3)

19. (2)

20. (2)

21. (1)

22. (1)

23. (4)

24. (4)

25. (3)

26. $A = 0.6,\ B = 0.4\pi$

27. $y = 2\sin x + 3$

28. $y = 3\cos x - 2$

29. July 15; 6600

30. $y = \pm 2\sin\dfrac{1}{2}x + 3$

31. *a.* See graph   *b.* (3)

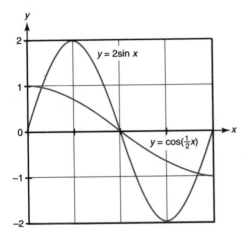

32. *a.* See graph   *b.* π

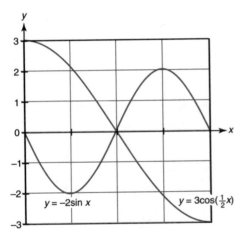

33. *a.* See graph.   *b.* $\pm\dfrac{\pi}{2}$

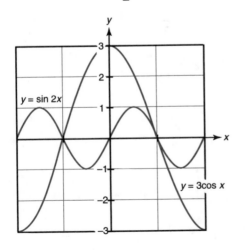

34. *a.* 3.0   *b.* 20   *c.* See graph.

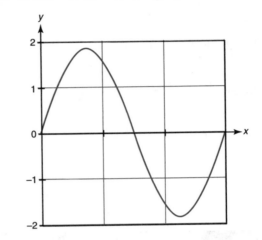

35. *a.* 40 centimeters   *b.* 2   *c.* See graph.

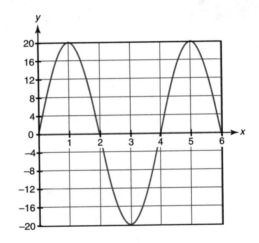

36. *a.* $d = 8\sin\dfrac{\pi}{3}t$ *b.* See graph.

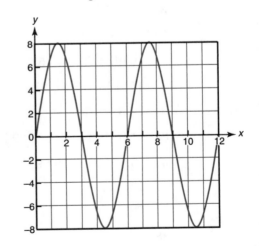

37. 8

38. 4.2

39. $A = 2000; B = \dfrac{\pi}{3}; D = 4000$

40. $A = 1.5; B = \dfrac{\pi}{6}; D = 6.5$

41. *a.* See graph.

   b. One

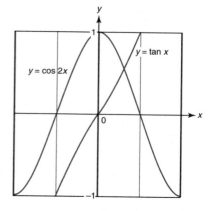

ANSWERS TO CHECKUP EXERCISES

1. (1)
2. (3)
3. (3)
4. (3)
5. (3)
6. (1)
7. (1)
8. (1)
9. (3)
10. (3)
11. (2)
12. (3)
13. (1)
14. (4)
15. (4)
16. (3)
17. (1)
18. (4)
19. (4)
20. (4)

21. $-\dfrac{1}{9}$

*22. See Selected Solutions.

23. $45°, 135°, 225°, 315°$

24. $0°, 90°, 180°$

25. $210°, 330°$

26. $0°, 120°, 180°, 300°$

27. $90°, 135°, 270°, 315°$

28. $120°, 240°$

29. 43

30. $120°, 240°$

31. 228.6, 311.4

32. 194.5, 270, 345.5

33. 45, 108.4, 225, 288.4

34. 78.5, 281.5

35. 61.3, 140.5, 241.3, 320.5

36. 53.6, 126.4, 187.9, 352.1

37. $\dfrac{13}{85}$

38. $\dfrac{36}{85}$

39. $\dfrac{4 - 3\sqrt{3}}{3 - 4\sqrt{3}}$

40. $\dfrac{10}{\sqrt{2}}$

41. 0.5

42. 66.25

43. 90, 194.5, 270, 345.5

44. 41.4, 180, 318.6

45. 41.8, 138.2, 210, 330

46. 60, 104.5, 255.5, 300

47. $-\dfrac{3}{5}$

48. $\dfrac{1}{\sqrt{15}}$

49. $\dfrac{10}{9}$

50. $\dfrac{20}{29}$

*51–60. See Selected Solutions.

SELECTED SOLUTIONS

22. $\dfrac{\cot^2\theta-1}{2\cot\theta}=\dfrac{\dfrac{1}{\tan^2\theta}-1}{\dfrac{2}{\tan\theta}}=\dfrac{1-\tan^2\theta}{2\tan\theta}=\dfrac{1}{\tan 2\theta}=\cot 2\theta$

51. $\left(\cot\theta+\csc\theta\right)\left(1-\cos\theta\right)=\left(\dfrac{\cos\theta+1}{\sin\theta}\right)\left(1-\cos\theta\right)=\dfrac{1-\cos^2\theta}{\sin\theta}=\dfrac{\sin^2\theta}{\sin\theta}=\sin\theta$

52. $\dfrac{\tan\theta}{\cot\theta}+1=\tan^2\theta+1=\sec^2\theta$

53. $\dfrac{\sin\theta}{\sin^2\theta+\cos 2\theta}=\dfrac{\sin\theta}{\sin^2\theta+1-2\sin^2\theta}=\dfrac{\sin\theta}{1-\sin^2\theta}=\dfrac{\sin\theta}{\cos^2\theta}=\dfrac{\sin\theta}{\cos\theta}\cdot\dfrac{1}{\cos\theta}=\tan\theta\sec\theta=\dfrac{\sec\theta}{\cot\theta}$

54. $\dfrac{\csc x\left(\cos x+\sin x\right)^2}{1+\sin 2x}=\dfrac{\csc x\left(\cos^2 x+2\sin x\cos x+\sin^2 x\right)}{1+\sin 2x}=\dfrac{\csc x\left(1+\sin 2x\right)}{\left(1+\sin 2x\right)}=\csc x.$

   On the right side, $\dfrac{\cot x}{\cos x}=\dfrac{\cos x}{\sin x\cdot\cos x}=\dfrac{1}{\sin x}=\csc x.$

55. $\dfrac{\sin 2A}{1+\cos 2A}=\dfrac{2\sin A\cos A}{1+2\cos^2 A-1}=\dfrac{\sin A\cos A}{\cos^2 A}=\dfrac{\sin A}{\cos A}=\tan A$

56. $\dfrac{1+\csc A}{\sec A}=\left(\dfrac{1+\dfrac{1}{\sin A}}{\dfrac{1}{\cos A}}\right)\cdot\left(\dfrac{\sin A\cos A}{\sin A\cos A}\right)=\dfrac{\sin A\cos A+\cos A}{\sin A}=\dfrac{\sin A\cos A}{\sin A}+\dfrac{\cos A}{\sin A}=\cos A+\cot A$

57. $\cot A-\dfrac{\cos 2A}{\sin A\cos A}=\dfrac{\cos A}{\sin A}-\dfrac{2\cos^2 A-1}{\sin A\cos A}=\dfrac{\cos^2 A-\left(2\cos^2 A-1\right)}{\sin A\cos A}$

   $=\dfrac{1-\cos^2 A}{\sin A\cos A}=\dfrac{\sin^2 A}{\sin A\cos A}=\dfrac{\sin A}{\cos A}=\tan A$

58. $\dfrac{\sec A\sec B}{\tan A-\tan B}=\left(\dfrac{\dfrac{1}{\cos A\cos B}}{\dfrac{\sin A}{\cos A}-\dfrac{\sin B}{\cos B}}\right)\cdot\left(\dfrac{\cos A\cos B}{\cos A\cos B}\right)$

   $=\dfrac{1}{\sin A\cos B-\sin B\cos A}=\dfrac{1}{\sin\left(A-B\right)}=\csc\left(A-B\right)$

59. $\dfrac{\tan\theta - \cot\theta}{\tan\theta + \cot\theta} = \dfrac{\tan\theta - \dfrac{1}{\tan\theta}}{\tan\theta + \dfrac{1}{\tan\theta}} = \dfrac{\tan^2\theta - 1}{\tan^2\theta + 1} = \dfrac{\tan^2\theta - 1}{\sec^2\theta} = \cos^2\theta\left(\tan^2\theta - 1\right)$

$\qquad = \cos^2\theta\left(\dfrac{\sin^2\theta}{\cos^2\theta} - 1\right) = \sin^2\theta - \cos^2\theta = \sin^2\theta - \left(1 - \sin^2\theta\right)$

$\qquad = 2\sin^2\theta - 1$

60. $\dfrac{\sin A}{1 - \cos A} + \dfrac{\sin A}{1 + \cos A} = \dfrac{\sin A\left(1 + \cos A\right) + \sin A\left(1 - \cos A\right)}{\left(1 - \cos A\right)\left(1 + \cos A\right)}$

$\qquad = \dfrac{2\sin A}{1 - \cos^2 A} = \dfrac{2\sin A}{\sin^2 A} = \dfrac{2}{\sin A} = 2\csc A$

Chapter 12	ANSWERS TO CHECKUP EXERCISES

1. (3)	12. (2)	23. 15.1	34. 16.7
2. (2)	13. (3)	24. 254.7	
3. (4)	14. (4)	25. 4.6	35. $-\dfrac{2}{7}$
4. (4)	15. (4)	26. 0.15 hour	36. 157.4
5. (3)	16. 25	27. 2700	*37. 6246
6. (1)	17. 1.15	28. 66.6	*38. 955
7. (3)	18. 65.27	*29. 10.6	39. 105.6
8. (4)	*19. 49.8, 65.1, 65.1	30. 109.3	*40. 60.8
9. (3)	*20. 23 feet, 5 inches	*31. 42.6	*41. 62
10. (3)	21. 796	32. 451.2	42. 28.3
11. (4)	*22. 585.9	33. 41	*43. 32.9

SELECTED SOLUTIONS

19. If, in $\triangle ABC$, $AC = BC = 12$, then

$$\frac{1}{2}(12)(12)\sin A = 55, \text{ so } \sin A = \frac{55}{72} \text{ and } \angle A \approx 49.8°.$$

The sum of the measures of the remaining angles is $180° - 49.8° = 130.2°$. Because these angles are opposite the congruent sides, they have equal measures. Hence, each of the base angles measures $\dfrac{130.2°}{2} = 65.1°$. The degree measures of the three angles of the triangle are 49.8, 65.1, and 65.1.

20. Let x represent the unknown distance. In the obtuse triangle, use the Law of Sines to find x:

$$\frac{x}{\sin 1.3°} = \frac{54}{\sin 3°}, \text{ so } x = \frac{54 \sin 1.3°}{\sin 3°} = 23.4 \text{ feet} \approx 23 \text{ feet, 5 inches.}$$

22. In $\triangle ABC$, $\angle A = 26°40'$, $\angle B = 38°10'$, and $\angle C = 180° - \left(38°10' + 26°40'\right) = 115°10'$. According to the Law of Sines,

$$\frac{AB}{\sin 115°10'} = \frac{400}{\sin 38°10'}, \text{ so } AB = \frac{400 \sin 115°10'}{\sin 38°10'} \approx 585.9.$$

29. Draw isosceles trapezoid $ABCD$, with $AB = DC = 5$, $BC = 8$, and $\angle BAD = 73°30'$. Then

$$\angle ABC = 180° - 73.5° = 106.5°.$$

In $\triangle ABC$, use the Law of Cosines to find AC:

$$AC = \sqrt{5^2 + 8^2 - 2(5)(8)\cos 106.5°} \approx 10.6.$$

31. Draw $\triangle FHP$, where the pitcher is at point P, home plate corresponds to point H, and first base corresponds to point F. Then

$$FH = 60, \ HP = 46, \text{ and } \angle PHF = \frac{1}{2} \times 90° = 45°.$$

Use the Law of Cosines to find FP:

$$FP = \sqrt{46^2 + 60^2 - 2(46)(60)\cos 45°} \approx 42.6.$$

37. Draw $\triangle BCJ$, where B, C, and J represent the locations of the balloon, Carmen, and Jamal, respectively. Drop a perpendicular segment from B to $\overline{CJ}$, intersecting $\overline{CJ}$ at H. Thus, BH represents the unknown height of the balloon. In $\triangle BCJ$, $\angle CBJ = 180° - \left(60° + 75°\right) = 45°$. According to the Law of Sines,

$$\frac{BC}{\sin 75°} = \frac{5280}{\sin 45°}, \text{ so } BC = \frac{5280 \sin 75°}{\sin 45°} \approx 7212.6.$$

In right triangle BHC,

$$\sin 60° = \frac{BH}{7212.6}, \text{ so } BH = \sin 60° \times 7212.6 = 6246.3 \approx 6246.$$

38. Draw $\triangle ABC$ with $AB = AC = x$ and $\angle A = 53°10'$. Because the area of the triangle is $43,560$ square feet,

$$\frac{1}{2}x^2 \sin 53°10' = 43,560, \text{ so } x = \sqrt{\frac{2 \times 43,560}{\sin 53°10'}} \approx 329.9 \text{ feet.}$$

According to the Law of Cosines,

$$BC = \sqrt{(329.9)^2 + (329.9)^2 - 2(329.9)\cos 53°10'} \approx 295.3 \text{ feet.}$$

The perimeter of $\triangle ABC$ is $329.9 + 329.9 + 295.3 = 955.1 \approx 955$.

40. Represent the forces by parallelogram $ABCD$, with $AB = CD = 30$ and $AD = BC = 40$. Draw the resultant force AC in such a way that $\angle BAC = 35.2°$. Let $\angle CAD = \angle ACB = x$. In $\triangle ABC$, use the Law of Sines to find x:

$$\frac{40}{\sin 35.2°} = \frac{30}{\sin x}, \text{ so } x = \sin^{-1}\left(\frac{30\sin 35.2°}{40}\right) \approx 60.8.$$

41. Represent the forces by parallelogram $ABCD$, with the car at point A. Let $AB = CD = 2000$ and $AD = BC = 1500$. You need to find $\angle BAD$. Draw the resultant force AC, and set $AC = 3000$. In $\triangle ABC$, use the Law of Cosines to find $\angle B$ (the supplement of $\angle BAD$):

$$3^2 = (1.5)^2 + 2^2 - 2(1.5)(2)\cos B, \text{ so } \angle B = \cos^{-1}\left(\frac{-2.75}{6}\right) \approx 117.3.$$

Thus, $\angle BAD = 180 - 117.3 = 62.7$. In order that $AC \geq 3,000$, round $\angle BAD$ *down* to 62.

43. In $\triangle PQR$, $PQ = 6, PR = 11$, and $QR = 13$. Use the Law of Cosines to find an angle of the triangle, say $\angle P$:

$$13^2 = 6^2 + 11^2 - 2(6)(11)\cos P, \text{ so } \angle P = \cos^{-1}\left(-\frac{1}{11}\right) \approx 84.8°.$$

The area of $\triangle PQR = \frac{1}{2}\times(6\times 11)\times\sin 84.8° = 32.9$.

Chapter 13 — ANSWERS TO CHECKUP EXERCISES

1. $6i$

2. $-\dfrac{\sqrt{3}}{6} + \dfrac{1}{6}i$

3. $-3\sqrt{2} - 3\sqrt{2}\,i$

4. $2 - 2\sqrt{3}\,i$

5. $-\dfrac{9}{4}$

6. $(x-3)^2 + y^2 = 9$

7. $x^2 + (y+2)^2 = 4$

8. $x = 2$

9. $(x+1)^2 + \left(y - \dfrac{1}{2}\right)^2 = \dfrac{5}{4}$

10. $r = 8\cos\theta$

11. $r^2 = \dfrac{12}{4 - \sin^2\theta}$

12. $r = 10\sin\theta - 4\cos\theta$

13. $r^2 = 2\sec\theta\csc\theta$

14. $y = x^2 + 2x$

15. $x = \dfrac{2}{3}y^2$

16. $y = x^2$

17. $y = e^{-2x}$

18. $(x-2)^2 + (y+1)^2 = 16$

19. $\dfrac{x^2}{9} + \dfrac{y^2}{4} = 1$

20. $\dfrac{(x+3)^2}{4} - \dfrac{(y-1)^2}{9} = 1$

21. $y = \dfrac{1}{2}x^2 - 1$

22. $x = 4t - 1, y = \sqrt{4t+3};\ t \geq -\dfrac{3}{4}$

23. $x = (140\cos\theta)t$

$y = -16t^2 + (140\sin\theta)t + 3$

*24. 22.7

*25 a. See Selected Solutions.   b. $y - 2 = -\dfrac{1}{2}(x - 1)$

26. $256\left(\cos\dfrac{\pi}{5} + i\sin\dfrac{\pi}{5}\right)$

27. $243\left(\cos\dfrac{2\pi}{9} + i\sin\dfrac{2\pi}{9}\right)$

28. 18.5

29. $-243i$

30. $\dfrac{1}{64}$

31. -64

32. $-16 - 16\sqrt{3}\,i$

33. $-1024i$

34. -64

35. $4\operatorname{cis}110°,\, 4\operatorname{cis}230°,\, 4\operatorname{cis}350°$

36. $6\sqrt{2}\operatorname{cis}107.5°,\, 6\sqrt{2}\operatorname{cis}287.5°$

37. $3\operatorname{cis}45°,\, 3\operatorname{cis}135°,\, 3\operatorname{cis}225°,\, 3\operatorname{cis}315°$

38. $3\operatorname{cis}24°,\, 3\operatorname{cis}96°,\, 3\operatorname{cis}168°,\, 3\operatorname{cis}240°,\, 3\operatorname{cis}312°$

39. $2\operatorname{cis}45°,\, 2\operatorname{cis}105°,\, 2\operatorname{cis}165°,\, 2\operatorname{cis}225°,\, 2\operatorname{cis}285°,\, 2\operatorname{cis}345°$

40. $\sqrt[3]{5}\operatorname{cis}53.13°,\, \sqrt[3]{5}\operatorname{cis}137.71°,\, \sqrt[3]{5}\operatorname{cis}257.71°$

41. $2.96 + 0.47i,\, 0.47 + 2.96i,\, -2.67 + 1.36i,\, -2.12 - 2.12i,\, 1.36 - 2.67i$

42. $3.88 + 0.95i,\, -2.77 + 2.89i,\, -1.12 - 3.84i$

43. a. $2i,\, -\sqrt{3} - i,\, \sqrt{3} - i$

44. a. $-3,\, \dfrac{3}{2} \pm \dfrac{3\sqrt{3}}{2}i$

45. $\pm 1,\, \pm i$

SELECTED SOLUTIONS

24. Because $x = 420$ and $v_0 = 140$, $t = \dfrac{420}{140\cos\theta} = \dfrac{3}{\cos\theta}$. Thus:

$$y = -16\left(\dfrac{3}{\cos\theta}\right)^2 + 140\sin\theta\left(\dfrac{3}{\cos\theta}\right) + 3 > 9\ .$$

Using a graphing calculator, set $Y1 = -\dfrac{144}{\cos^2\theta} + 420\tan\theta + 3$. Using a step value of 0.1, create a table. Scroll down to the smallest value of θ, correct to the nearest tenth of a degree, that makes $Y1 > 9$. The answer is 22.7.

25. a. As $t = \dfrac{x - x_A}{x_B - x_A}$, $y = y_A + \dfrac{x - x_A}{x_B - x_A}(y_B - y_A)$, so

$$y - y_A = \dfrac{y_B - y_A}{x_B - x_A}(x - x_A) = m(x - x_A).$$

Chapter 14   ANSWERS TO CHECKUP EXERCISES

1. $(0,0)$; $V(0,\pm\sqrt{13})$; $F(0,\pm3)$

2. $(0,0)$; $V(\pm7,0)$; $F(\pm\sqrt{74},0)$

3. $(0,0)$; $V(\pm4,0)$; $F(\pm\sqrt{7},0)$

4. $V(0,0)$; $F\left(0,-\dfrac{1}{8}\right)$

5. $V\left(\dfrac{1}{2},0\right)$; $F(2,0)$

6. $(2,-1)$; $V(2,5)$, $V'(2,-7)$; $F(2,-1\pm2\sqrt{5})$

7. $(1,4)$

8. $V(4,-3)$; $F\left(\dfrac{33}{8},-3\right)$

9. $(-2,5)$; $V(2,5)$, $V'(-6,5)$; $F(-2\pm\sqrt{97},5)$

10. $V(-4,3)$; $F(-2,3)$

11. $(4,-2)$; $V(4,4)$, $V'(4,-8)$; $F(4,-2\pm3\sqrt{2})$

12. $(4,2)$; $V(4,8)$, $V'(4,-4)$; $F'(4,2\pm2\sqrt{34})$

13. $\dfrac{x^2}{11}+\dfrac{y^2}{4}=1$

14. $\dfrac{(x+4)^2}{9}+\dfrac{(y-1)^2}{36}=1$

15. $\dfrac{(x+1)^2}{169}+\dfrac{(y-2)^2}{25}=1$

16. $\dfrac{(x-2)^2}{16}+\dfrac{\left(y-\dfrac{\sqrt{7}}{4}\right)^2}{9}=1$

17. $\dfrac{(x+1)^2}{33}+\dfrac{(y+1)^2}{49}=1$

18. $\dfrac{(x+1)^2}{144}-\dfrac{(y-3)^2}{25}=1$

19. $\dfrac{(y-4)^2}{16}-\dfrac{(x+2)^2}{9}=1$

20. $\dfrac{(x-4)^2}{25}-\dfrac{(y-4)^2}{75}=1$

21. $\dfrac{y^2}{81}-\dfrac{x^2}{56.25}=1$

22. $\dfrac{(x+1)^2}{9}-\dfrac{(y+2)^2}{4}=1$

23. $(x+1)^2=-8(y-3)$

24. $(y+2)^2=16(x-4)$

25. $(x-3)^2=-8(y-2)$

26. $(y-5)^2=24(x-4)$

27. $(x+4)^2=-8(y+1)$

28. $(x-3)^2=-3(y-12)$

29. $F(-2\pm4\sqrt{5},-1)$

30. $F(3,1\pm\sqrt{10})$

*31. 1 foot from the center of the base of the dish (the vertex of the parabola in the coordinate plane) and along the axis of symmetry.

32. Hyperbola

33. Ellipse

34. Parabola

35. Hyperbola

36. Ellipse

37. Ellipse

38. $Y = \dfrac{\left(\sqrt{3}\,x - 1\right) \pm \sqrt{-x^2 + \left(8 - 2\sqrt{3}\right)x + 9}}{2}$

39. $Y = \dfrac{\left(8x + \dfrac{2}{\sqrt{5}}\right) \pm \sqrt{96x^2 + \left(\dfrac{72}{\sqrt{5}}\right)x + \dfrac{164}{5}}}{-4}$

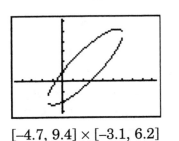

[−4.7, 9.4] × [−3.1, 6.2]   [−4.7, 4.7] × [−3.1, 3.1]

40. $\left(x'\right)^2 - \dfrac{\left(y'\right)^2}{1/4} = 1$

45. $\dfrac{\left(x'\right)^2}{1/4} - \dfrac{\left(y'\right)^2}{1/9} = 1$

41. $\left(y'\right)^2 = 3\left(x' + 1\right)$

46. $\left(2,1\right); \left(3,2\right)$

42. $\left(x'\right)^2 + \dfrac{\left(y' - 1\right)^2}{4} = 1$

47. $\left(\pm 3, \mp 7\right); \left(\pm \dfrac{1}{\sqrt{3}}, \mp \dfrac{2}{\sqrt{3}}\right)$

43. $\left(x' - 2\right)^2 = \dfrac{1}{2}\left(y' + 3\right)$

48. $\left(6, -3\right); \left(2, 1\right)$

44. $\left(x' - 4\right)^2 + \dfrac{\left(y'\right)^2}{2} = 1$

49. $\left(\pm \dfrac{1}{\sqrt{3}}, \pm \dfrac{2}{\sqrt{3}}\right); \left(\pm 3, \pm 7\right)$

50. Hyperbola: $\dfrac{\left(x - 3.6\right)^2}{5.76} - \dfrac{y^2}{7.2} = 1$; foci: $\left(0,0\right), \left(7.2,0\right)$; vertices: $\left(1.2,0\right), \left(6,0\right)$

51. Parabola: $x^2 = 12\left(y + 3\right)$; focus: $(0,0)$; vertex: $(0,-3)$

52. Ellipse: $\dfrac{\left(x + 8.4\right)^2}{158.76} + \dfrac{y^2}{88.2} = 1$; foci: $\left(0,0\right), \left(-16.8,0\right)$; vertices: $\left(-21.0,0\right), \left(4.2,0\right)$

SELECTED SOLUTIONS

31. Consider a two-dimensional cross section of the parabolic surface of the dish. Sketch its graph so that its vertex is at the origin, the maximum width of the parabola is 6 units, and its highest point is 2.25 units above the vertex. The equation of this parabola has the form $x^2 = 4py$. Since $(3, 2.25)$ is a point on the graph, find p by substituting 3 for x and 2.25 for y:

$$3^2 = 4p\left(2.25\right), \text{ so } 9 = 9p \text{ and } p = 1.$$

Because $p = 1$, the focus is located 1 unit above the vertex at $F\left(0, 1\right)$. Hence the receiver should be located 1 foot from the center of the base of the dish (the vertex of the parabola in the coordinate plane) and along the axis of symmetry.

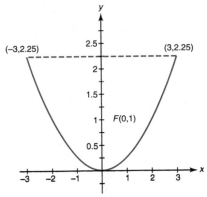

Chapter 15   ANSWERS TO CHECKUP EXERCISES

1. (3)

2. (4)

3. (1)

4. (1)

5. (2)

6. (4)

7. (1)

8. (3)

9. (2)

10. (1)

11. (3)

12. (2)

13. (4)

14. (1)

15. (3)

16. 4694

17. $\dfrac{4969}{3}$

18. 7

19. 3

20. 10

21. 98

22. 2540

23. $\dfrac{2343}{1024}$

24. 3

25. 147

26. $\dfrac{407}{999}$

27. 2

28. $\dfrac{1640}{6561}$

29. 2296

30. 7,980,000

31. 4896

32. 6475

33. $\dfrac{64}{2-\sqrt{2}}$

34. 23.99609375 feet

35. 24 feet

36. $\dfrac{4096}{27}$

37. $\dfrac{781}{3}$

38. $\dfrac{781}{432}$

39. $\dfrac{64}{27}$

40. 18

41. $a.$ 24 $b.$ 18

42. 33

43. 100

44. 4081; 175

45. $2-51i$

*46–56. See "Selected Solutions."

SELECTED SOLUTIONS

46. $P_1 : 2 = 1^2 + 1.$

$$P_k : \; 2+4+6+\cdots+2k \; = k^2 + k$$
$$2+4+6+\cdots+2k+2(k+1) = k^2 + k + 2(k+1)$$
$$= \left(k^2 + 2k + 1\right) + (k+1)$$
$$= (k+1)^2 + (k+1)$$
$$= P_{k+1}$$

47. $P_1 : 1 = 1\left(2 \cdot 1 - 1\right).$

$$P_k : \qquad 1 + 5 + 9 + \cdots + \left(4k - 3\right) = k\left(2k - 1\right)$$

$$1 + 5 + 9 + \cdots + \left(4k - 3\right) + \left[4\left(k+1\right) - 3\right] = k\left(2k-1\right) + \left[4\left(k+1\right) - 3\right]$$

$$= k\left(2k-1\right) + \left[4k+1\right]$$

$$= 2k^2 + 3k + 1$$

$$= \left(k+1\right)\left(2k+1\right)$$

$$= \left(k+1\right)\left(2\left(k+1\right) - 1\right)$$

$$= P_{k+1}$$

48. $P_1 : 2 = 3^1 - 1.$

$$P_k : \qquad 2 + 6 + 18 + \cdots + 2 \cdot 3^{k-1} = 3^k - 1$$

$$2 + 6 + 18 + \cdots + 2 \cdot 3^{k-1} + \left(2 \cdot 3^k\right) = 3^k - 1 + \left(2 \cdot 3^k\right)$$

$$= 3 \cdot 3^k - 1$$

$$= 3^{k+1} - 1$$

$$= P_{k+1}$$

49. $P_1 : 1^2 = \dfrac{1\left(4 \cdot 1^2 - 1\right)}{3} = \dfrac{3}{3} = 1.$

$$P_k : \qquad 1^2 + 3^2 + 5^2 + \cdots + \left(2k-1\right)^2 = \dfrac{k\left(4k^2 - 1\right)}{3}$$

$$1^2 + 3^2 + 5^2 + \cdots + \left(2k-1\right)^2 + \left(2k+1\right)^2 = \dfrac{k\left(4k^2 - 1\right)}{3} + \left(2k+1\right)^2$$

$$= \dfrac{k\left(2k+1\right)\left(2k-1\right)}{3} + \left(2k+1\right)^2$$

$$= \dfrac{k\left(2k+1\right)\left(2k-1\right) + 3\left(2k+1\right)^2}{3}$$

$$= \dfrac{\left(2k+1\right)\left[k\left(2k-1\right) + 3\left(2k+1\right)\right]}{3}$$

$$= \dfrac{\left(2k+1\right)\left(2k^2 + 5k + 3\right)}{3}$$

$$= \dfrac{\left(2k+1\right)\left(2k+3\right)\left(k+1\right)}{3}$$

$$= \dfrac{\left(k+1\right)\left(2k+1\right)\left(2k+3\right)}{3}$$

$$= \dfrac{\left(k+1\right)\left(4k^2 + 8k + 3\right)}{3}$$

$$= \dfrac{\left(k+1\right)\left[4\left(k+1\right)^2 - 1\right]}{3}$$

$$= P_{k+1}$$

50. $P_1 : 3 = \dfrac{3\left(3^1 - 1\right)}{2} = \dfrac{6}{2}.$

$$P_k : \qquad 3 + 3^2 + 3^3 + \cdots + 3^k = \frac{3\left(3^k - 1\right)}{2}$$

$$3 + 3^2 + 3^3 + \cdots + 3^k + 3^{k+1} = \frac{3\left(3^k - 1\right)}{2} + 3^{k+1}$$

$$= \frac{3\left(3^k - 1\right) + 2 \cdot 3^{k+1}}{2}$$

$$= \frac{3^{k+1} - 3 + 2 \cdot 3^{k+1}}{2}$$

$$= \frac{3 \cdot 3^{k+1} - 3}{2}$$

$$= \frac{3\left(3^{k+1} - 1\right)}{2}$$

$$= P_{k+1}$$

51. $P_1 : \dfrac{1}{1 \cdot 2} = \dfrac{1}{1+1} = \dfrac{1}{2}.$

$$P_k : \qquad \frac{1}{1 \cdot 2} + \frac{1}{2 \cdot 3} + \frac{1}{3 \cdot 4} + \cdots + \frac{1}{k\left(k+1\right)} = \frac{k}{k+1}$$

$$\frac{1}{1 \cdot 2} + \frac{1}{2 \cdot 3} + \frac{1}{3 \cdot 4} + \cdots + \frac{1}{k\left(k+1\right)} + \frac{1}{\left(k+1\right)\left(k+2\right)} = \frac{k}{k+1} + \frac{1}{\left(k+1\right)\left(k+2\right)}$$

$$= \frac{k\left(k+2\right) + 1}{\left(k+1\right)\left(k+2\right)}$$

$$= \frac{k^2 + 2k + 1}{\left(k+1\right)\left(k+2\right)}$$

$$= \frac{\left(\cancel{k+1}\right)\left(k+1\right)}{\left(\cancel{k+1}\right)\left(k+2\right)}$$

$$= \frac{\left(k+1\right)}{\left(k+2\right)}$$

$$= P_{k+1}$$

52. $P_1 : \dfrac{5^1 - 1}{4} = 1$, a whole number. Assume that $P_k : 5^k - 1 = 4p$, where p is a whole number.

$$5\left(5^k - 1\right) = 5\left(4p\right)$$

$$5^{k+1} - 5 = 20p$$

$$5^{k+1} - 4 - 1 = 20p$$

$$P_{k+1} : \quad 5^{k+1} - 1 = 20p + 4 = 4\left(5p + 1\right) \quad \leftarrow \text{Divisible by 4}$$

53. $P_1 : \dfrac{3^1 - 1}{2} = 1$, a whole number. Assume that $P_k : 3^k - 1 = 2p$, where p is a whole number.

$$3\left(3^k - 1\right) = 3\left(2p\right)$$
$$3^{k+1} - 3 = 6p$$
$$3^{k+1} - 2 - 1 = 6p$$
$$P_{k+1} : \quad 3^{k+1} - 1 = 6p + 2 = 2\left(3p + 1\right) \quad \leftarrow \text{Divisible by } 2$$

54. $P_1 : \dfrac{1^3 + 2 \cdot 1}{3} = 1$, a whole number. Assume that $P_k : k^3 + 2k = 3p$, where p is a whole number.
You need to prove that, $P_{k+1} : \left(k + 1\right)^3 + 2\left(k + 1\right)$ is divisible by 3.

$$\left(k^3 + 2k\right) + 3\left(k^2 + k + 1\right) = 3p + 3\left(k^2 + k + 1\right)$$
$$\left(k^3 + 3k^2 + 3k + 1\right) + \left(2k + 2\right) = 3\left(p + k^2 + k + 1\right)$$
$$P_{k+1} : \quad \left(k + 1\right)^3 + 2\left(k + 1\right) = 3\left(p + k^2 + k + 1\right) \quad \leftarrow \text{Divisible by } 3$$

55. $P_1 : 2 \cdot 3 + 1 = 7 < 2^3 \left(= 8\right)$. Assume that $P_k : 2k + 1 < 2^k$, where $k > 3$.
You need to prove $P_{k+1} : 2\left(k + 1\right) + 1 < 2^{k+1}$.

$$2\left(2k + 1\right) < 2\left(2^k\right), \text{ so } 4k + 2 < 2^{k+1}.$$

Because $k > 3$,

$$2\left(k + 1\right) + 1 < \underbrace{2\left(k + 1\right) + 2k}_{4k + 2} < 2^{k+1}.$$

56. $P_1 : 2^5 > 5^2$ because 32 > 25. Assume that $P_k : 2^k > k^2$, where $k \geq 5$.
You need to prove $P_{k+1} : 2^{k+1} > \left(k + 1\right)^2$ or, equivalently, $2^{k+1} > k^2 + 2k + 1$.

- $2 \cdot 2^k > 2 \cdot k^2$, so $2^{k+1} > 2k^2$ or $2^{k+1} > k^2 + k^2$.

- For $k \geq 5$, $k^2 = k \cdot k > \underbrace{3 \cdot k}_{2k+k} > 2k + 1$, so $k^2 + k^2 > k^2 + 2k + 1$.

Hence, $2^{k+1} > k^2 + 2k + 1$.

INDEX